New Challenges in Offshore Geotechnical Engineering Developments

New Challenges in Offshore Geotechnical Engineering Developments

Guest Editors

Pan Hu
Dong-Sheng Jeng

MDPI

Basel • Beijing • Wuhan • Barcelona • Belgrade • Novi Sad • Cluj • Manchester

Guest Editors

Pan Hu
Western Sydney University
Penrith
Australia

Dong-Sheng Jeng
Griffith University
Southport
Australia

Editorial Office
MDPI AG
Grosspeteranlage 5
4052 Basel, Switzerland

This is a reprint of the Special Issue, published open access by the journal *Journal of Marine Science and Engineering* (ISSN 2077-1312), freely accessible at: https://www.mdpi.com/journal/jmse/special_issues/new_challenges_in_offshore_geotechnical_engineering_developments.

For citation purposes, cite each article independently as indicated on the article page online and as indicated below:

Lastname, A.A.; Lastname, B.B. Article Title. *Journal Name* **Year**, *Volume Number*, Page Range.

ISBN 978-3-7258-1597-5 (Hbk)
ISBN 978-3-7258-1598-2 (PDF)
https://doi.org/10.3390/books978-3-7258-1598-2

Contents

Journal of
Marine Science and Engineering

MDPI

Article

Application of ANN in Construction: Comprehensive Study on Identifying Optimal Modifier and Dosage for Stabilizing Marine Clay of Qingdao Coastal Region of China

Qirui Bo [1], Junwei Liu [1,*], Wenchang Shang [1], Ankit Garg [2], Xiaoru Jia [1] and Kaiyue Sun [1]

1 School of Civil Engineering, Qingdao University of Technology, Qingdao 266525, China; 18254690329@163.com (Q.B.); 18766226632@163.com (W.S.); 15621123689@163.com (X.J.); qutsky@163.com (K.S.)

2 MOE Key Laboratory of Intelligent Manufacturing Technology, College of Engineering, Shantou University, Shantou 515063, China; ankit@stu.edu.cn

* Correspondence: liujunwei@qut.edu.cn; Tel.: +86-150-6619-8828

Abstract: Nowadays, the use of new compound chemical stabilizers to treat marine clay has gained significant attention. However, the complex non-linear relationship between the influencing factors and the unconfined compressive strength of chemically treated marine clay is not clear. In order to study the influence of various factors (dosage, type of stabilizer, curing age) on the unconfined compressive strength of solidified soil during chemical treatment, experiments were performed to determine the unconfined compressive strength of soft marine clay modified with various types of stabilizers. Further, an artificial neural network (ANN) model was used to establish a prediction model based on the unconfined compressive strength test data and to verify the performance. Sensitivity and optimization analyses were further conducted to explore the relative significance of parameters as well as the optimal dosage amount. Research has found that when the content of aluminate cement is 89.5% and the content of curing agent is 30%, the unconfined compressive strength significantly increases after 28 days of solidification, and the change in quicklime content has the greatest effect on the improvement in the unconfined compressive strength. The influence of modifiers on the unconfined compressive strength is in the order: potassium hydroxide > kingsilica > quick lime > bassanite. The values of each factor were obtained when the unconfined compressive strength was the maximum, which provided support for the optimization of the treatment scheme. The analysis of chemical treatment is no longer limited to the linear relationship according to the test results, which proves the feasibility of non-linear relationship analysis based on the artificial neural network.

Keywords: marine clay; chemical treating; artificial neural network; optimization analysis; sensitivity analysis

Citation: Bo, Q.; Liu, J.; Shang, W.; Garg, A.; Jia, X.; Sun, K. Application of ANN in Construction: Comprehensive Study on Identifying Optimal Modifier and Dosage for Stabilizing Marine Clay of Qingdao Coastal Region of China. *J. Mar. Sci. Eng.* **2024**, *12*, 465. https://doi.org/10.3390/jmse12030465

Academic Editor: Anabela Oliveira

Received: 7 February 2024
Revised: 4 March 2024
Accepted: 5 March 2024
Published: 8 March 2024

1. Introduction

China's coastal area has a complex topography and rich landforms, among which the beach area is very broad. Offshore wind farms are a characteristic of China. For example, offshore wind farms in coastal provinces such as Shandong, Zhejiang, and Jiangsu have been established. The relevant investigation and statistics show that it is of great strategic significance to make full use of resources for wind farm construction and rational and orderly development and utilization. However, most of the offshore wind turbines built in the intertidal zone, which are located in silt sites and have the characteristics of high moisture content, low strength, and a large pore ratio, cannot meet the requirements of the superstructure in terms of strength, deformation, and stability [1–3]. These properties have caused great obstacles to the construction of projects. Therefore, before the construction of a project, it is necessary to treat the silty soil to improve its strength and meet the construction requirements. At present, the chemical curing method can effectively solve the adverse factors of beach silt, which adds an inorganic chemical stabilizer to the beach silt to improve its performance. The traditional solidifying materials for soil are all solid inorganic binders. Good results have been achieved by using lime, cement, and fly ash to improve the soil [4,5]. However, the traditional single chemical stabilizer can no longer meet the engineering needs of beach silt, and the development of a new composite stabilizer has gained significant attention recently. At present, there are abundant research results on chemical curing:

Chew et al. and Porbaha et al. found that the water content of silt would decrease rapidly with the addition of cement, but then it would decrease at a slower rate [6,7]. The larger the curing age, the lower the water content. Lorenzo and Bergado found that the density of solidified soil would increase with an increase in cement content and also curing age [8]. The specific gravity of solidified soil decreases with the addition of cement and has no significant relationship with water content and curing age. Horpibulsuk et al. found that water content not only affects the amount of hydration products during chemical reactions in solidified soil but also has a greater impact on the porosity of 0.1–1 μm [9].

Wang Lifeng [10] used silica nanopowder as an external admixture to be mixed into hydraulic soil and carried out unconfined compressive strength tests on it. The results showed that nanosilica powder can improve the macroscopic mechanical properties of hydraulic soil. Ding Jianwen et al. [11], using the traditional cement curing treatment method based on the use of mixed cement and phosphogypsum joint curing treatment of dredged silt with high water content, reported the results of indoor experiments showing that the effect of phosphogypsum on silt curing soil enhancement is significant, and the amount of this dosage increases with the increase in the initial water content of silt.

Sharma et al. found that the microstructure properties of solidified soil are closely related to the curing scheme [12]. The addition of cement to the stabilizer has a greater effect on the properties of the improved soil than lime. The study showed that the compressive strength of soil samples cured for 28 days was nearly four to six times higher than that of untreated soil samples, reflecting the effect of curing age on strength. Dahal et al. found the maximum strength of the soil by using ordinary Portland cement with different contents [13]. The authors used statistical tools to establish linear relationships between these parameters.

Liu et al. studied the strength characteristics of steel slag, cement, and metakaolin composite (SCM composite)-stabilized soil under different clay contents, water contents, and curing time [14]. SCM composites can effectively improve the strength of solidified soil, and SCM-stabilized soil exhibits similar characteristics as soil-cement. Gong et al. found an efficient stabilizer that could reduce environmental damage through experiments using an eco-friendly stabilizer to solidify soft soil in Nansha [15]. The main component was still cement, but a multi-component stabilizer (18% cement, 3% lime, 4% gypsum, 3% expansive soil, and 0.8% sodium hydroxide) was found to produce the highest strength and curing efficiency.

At present, the research results are still mainly focused on the analysis of the linear relationship between different components and the curing effect in the process of chemical curing according to the test results, and there are few studies on complex non-linear fitting in chemical curing. In recent years, with the rapid development of artificial intelligence technology, machine learning has become an important tool for scientific research and analysis. The dataset for artificial neural network (ANN) modeling is the basis for neural network training, which determines the information that the network can learn and the tasks it can perform. A dataset is usually a set of labeled data that is used to train a neural network to recognize patterns, predict outcomes, or perform classification. Artificial neural network (ANN) dataset parameters mainly include the size, quality, distribution, and labeling quality of the data. Of course, there are some limitations of datasets, such as sample bias, data sparsity, and data imbalance. In this study, an artificial neural network (ANN) model was used, which has the advantages of flexibility in optimization, sensitivity analysis, and risk analysis [16] to solve the fitting of a non-linear relationship between curing strength and different influencing factors [17]. Also, the model analysis was carried out on the results of unconfined compressive strength testing of a set of curing schemes with optimal mix ratio, and the complex non-linear relationship between the dosage of stabilizer components and the unconfined compressive strength was obtained. Further, optimization and sensitivity analyses were conducted to understand the optimum dose and also the ranking of parameters.

2. Materials and Methods

2.1. Sampling and Basic Properties of Marine Clay

The marine clay utilized in this study was collected at the depth of 1–2 m from the vicinity of Hongshiya, Huangdao District, Qingdao City. The geographical location is shown in Figure 1a, and the site is shown in Figure 1b. The surface color of the marine clay was dark brown, while the interior color was black, with fine mud, and an odor. The marine clay soil samples were collected and sealed in plastic buckets. The original state of the marine clay and sampling area of the silt soil sample are shown in Figure 1c,d.

Figure 1. Sampling details. (**a**) Geographical location (Huangdao District, Qingdao City) of sampling points. (**b**) Scene of sampling. (**c**) Original state of marine clay. (**d**) Sampling and preservation of marine clay.

According to the Chinese standard for geotechnical testing (GBT 50123-2019 [18]) and soil testing (NY/T 1121.16-2006 [19]), a series of laboratory geotechnical tests were carried out to determine the natural density, water content, optimum moisture content, pH value, total water-soluble salt content, and other physical and mechanical properties of the marine clay. The maximum dry density of the soil sample was 1.64 g/cm^3, and the optimum water content was 19.8%, which is shown in Figure 2. The physical and mechanical indexes of tidal silt are shown in Table 1.

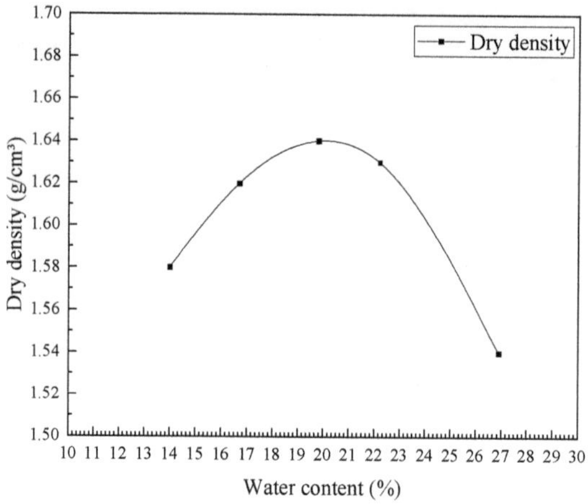

Figure 2. Compaction curve.

Table 1. Basic geotechnical parameters of beach silt soil.

Index	Value
Natural Density, ρ (g/cm^3)	1.68
Moisture Content, ω (%)	120.1
Specific gravity, G_s	2.73
Void ratio, e	1.65
Liquid limit, w_l (%)	83
Plastic limit, w_p (%)	36
pH	8.25
Soil soluble salts, (g/kg)	29.6

2.2. Experimental Details

Chemical treatment of marine clay is achieved by adding a stabilizer to the marine clay, which has the characteristics of high strength, low compressibility, and low permeability [20–23]. Firstly, this section describes the soil solidification treatment measures and the selection of suitable main solidifying agents and additives based on the countermeasure analysis. Secondly, the orthogonal design principle was used to divide different proportion groups of stabilizers. Curing age and dosage of stabilizer were taken as influencing factors for analyzing variations in strength. Marine clay was treated with different proportions of composite stabilizer, the test block was prepared, and the maintenance was carried out. Subsequently, unconfined compressive strength tests were carried out to evaluate the strength of solidified soil blocks under different influencing factors. Finally, the best mix ratio of marine clay stabilizer was obtained, which satisfied the construction requirements of strong stability, high strength, strong bearing capacity, and good compressibility.

2.2.1. Main Stabilizer and Admixture

The main stabilizer used in this study was alumina cement(Zhengzhou Kanghui Refractories Co., Zhengzhou, China). Admixtures included bassanite(Jinan Shengteng Chemical Co., Jinan, China), kingsilica(Shenzhen Haiyang Powder Technology Co., Shenzhen, China), quicklime(Tianjin Zhiyuan Chemical Reagent Co., Tianjin, China), and potassium hydroxide(Jinan Xiaotest Chemical Co., Jinan, China). The components of the stabilizer and their main chemical constituents are shown in Table 2. The composition diagram of the stabilizer is shown in Figure 3.

Table 2. Stabilizer components and their main chemical composition.

Stabilizer Component	Main Chemical Composition
Alumina cement	$3CaO{\cdot}Al_2O_3$
Bassanite	$CaSO_4{\cdot}0.5H_2O$
Kingsilica	SiO_2
Quick lime	CaO
Potassium hydroxide	KOH

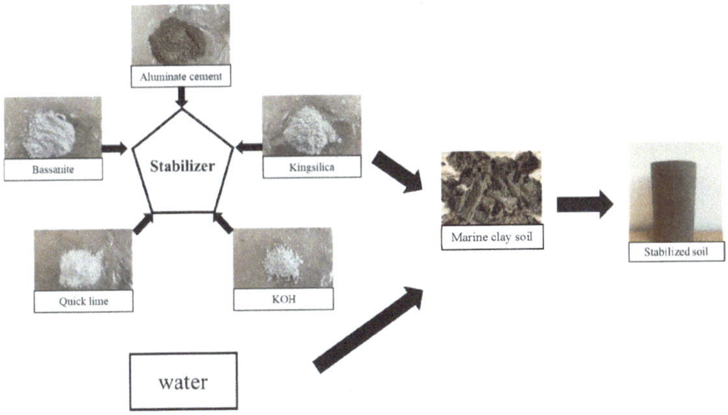

Figure 3. Schematic diagram of the solidified soil production process.

2.2.2. Unconfined Compressive Strength Tests

Water and marine clay were mixed in a ratio of 0.28 by weight. After the mixture reached equilibrium, the main stabilizer and external additives were added to it and uniformly allowed to mix. Firstly, 50 mm diameter and 100 mm high concrete energized volume molds were prepared, and the cured soil was divided into 5–8 layers using the layered manual compaction and molding method, loaded into the molds layer by layer, and manually pounded using a percussion hammer. The prepared specimens were wrapped with cling film and then put into a sealed bag for maintenance to reduce the evaporation of water in the specimens and labeled as a mark. After sample preparation, the mold was removed after 24 h maintenance under natural conditions. The demolded test pieces were wrapped in plastic and placed in sealed bags to prevent any loss of moisture through evaporation. Considering that the basic construction in the intertidal zone needs to be combined with the time of ocean tidal fluctuation, the maintenance age of chemically treated marine clays was increased by 6 h in addition to the conventional 3 or 7 days. The test block was maintained under 95% humidity and 20 ± 5 °C until it reached the maintenance age. The sample was then tested for unconfined compressive strength. The process placed the specimen block on the lower platen and operated the motor to move the key up and down so that the upper pressurized plate just touched the specimen. The data were zeroed manually in the computer terminal, the loading rate of 1 mm/min was entered to start pressurization, and the computer was used to collect the stress–strain relationship data. Similarly, samples under different curing ages and stabilizer dosage factors were tested for compressive strength.

In order to minimize expenditure and also time for conducting the number of tests, an orthogonal test was used for designing the experimental scheme. The test plan based on the orthogonal design approach is summarized in Table 3.

Table 3. Testing plan.

Number	Aluminate Cement Content (%)	Bassanite Content (%)	Kingsilica Content (%)	KOH Content (%)	Quick Lime Content (%)	Solidified Agent Content (%)
0	100	0	0	0	0	
1	95.5	2	1	0.5	1	
2	93.5	2	1.5	1	2	
3	91.5	2	2	1.5	3	
4	91	4	1	1	3	10%
5	92	4	1.5	1.5	1	20%, 30%
6	91.5	4	2	0.5	2	
7	89.5	6	1	1.5	2	
8	89	6	1.5	0.5	3	
9	90	6	2	1	1	

Note: The figures in the table are the mass ratio of each material quality to the solidified agent quality.

The data processing of the unconfined compressive strength tests was performed according to the following formula. The axial strain of the specimens follows the formula:

$$\varepsilon = \frac{\Delta H}{H_0} \times 100 \quad (1)$$

In the formula: ε indicates the strain (%) produced by compression of the test block; ΔH represents the compression (cm); and H_0 indicates the height (cm) of the test block before compression. The average cross-sectional area of the sample follows the formula:

$$A_a = \frac{A_0}{1 - 0.01\varepsilon} \quad (2)$$

In the formula: A_a indicates the corrected area of the sample; and A_0 is the cross-sectional area before compression of the test block (cm^2). The axial stresses on the specimens follow the formula:

$$\sigma = \frac{P}{A_a} = \frac{P}{A_0} \times \left(1 - \frac{\varepsilon}{100}\right) \times 10 \quad (3)$$

In Formula (3): P represents the load produced by the compression of the test block (N).

2.3. Development of Artificial Neural Network Model (ANN)

The artificial neural network model used in this experiment was multi-layer. They were the input layer, hidden layer, and output layer. The input layer consisted of seven nodes, which were the seven factors influencing the unconfined compressive strength of chemically treated marine clay, including the alumina cement content, bassanite content, kingsilica content, potassium hydroxide content, quick lime content, stabilizer content, and curing age. Hidden layers contain functions that transmit and process signals between neurons and their corresponding connections. The output layer had only one node, which was the unconfined compressive strength. The topological structure of the ANN model is shown in Figure 4. In the unconfined compressive strength test of silt-consolidated soil, a total of 90 sets of valid test data were obtained, of which 82 sets of data were used as training for the artificial neural network model, and 8 randomly selected sets were used for the prediction of the neural network model.

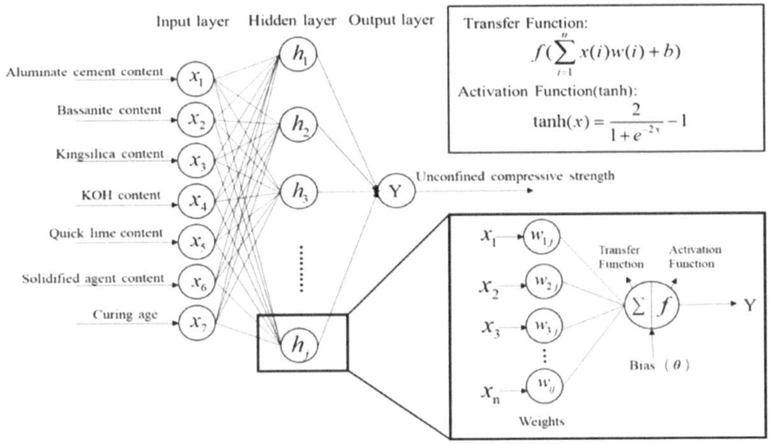

Figure 4. Schematic structure of artificial neural network model.

The operation flow of the ANN model is shown in Figure 5. Data were first normalized using the equation below:

$$X^* = \frac{X - X_{min}}{X_{max} - X_{min}} \quad (4)$$

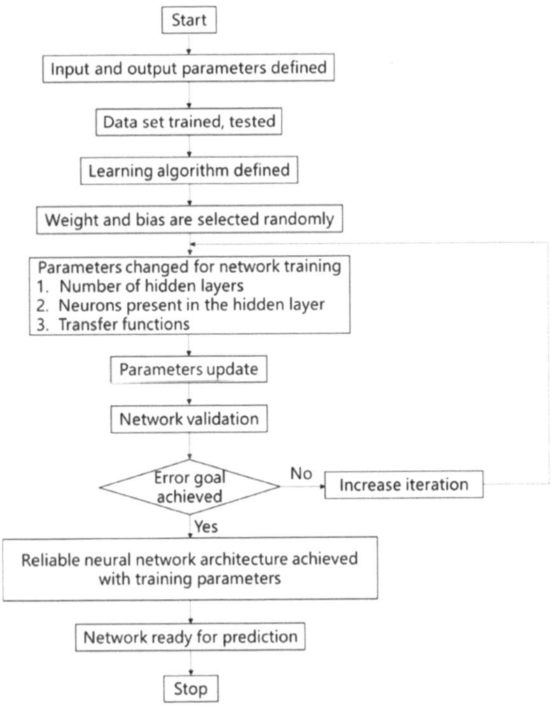

Figure 5. Operation flow chart of the artificial neural network model [24].

In Formula (4): X* is the value after normalization; X is the true value of the raw data; and Xmax and Xmin are the maximum and minimum values of the data, respectively. In this study, all unconfined compressive strength test data collected were divided into 70% training data, 15% validation data, and 15% detection data. In this study, the development of an artificial neural network model mainly used two network forms: multi-layer perceptron (MLP) and radial basis function (RBF). MLP is the most common form of network. Its network is compact and fast to execute, providing better results than other types of networks once trained. However, if these networks require repeated training, the model will run slowly. RBL networks are trained very quickly and are more sensitive to a small number of input variables.

However, RBL often shows relatively poor performance, which is not as effective as MLP networks [23]. In this study, the number of nodes in the hidden layer of the MPL network model selected in the artificial neural network model ranged from 4 to 12.

All data were fitted by the ANN model, and 10 models with a high fitting degree were finally selected. A group of models with MLP 7-8-1 having the highest fitting between the predicted unconfined compressive strength and the actual measured value was selected for subsequent analysis.

Parameter Setting for Optimization Analysis and Sensitivity Analysis

An artificial neural network can fit the proportion of influencing factors when the unconfined compressive strength reaches the maximum, according to the complex non-linear relationship already known. Based on this analysis, the highest strength of chemically treated marine clay can be obtained, and when the strength of chemically treated marine clay reaches the highest value, the amount of hardener added, the proportion of its components, and the age of the curing period can be determined. Through the analysis results, the existing chemical treatment schemes for marine clay can be optimized and improved. In the optimization process, we used the simplex algorithm, the most common and simplest optimization algorithm, to find the maximum value to consider extreme cases. Simplex search is a gradient-free optimization algorithm for minimizing or maximizing any function in a limited number of iterations [25]. A total of 450 iterations were performed for this test optimization analysis.

An artificial neural network can also carry out sensitivity analysis on different influencing factors of the input, i.e., quantify the proportion of influencing factors among output results by analyzing the degree of influencing different factors on the output, the unconfined compressive strength. Sensitivity analysis provides a general concept of output sensitivity by changing one input variable while leaving several other input variables unchanged. In this study, we individually changed all influencing factors to study the sensitivity of the unconfined compressive strength with other inputs unchanged.

3. Results and Discussion

3.1. *Effect of Chemical Treatment on Unconfined Compressive Strength (UCS) of Marine Clay*

In this experiment, two test variables, conservation age and curing agent dosage, were set up to explore the effects of both on the effectiveness of curing agent in curing beach silt. Three gradients of 10%, 20%, and 30% of curing agent dosage were designed for the preparation of test blocks for each proportion group; and three time periods of 6 h, 3 days, and 7 days were set for the routine maintenance of the test blocks. At the same curing agent dosage, 9 samples were tested at each age, totaling 81 samples.

It can be observed from Figure 6a,b that when the content of additives is low (i.e., 10% and 20%), the influence of each additive on the average unconfined compressive strength is minimal at different levels. However, the situation changes significantly when the stabilizer content reaches 30% (refer to Figure 6c). The unconfined compressive strength not only varies significantly with the level of each additive but also increases significantly at different curing ages. Taking 30% stabilizer as an example, the primary and secondary order of influence on the unconfined compressive strength at the curing age of 6 h is quick lime > potassium hydroxide > bassanite > kingsilica. When the curing period is 3 days, the primary and secondary order of influence on the unconfined compressive strength is potassium hydroxide > bassanite > quick lime > kingsilica. When the curing period is 7 days, the primary and secondary order of influence on the unconfined compressive strength is potassium hydroxide > kingsilica > quick lime > bassanite.

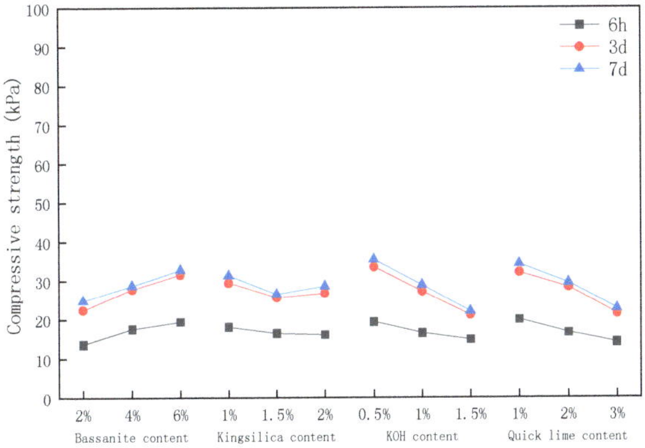

(a)

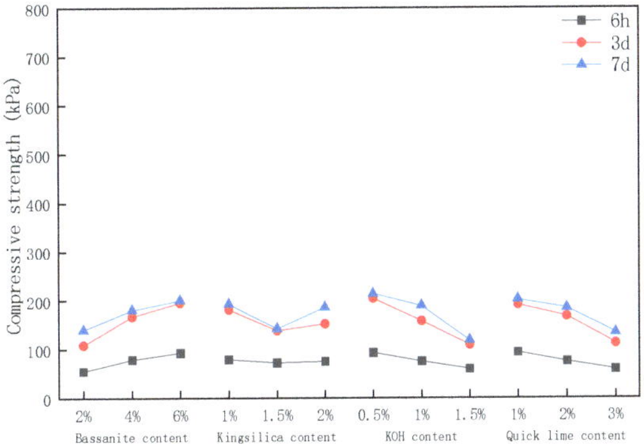

(b)

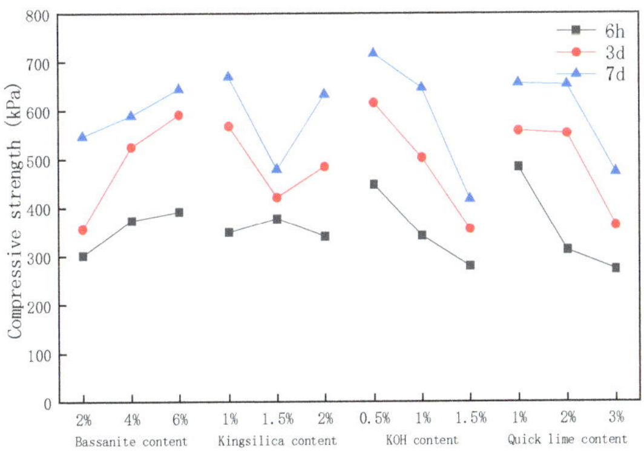

(c)

Figure 6. Average unconfined compressive strength of different external admixtures at different levels. (**a**) 10% Stabilizer Content. (**b**) 20% Stabilizer Content. (**c**) 30% Stabilizer content.

At the early stage of treated clay, the main influencing factors of strength are ion exchange of calcium oxide, pozzolanic reaction, and carbonation. It shows that the addition of lime contributes most to the early strength of treated clay. However, with an increase in curing age, there is not enough active aluminum oxide and silicon oxide in the soil to react with lime. In the later stage of solidified soil, its strength is more affected by potassium hydroxide, followed by gypsum and active silicon powder.

Stress–Strain Curve for Unconfined Compressive Strength Tests

The stress–strain curves of chemically treated marine clay samples were obtained through uniaxial compression tests. In order to illustrate stress–strain behavior, testing plan no. 7 was considered due to its superior performance among all other plans. The stress–strain relationship curves of marine clay treated under test plan no. 7 under different stabilizer dosages and curing ages were plotted to analyze the strength and deformation characteristics.

Loading failure of treated clay samples requires three deformation stages: elastic stage, plastic yield stage, and softening stage. The elastic stage is the initial straight-line part of the stress–strain curve, at which the linear relationship between stress and strain is the main stage of stress growth. The compression deformation of the particles in the treated clay is within the elastic range and there is no breakage. The plastic yield stage is the non-linear growth stage of the stress-strain curve. When the strength exceeds the elastic limit stress, the treated clay is gradually damaged, the slope of the stress-strain curve of the solidified soft soil is gradually reduced, the particles inside the sample are broken, and the voids between the soil particles are constantly compacted. At this stage, the effect of soil compaction on the strength increase of the sample is greater than that of particle destruction, and the plastic deformation of solidified soft soil cannot be recovered. Finally, the non-linear downward section of the stress–strain curve is called the softening stage. Cracks in treated clay specimens continue to develop and gradually penetrate, resulting in the failure of the specimens.

As observed from Figure 7, at 10% content, the treated clays at three ages have very low strength and basically show a soft clay state. With the increase in strain, the stress is slowly and gradually elevated without obvious damage peaks, and the elastic phase of the stress–strain curve is short and rapidly enters the plastic phase and the final softening phase. The stress is always below 50 kPa and can hardly withstand pressure. The results show that, at 10% content, the curing effect is not obvious and is unlikely to meet the engineering requirements. It can be observed from Figure 7b that the stress–strain curves for 20% content at the curing age of 3 d and 7 d have gone through three stages. The yield strength reaches about 200 kPa to 250 kPa. However, the elastic phase of the treated clays is still not evident in the specimens at the curing age of 6 h. The results show that the early strength of the stabilized soil is still low when the content of the stabilizer is 20%, which is unlikely to meet the engineering requirements. As observed from Figure 7c, when the stabilizer content is 30%, the treated clay samples can reach a certain strength under the curing age of 6 h, 3 d, and 7 d. The stress–strain curve also completely undergoes the elastic stage, plastic stage, and softening stage. With an increase in curing age, the yield strength increases continuously, up to 830 kPa. The results show that at this optimum mixing ratio, the stabilizer content should not be less than 30% to achieve the full curing effect.

Mixed materials often show non-linear deformation, and the deformation coefficient E50 is often used to indicate the ability of mixed materials to resist elastic-plastic deformation. The definition of the formula is shown in Equation (5):

$$E_{50} = \frac{\sigma_{1/2}}{1/2\varepsilon_f} \tag{5}$$

In the formula: $\sigma_{1/2}$ indicates the strain to destruction in the stress–strain curve of the unconfined compressive strength (%); and ε_f indicates the stress corresponding to half of the breaking strain in the stress–strain curve for the unconfined compressive strength.

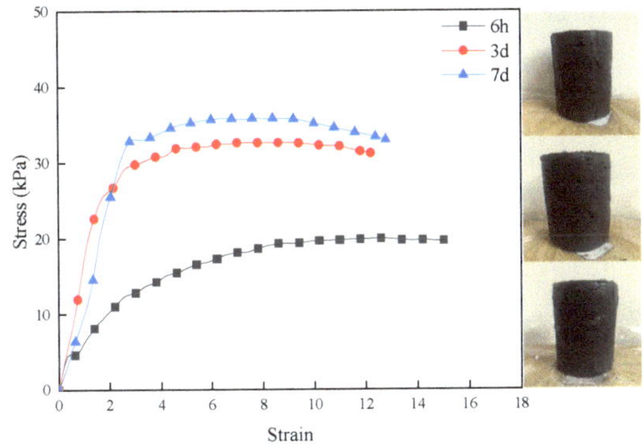

(a)

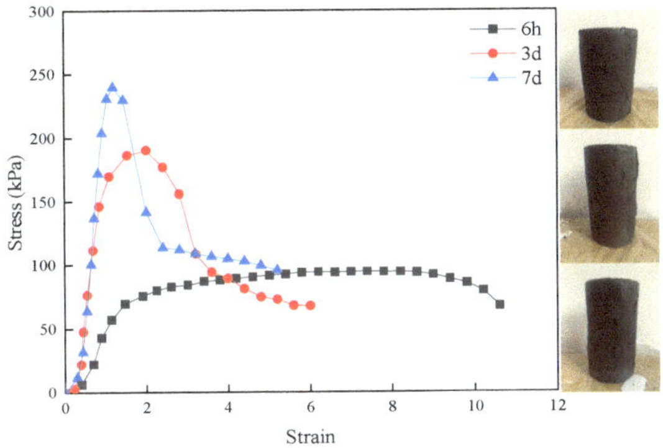

(b)

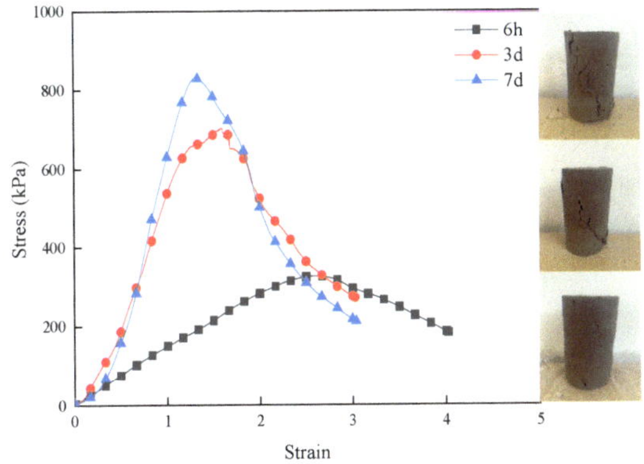

(c)

Figure 7. Stress–strain curves of cured soil at different ages with various stabilizer contents. (**a**) 10% Stabilizer content. (**b**) 20% Stabilizer content. (**c**) 30% Stabilizer content.

The deformation coefficients of each group of cured soils at different maintenance ages and curing agent dosage conditions were calculated for comparison, as shown in Figure 8.

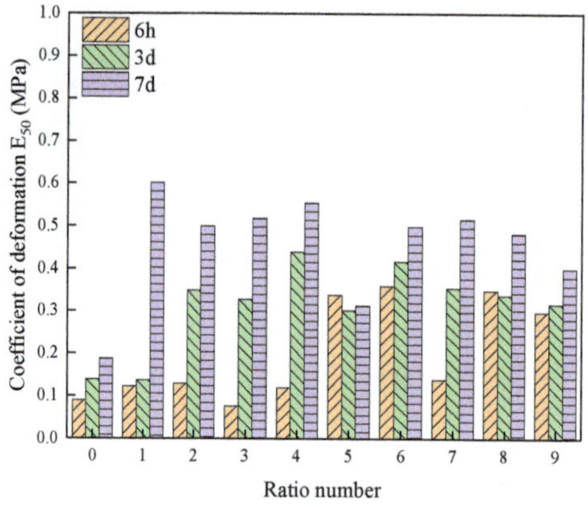

(**a**)

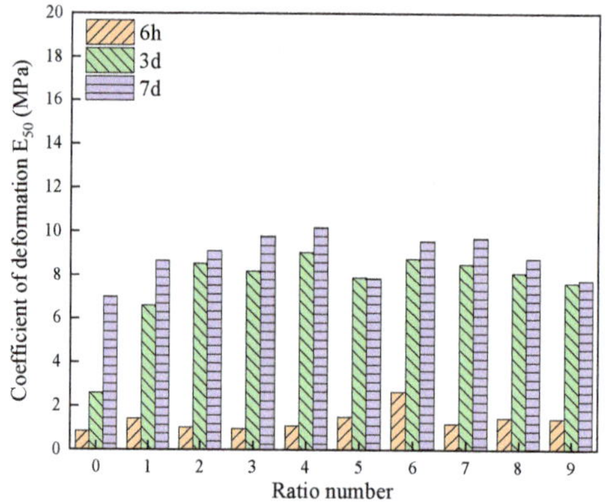

(**b**)

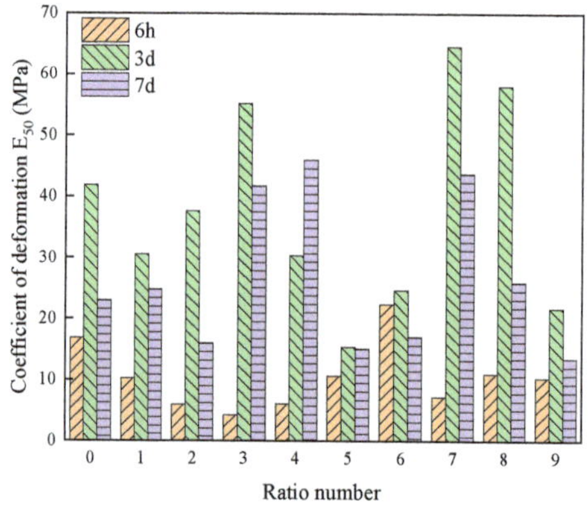

(**c**)

Figure 8. Deformation coefficients of cured soils under different maintenance age conditions. (**a**) 10% Stabilizer content. (**b**) 20% Stabilizer content. (**c**) 30% Stabilizer content.

As can be seen from Figure 8, when the dosage of curing agent is less than or equal to 20%, the deformation coefficient increases with the increase in maintenance age, and the larger the dosage, the more significant the deformation coefficient enhancement effect with the increase in maintenance age, i.e., the stronger the deformation resistance. When the dosage of curing agent is more than 20%, the deformation coefficient increases and then decreases with the increase in maintenance age, and then there is a certain enhancement in the late stage of maintenance, but lower than the deformation coefficient in the early stage, which indicates that the increase in the dosage of curing agent weakens the deformation resistance although the strength will be significantly increased.

3.2. *Prediction of Unconfined Compressive Strength Based on the ANN Model*

3.2.1. Performance Analysis of Selected Neural Network Model

All original data of the ANN model were fitted, and a group of models with MLP 7-8-1 having the highest fitting between the predicted and measured values was selected for subsequent analysis. The model had 7 input layer nodes, 8 hidden layer nodes, and only 1 output layer node. The performance analysis, optimization analysis, and sensitivity analysis of the above models were used to simulate the unconfined compressive strength of the chemically treated marine clay.

Detailed descriptions of the selected ANN models are summarized in Table 4. The results of artificial neural network training are shown in Figure 8.

Table 4. Detailed description of selected ANN model for UCS tests.

Index	Value
Net. name	MLP 7-8-1
Training perf.	0.984256
Test perf.	0.993253
Validation perf.	0.996300
Training error	0.000947
Test error	0.001037
Validation error	0.012440
Algorithm	BFGS 54
Error function	SOS
Hidden activation	Exponential
Output activation	Tanh

The numbers in rows 2–4 and 5–7 in Table 4 are the fitted regression values and errors of the model in the training, testing, and validation phases, respectively, and they represent the correlation between the output value and the target expectation. When the fitted regression value is closer to 1 and the error tends to 0, it means that the fitting effect is better. As can be seen from the data in Table 4, the model MLP 7-8-1 has a good fit.

It can be observed from Figure 9 that the measured and predicted values are similar to each other, with an R^2 value of 0.992. It should be noted that only the normalized data of each factor is given as input. The ANN model can predict the value of unconfined compressive strength through the existing non-linear relationship. This method can help to replace unnecessary laboratory testing, thereby saving time and expenditure for engineers. It can also achieve the selection and optimization of stabilizers.

In order to verify the authenticity of the predicted results by fitting an artificial neural network, the data which were not involved in model training were randomly selected from the original data. Predicted values from the ANN model were compared with these data. The comparison between the experimental measurements and the network predictions is shown in Figure 10. It can be observed that the predicted value of the model and the measured value of the sample data are very close to each other. An R^2 value of 0.99 is observed.

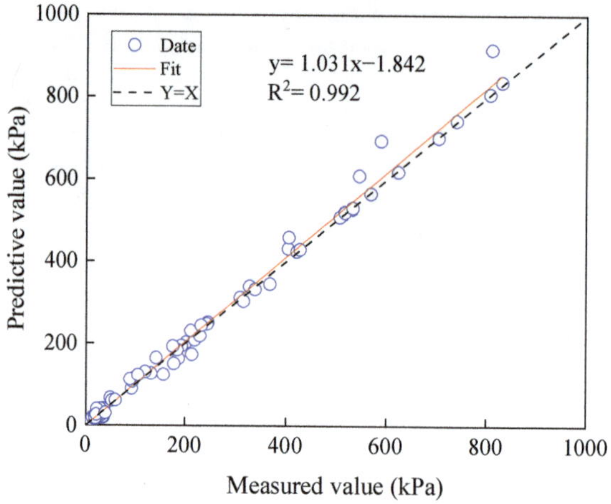

Figure 9. Comparison between measured values and network-predicted values of sample data.

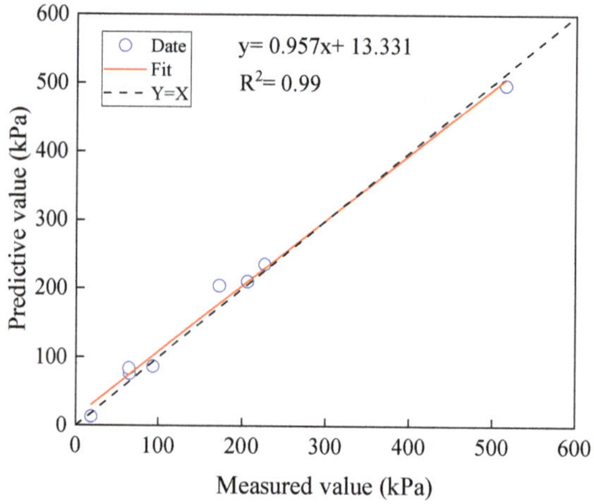

Figure 10. Comparison between measured values (random data not selected for training) and network-predicted values of sample data.

3.2.2. Influence of Different Parameters on Unconfined Compressive Strength

The complex non-linear dependence of one parameter with respect to several parameters can be well displayed in three-dimensional images. As can be seen from Figure 11a,c,e, when the content of potassium hydroxide is about 0.2% and 1.6%, respectively, and the content of gypsum hemihydrate is around 4–6% (Figure 11a); the content of potassium hydroxide is about 0.2% and 1.6%, respectively, and the content of activated silica micronutrient powder is about 2.2% and 1.6%, respectively (Figure 11c); and the content of potassium hydroxide is about 0.6% and 1.6%, respectively, and the lime content is about 0.5% and 3%, respectively (Figure 11e), these parameters together with each other promote the strength of cured soil. With the appropriate amount of potassium hydroxide into the cured soil to improve the pH value, the alkaline environment helps the volcanic ash reaction, and the increase in the content of potassium ions promotes the adsorption and exchange of calcium ions, so that the soil particles flocculate, promoting the dispersion of small soil particles to become larger soil agglomerates, which strengthens the cured soil. But when the content of potassium hydroxide is too much, to a certain extent "robbing" the calcium ions, it reduces the generation of a hydration gel and calcium hydroxide is rapidly consumed without timely replenishment; thus, the strength of the cured soil is weakened to a certain degree.

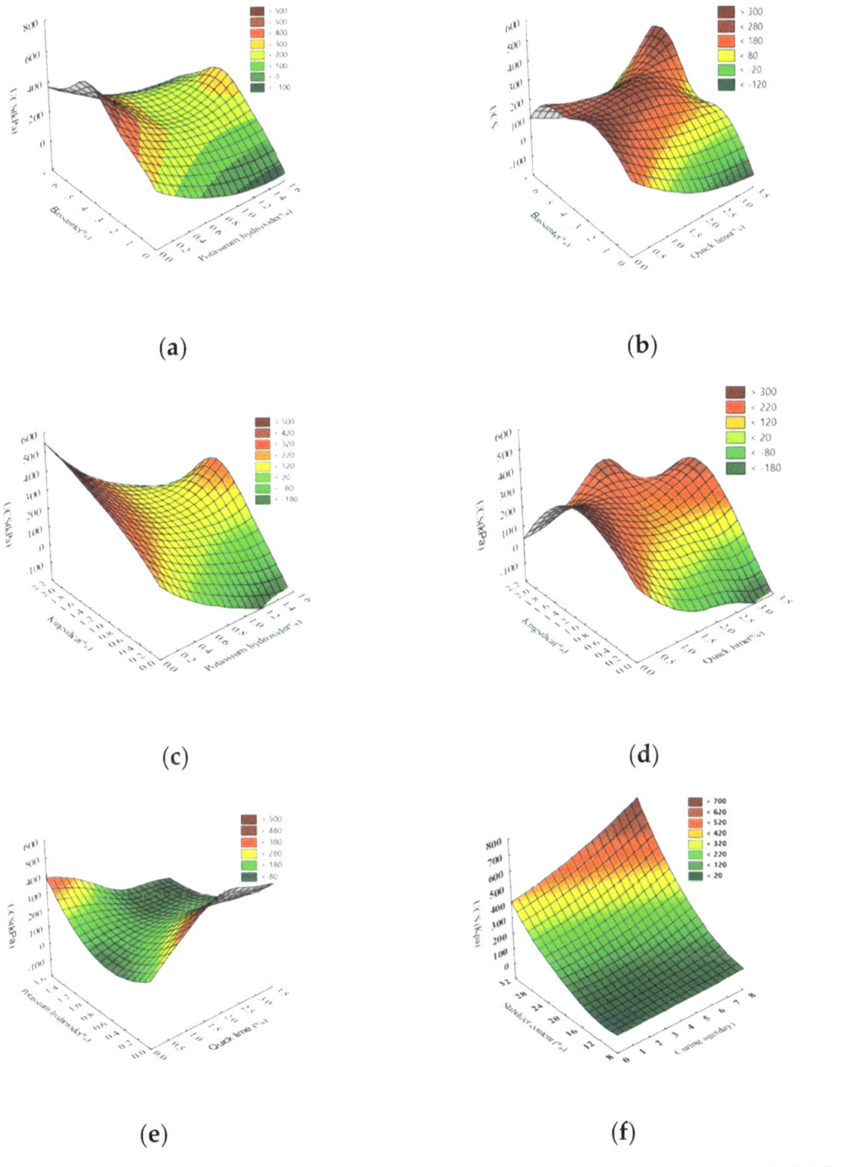

Figure 11. Influence of different influencing factors on unconfined compressive strength. (**a**) Influence of bassanite and potassium hydroxide on unconfined compressive strength. (**b**) Influence of bassanite and quick lime on unconfined compressive strength. (**c**) Influence of kingsilica and potassium hydroxide on unconfined compressive strength. (**d**) Influence of kingsilica and quick lime on unconfined compressive strength. (**e**) Influence of potassium hydroxide and quick lime on unconfined compressive strength. (**f**) Influence of stabilizer content and curing age on unconfined compressive strength.

As can be seen from Figure 11b,d, the interaction of quicklime with hemihydrate gypsum and activated silica micropowder has a significant effect on the unconfined compressive strength of silt-consolidated soil, which is due to the fact that after the calcium oxide, a constituent of quicklime, is dissolved in water, the reaction generates a large amount of $Ca(OH)_2$ that promotes the fracture of the Al–O bond, the Si–O bond, etc., in the hemihydrate gypsum and activated silica micropowder and further reacts to generate the gel substance, i.e., calcium silicate and hydrated calcium aluminate. When the lime content is low, the strength of the cured soil gradually increases with the increase in the content of gypsum hemihydrate and activated silica fume, and then the lime is used as the raw material for the reaction to continue to help the chemical reaction within the cured soil. When the lime content is higher, the strength of cured soil appears to be reduced. Analyzing the reason, too much calcium hydroxide produced by the hydration reaction tends to be saturated, and after saturation, it precipitates in the form of crystals and constitutes a loose structure, which weakens the enhancement of the strength of cured soil. When the contents of activated silica micropowder and semi-anhydrous gypsum exceed a certain range, the chemical reaction "place" within the cured soil is occupied due to the excessive output of hydrated gel material, weakening the degree of reaction, and at the same time, the generation of excessive calcium alumina destroys the internal structure, thus leading to the phenomenon of decreasing the strength of the cured soil.

From Figure 11f, it can be seen that the interaction of stabilizer content and curing age has a significant effect on the unconfined compressive strength of silt-cured soil, but the effect of the stabilizer content on the unconfined compressive strength is more obvious.

3.3. *Optimization of Chemical Treatment Technology for Improving the Strength of Marine Clay*

3.3.1. Sensitivity Analysis

ANN was used for sensitivity analysis to analyze the relative significance of different factors on the unconfined compressive strength [26–29]. The results of the sensitivity analysis and optimization analysis of ANN are summarized in Table 5.

Table 5. Sensitivity analysis and optimization analysis of ANN.

Variable	Sensitivity Analysis	Optimization Analysis
UCS	-	830 kPa
Aluminate cement content	4.032583	90.06%
Bassanite content	5.186851	3.49%
Kingsilica content	2.998604	2.00%
KOH content	5.807388	0.3%
Quick lime content	30.29768	1.08%
Curing age	2.136436	7 days
Stabilizer content	15.61476	30%

From the sensitivity analysis, it can be observed that the lime content has the greatest influence on the unconfined compressive strength, followed by the stabilizer content. The curing age has the least effect on the strength of treated clay. Due to the large amount of water in sludge, the chemical reaction between lime and pore water in sludge results in hydration as well as physical reactions (ion exchange reaction and calcium hydroxide crystallization reaction). Lime continuously absorbs water to form $Ca[OH]_2{\cdot}nH_2O$, which forms crystals that combine and bond with soil particles. Calcium hydroxide reacts with carbon dioxide in the air to produce calcium carbonate. Lime enhances the water stability of the stabilized soil and significantly increases its strength. The change in lime content has a great influence on UCS, which is consistent with the results of the range analysis of the unconfined compressive strength tests. The sensitivity analysis of the ANN model very well showed the influence of each component of the chemical stabilizer. According to the analysis, the proportioning scheme of the stabilizer can be adjusted accordingly to improve the curing effect.

3.3.2. Optimization Analysis

ANN-based models can be optimized and analyzed to find the parameters that can achieve the specific expected results, such as obtaining the maximum unconfined compressive strength in extreme cases and analyzing the values of various influencing factors in this case. The results of the optimal scheme are close to the results of the unconfined compressive strength tests. The results are summarized in Table 5. The maximum convergence diagram of the output results is shown in Figure 12.

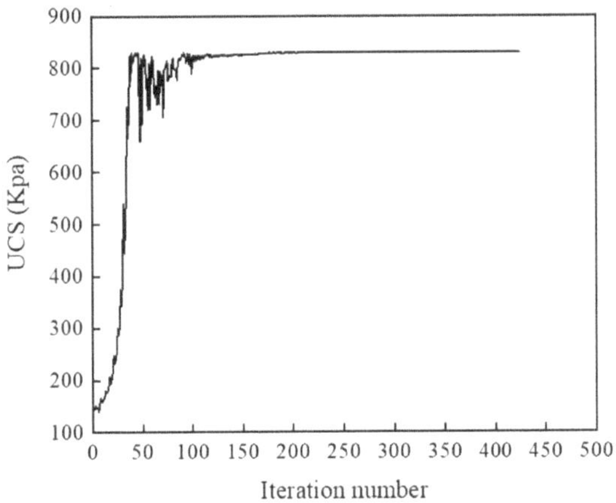

Figure 12. Convergence of maximum output results of unconfined compressive strength tests.

When the number of iterations of the ANN model reaches about 150 times, the unconfined compressive strength converges to the maximum value, which is 830 kPa. When the unconfined compressive strength reaches the maximum value, aluminate cement plays the most important role, with the specific gravity reaching 90.06%. The influence of stabilizer dosage is second, with a specific gravity of up to 30%. It shows that the maximum strength comes from aluminate cement, and the amount of stabilizer is also an important factor to improve the strength of cured soil. On the premise of economy, the amount of stabilizer at 30% tends to achieve the best curing effect. According to this optimization analysis, the dosage and curing age of each component of the stabilizer can be conveniently optimized and reasonably distributed according to the different effects of each factor.

The trend of influence of various factors on the unconfined compressive strength of marine clay can be optimized and analyzed by using the selected ANN model. Figure 13 shows the response curve of the unconfined compressive strength to various influencing factors. The model obtains the maximum unconfined compressive strength by optimizing and analyzing the complex non-linear relationship. At this time, the values of each influencing factor are determined when the maximum UCS is reached and are shown in the diagram.

As can be seen from Figure 13, when only the effect of bassanite is considered, the unconfined compressive strength will first increase and then decrease with an increase in bassanite content. When the content of bassanite reaches about 2%, there is a critical value of UCS, which reaches a maximum of about 828 kPa. It shows that when other factors are set as the current setting values, a small amount of bassanite can improve UCS, while the excessive addition of bassanite will inhibit the curing effect of the stabilizer on beach mud. This is because gypsum reacts with calcium aluminate hydrate in cement hydration products to produce high sulfur-type calcium sulphate aluminate hydrate, i.e., alunite. Aluminum alunite is characterized by volume expansion and can fill the pores of cured soil, thus improving its strength. But excessive addition of gypsum decreases the strength of the stabilized soil because alunite has no cementing effect and only a filling effect [30].

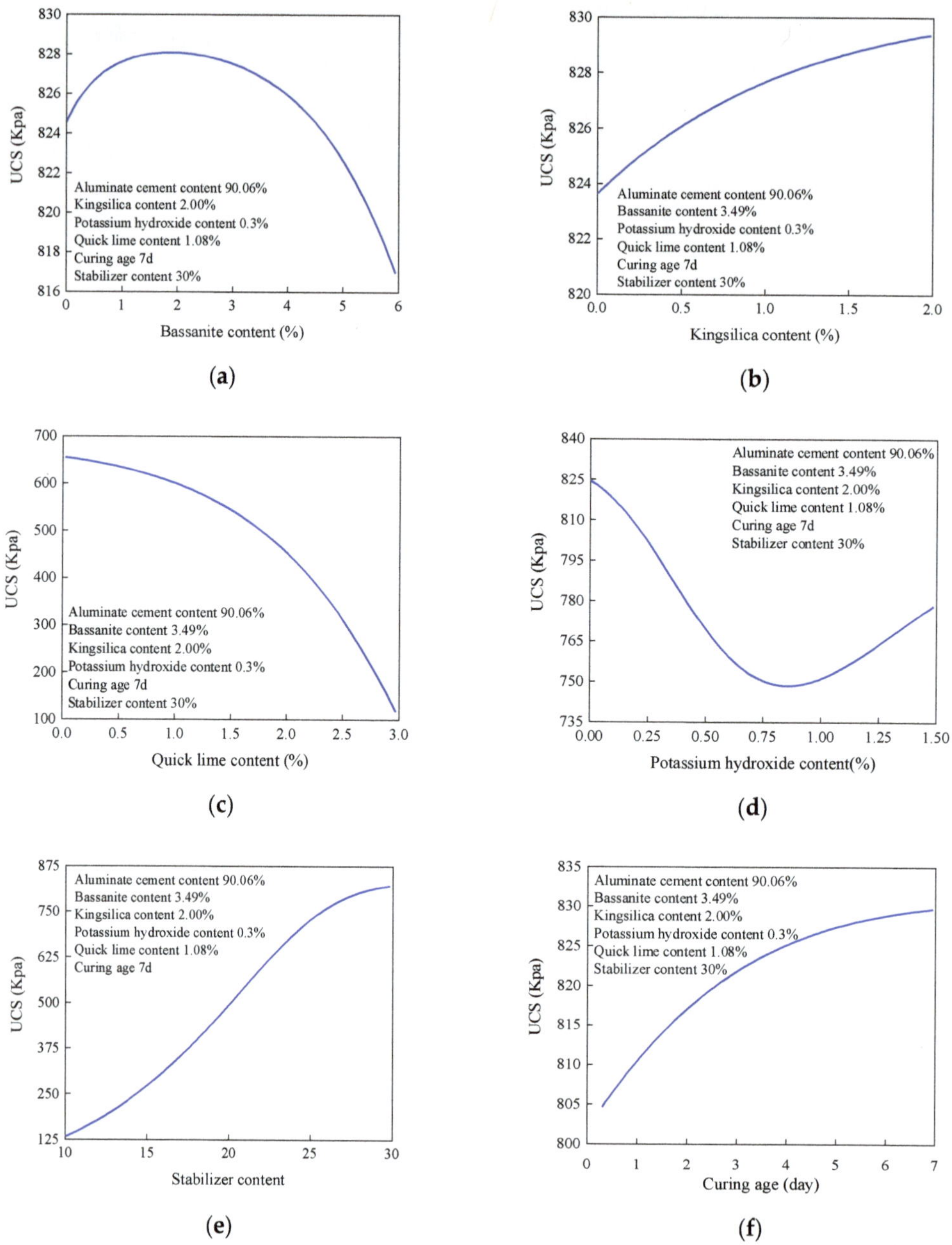

Figure 13. Influence of various influencing factors on unconfined compressive strength. (**a**) Influence of bassanite content on unconfined compressive strength. (**b**) Influence of kingsilica content on unconfined compressive strength. (**c**) Influence of quick lime content on unconfined compressive strength. (**d**) Influence of potassium hydroxide content on unconfined compressive strength. (**e**) Influence of stabilizer content on unconfined compressive strength. (**f**) Influence of curing age on unconfined compressive strength.

It can be observed from Figure 13b that the UCS of treated marine clay will increase with increasing content of kingsilica. It shows that the unconfined compressive strength of the stabilized soil can be improved by adding kingsilica when other influencing factors are determined. Kingsilica can react with CH in cement hydration products to form CSH gel. This reduces the content of CH to a certain extent and improves the degree of cement paste. CSH gel can also continuously fill the pores between cement particles, reduce the water–cement ratio, and thus improve the strength of cured soil.

As observed from Figure 13c, in the current analysis of the optimization results, UCS decreases with the increase in quick lime content when other influencing factors are determined by optimization analysis. It shows that the curing effect of the stabilizer will be counteracted by adding lime in this case. The main functions of lime addition in marine clay are ion exchange, pozzolanic reaction, and carbonation. However, when the amount of lime is large, there is not enough active aluminum oxide and silica in the clay to react with lime chemically, which results in relatively fewer stable minerals in the treated marine clay, thus reducing the strength of the chemically treated marine clay [30].

It can be seen from Figure 13d,e,f that the addition of potassium hydroxide will make the UCS decrease first and then increase, and there is a minimum value for UCS. When the content of potassium hydroxide increases to about 0.8%, the UCS decreases to the lowest strength, 750 kPa. It will then increase with the addition of potassium hydroxide. The dosage of stabilizer and curing age all promote UCS, and there is a positive correlation between UCS, stabilizer dosage, and curing age. It is not difficult to find that the more stabilizer is added, the harder the marine clay will be, and the longer the treated clay is cured, the stronger it will be.

4. Conclusions

This study conducts comprehensive experiments and ANN-based modeling for analyzing the effects of stabilizer type, curing age, and content on the compressive strength of marine clay of the Qingdao region of China. A series of unconfined compressive strength tests, that were designed based on an orthogonal approach, was conducted first, followed by the development of ANN-based modeling. Further, optimization and sensitivity analyses were conducted to explore the influence of different parameters on the compressive strength of treated marine clay. The following conclusions can be drawn based on the study:

1. Through the unconfined compressive strength test and range analysis, it can be found that there is a complex non-linear relationship between the influencing factors and the unconfined compressive strength under different stabilizer dosages and curing ages. The increment of stabilizer content from 10% to 30% shows a substantial increment in the unconfined compressive strength of the admixed soil after a 28 d curing period when the aluminate cement content is 89.5%, in which the primary and secondary order of influence on the unconfined compressive strength is potassium hydroxide > kingsilica > quick lime > bassanite.
2. By analyzing the stress–strain behavior of soil samples, it is found that the strength increases continuously with an increase in stabilizer content and the extension of curing age, showing a trend of transition from plasticity to brittleness.
3. The forecasting model is established using an artificial neural network (ANN), which verifies the validity and high performance of the model. It is proved that ANN-based modeling can be used as a tool to analyze complex non-linear relationship problems under multiple factors.
4. Sensitivity analysis results based on the ANN model show that the unconfined compressive strength (UCS) is most sensitive to the change in quick lime content. The influence of quick lime should be given priority in the formulation design of stabilizers. The results are in agreement with the range analysis. It further proves that the artificial neural network has high performance and good forecasting ability.
5. The results of optimization analysis based on the ANN model show that the maximum UCS can reach about 830 kPa. When the maximum UCS is reached, the model fits the values of various factors, similar to the results of the unconfined compressive strength tests, and the curing scheme can be optimized according to the results of the model.

6. The influence of various influencing factors on the unconfined compressive strength of solidified soil can be optimized using the ANN model, and the influence trend of a single variable on UCS can be obtained. The results show that there is a critical value for the influence of the amount of bassanite and potassium hydroxide, which can be used for further optimize the design of stabilizer components.

Author Contributions: Conceptualization, writing—original draft preparation and data curation, Q.B.; conceptualization, methodology and funding acquisition, J.L.; investigation, Q.B. and K.S.; writing—review and editing, A.G., W.S. and X.J.; visualization, J.L.; supervision and project administration, J.L. and A.G. All authors have read and agreed to the published version of the manuscript.

Funding: This study was funded by Shandong Provincial Natural Science Foundation (Grant No. ZR2021YQ31), Joint Funds of the National Natural Science Foundation of China (Grant No. U2006225), Taishan Scholars Program (Grant No. tsqn202211176), and National Nature Science Foundation of China (Grant No. 42277135).

Institutional Review Board Statement: Not applicable.

Informed Consent Statement: Not applicable.

Data Availability Statement: Data are contained within the article.

Conflicts of Interest: All authors declare that the research was conducted in the absence of any commercial or financial relationships that could be construed as a potential conflicts of interest. In the collection, analyses, and interpretation of data, the Joint Funds of the National Natural Science Foundation of China provided assistance; in the writing of the manuscript, Shandong Provincial Natural Science Foundation, National Nature Science Foundation of China, and Taishan Scholars Program provided assistance.

References

1. Zhang, P.; Guo, Y.; Liu, Y.; Ding, H. Experimental study on installation of hybrid bucket foundations for offshore wind turbines in silty clay. *Ocean Eng.* **2016**, *114*, 87–100. [CrossRef]
2. Ding, H.; Hu, R.; Zhang, P.; Le, C. Load bearing behaviors of composite bucket foundations for offshore wind turbines on layered soil under combined loading. *Ocean Eng.* **2020**, *198*, 106997. [CrossRef]
3. Zhang, J.; Wang, H. Development of offshore wind power and foundation technology for offshore wind turbines in China. *Ocean Eng.* **2022**, *266*, 113256. [CrossRef]
4. Sai, Z.; Wang, L.; Han, H.; Wu, W.; Sun, Z.; Wei, J.; Zhang, L.; Hu, G.; Wu, H. Mechanical and Self-Healing Performance of Yellow River Alluvial Silt Treated with Composite Flexible Curing Agent. *Coatings* **2022**, *12*, 870. [CrossRef]
5. Ning, B. Permeability and Mechanical Behaviors of Stabilized Silt by Inorganic Soil Stabilizer. Ph.D. Thesis, Zhengzhou University, Zhengzhou, China, 2020.
6. Chew, S.H.; Kamruzzaman, A.H.M.; Lee, F.H. Physicochemical and engineering behavior of cement treated clays. *J. Geotech. Geoenviron. Eng.* **2004**, *130*, 696–706. [CrossRef]
7. Porbaha, A.; Shibuya, S.; Kishida, T. State of the art in deep mixing technology. Part III: Geomaterial characterization. *Ground. Improv.* **2000**, *4*, 91–110. [CrossRef]
8. Lorenzo, G.A.; Bergado, D.T. Fundamental parameters of cement-admixed clay—New approach. *J. Geotech. Geoenviron. Eng.* **2004**, *130*, 1042–1050. [CrossRef]
9. Horpibulsuk, S.K.; Rachan, R.L.; Chinkulkijniwat, A.; Raksachon, Y.; Suddeepong, A. Analysis of strength development in cement-stabilized silty clay from microstructural considerations. *Constr. Build. Mater.* **2010**, *24*, 2011–2021. [CrossRef]
10. Wang, L.F. Analysis of microstructural characteristics and pores of nanometer silicon and cement-stabilized soils. *Bull. Sci. Technol.* **2012**, *28*, 31–35. [CrossRef]
11. Ding, J.W.; Zhang, S.; Hong, Z.S.; Liu, S.Y. Experimental study of solidification of dredged clays with high water content by adding cement and phosphogypsum synchronously. *Rock Soil Mech.* **2010**, *31*, 2817–2822. [CrossRef]
12. Sharma, L.; Sirdesai, N.; Sharma, K.; Singh, T. Experimental study to examine the independent roles of lime and cement on the stabilization of a mountain soil: A comparative study. *Appl. Clay Sci.* **2018**, *152*, 183–195. [CrossRef]
13. Dahal, B.K.; Zheng, J.-J.; Zhang, R.-J.; Song, D.-B. Enhancing the mechanical properties of marine clay using cement solidification. *Mar. Georesources Geotechnol.* **2019**, *37*, 755–764. [CrossRef]
14. Liu, L.; Zhou, A.; Deng, Y.; Cui, Y.; Yu, Z.; Yu, C. Strength performance of cement/slag-based stabilized soft clays. *Constr. Build. Mater.* **2019**, *211*, 909–918. [CrossRef]
15. Gong, X.; Niu, J.; Liang, S.; Feng, D.; Luo, Q. Solidification of Nansha soft clay using cement-based composite curing agents. *Adv. Cem. Res.* **2020**, *32*, 66–77. [CrossRef]

16. Kurugodu, H.; Bordoloi, S.; Hong, Y.; Garg, A.; Garg, A.; Sreedeep, S.; Gandomi, A. Genetic programming for soil-fiber composite assessment. *Adv. Eng. Softw.* **2018**, *122*, 50–61. [CrossRef]
17. Pacheco, V.L.; Bragagnolo, L.; Thomé, A. Artificial neural networks applied for solidified soils data prediction: A bibliometric and systematic review. *Eng. Comput.* **2021**, *38*, 3104–3131. [CrossRef]
18. *GB/T 50123-2019*; Standard for Geotechnical Testing Method. China Planning Press: Beijing, China, 2019. (In Chinese)
19. *NY/T 1121.16-2006*; Soil Testing Part 16: Determination of Total Soil Water-Soluble Salts. Ministry of Agriculture and Rural Affairs: Beijing, China, 2006. (In Chinese)
20. Nalbantoglu, Z.; Tuncer, E.R. Compressibility and hydraulic conductivity of a chemically treated expansive clay. *Can. Geotech. J.* **2001**, *38*, 154–160. [CrossRef]
21. Rout, P.K. Study of Compressibility and Hydraulic Conductivity Characteristics of a Chemically Treated Expansive Clay. *Int. J. Eng. Manag. Humanit. Soc. Sci. Paradig.* **2018**, *30*, 284–295.
22. Locat, J.; Trembaly, H.; Leroueil, S. Mechanical and hydraulic behaviour of a soft inorganic clay treated with lime. *Can. Geotech. J.* **1996**, *33*, 654–669. [CrossRef]
23. Hartono, E.; Wardani SP, R.; Muntohar, A.S. The behavior of the clay shale stabilized by dry and wet cement mixing method. *J. GeoEng.* **2021**, *16*, 81–90.
24. Garg, A.; Wani, I.; Kushvaha, V. Application of Artificial Intelligence for Predicting Erosion of Biochar Amended Soils. *Sustainability* **2022**, *14*, 684. [CrossRef]
25. Cai, W.; Kumar, H.; Huang, S.; Bordoloi, S.; Garg, A.; Lin, P.; Gopal, P. ANN model development for air permeability in biochar amended unsaturated soil. *Geotech. Geol. Eng.* **2020**, *38*, 1295–1309. [CrossRef]
26. Elangasinghe, M.A.; Singhal, N.; Dirks, K.N.; Salmond, J.A. Development of an ANN–based air pollution forecasting system with explicit knowledge through sensitivity analysis. *Atmospheric Pollut. Res.* **2014**, *5*, 696–708. [CrossRef]
27. Nourani, V.; Fard, M.S. Sensitivity analysis of the artificial neural network outputs in simulation of the evaporation process at different climatologic regimes. *Adv. Eng. Softw.* **2012**, *47*, 127–146. [CrossRef]
28. Golzar, F.; Nilsson, D.; Martin, V. Forecasting wastewater temperature based on artificial neural network (ANN) technique and monte carlo sensitivity analysis. *Sustainability* **2020**, *12*, 6386. [CrossRef]
29. Chen, S.; Ren, Y.; Friedrich, D.; Yu, Z.; Yu, J. Sensitivity analysis to reduce duplicated features in ANN training for district heat demand prediction. *Energy AI* **2020**, *2*, 100028. [CrossRef]
30. Jin, Y.; Zheng, X.; Cai, C.; Wang, G.; Sha, L. Study on the Strength and Curing Mechanism of Beach Silt Stabilized Soil Using Cement-fly Ash. *Bull. Sci. Technol.* **2015**, *31*, 132–136. (In Chinese) [CrossRef]

Journal of
Marine Science and Engineering

MDPI

Article

Influence of Penetration Rate on Full-Flow Penetrometer Resistance in Underconsolidated Clay

Huanhuan Qiao [1,2], Xuening Liu [1], Ruixian Zhou [1], Huan He [1,*], Peng Peng [3] and Zhen Jiang [1,*]

1 Institute of Geotechnical Engineering, School of Transportation, Southeast University, Nanjing 211189, China; 230189834@seu.edu.cn (H.Q.); 230199158@seu.edu.cn (X.L.); 220213349@seu.edu.cn (R.Z.)
2 School of Engineering, University of Warwick, Coventry CV4 7AL, UK
3 School of Civil Engineering and Transportation, Shenzhen University, Shenzhen 518060, China; pengpeng@szu.edu.cn
* Correspondence: h_he@seu.edu.cn (H.H.); 230209177@seu.edu.cn (Z.J.)

Abstract: In the past few years, offshore site investigations have extensively utilized full-flow penetrometers like the Ball and T-bar penetrometers to assess penetration resistance and subsequently analyze the strength characteristics of marine clay. The relationship between penetration rate and the measured resistance to penetration and shear strength in clays has been extensively documented through full-flow penetration tests. Although previous studies have shown empirical correlations between undrained shear strength and penetration resistance, the resistance factor utilized in these correlations is typically suggested for cohesive soils that are overconsolidated or normally consolidated, rather than underconsolidated soils. The effects of penetration rate undrained penetration resistances in underconsolidated marine clay are investigated in this study by considering the outcomes of variable rate penetration testing and twitch penetration testing using full-flow penetrometers in laboratory model tests. The discussion focuses on penetration resistances depending on the normalized velocity of the full-flow penetrometers (Ball and T-bar).

Keywords: penetration rate; full-flow penetrometer; marine clay; underconsolidated

Citation: Qiao, H.; Liu, X.; Zhou, R.; He, H.; Peng, P.; Jiang, Z. Influence of Penetration Rate on Full-Flow Penetrometer Resistance in Underconsolidated Clay. *J. Mar. Sci. Eng.* **2024**, *12*, 427. https://doi.org/10.3390/jmse12030427

Academic Editors: Dmitry A. Ruban, Pan Hu and Dong-Sheng Jeng

Received: 1 December 2023
Revised: 23 February 2024
Accepted: 23 February 2024
Published: 28 February 2024

1. Introduction

Over the past few decades, the scarcity of urban land resources has become more noticeable due to the growth and expansion of coastal cities in numerous nations [1]. Coastal reclamation has become an effective way to solve the problem of land resource shortage [2]. At present, the cumulative area of coastal reclamation in China has exceeded 11,000 km^2, resulting in the continuous emergence of a mass of coastal new cities' reclamation area. Among the regions of reclamation in Zhejiang Province, China, most of them are by hydraulic fill. The Wenzhou Oufei Project is the largest single reclamation project in China, with a total area of 326.67 km^2 [3]. The Wenzhou Oufei Project holds the distinction of being China's biggest reclamation project to date. The position of the view of the Wenzhou Oufei Project reclamation area and the photo of the dredger fill in the field are shown in Figure 1a,b, respectively. The coastal reclamation projects mainly rely on the reclamation of nearshore clay sediment as the dredger fill material, and generally use the preloading method for drainage and consolidation treatment [4,5]. Consequently, construction and building on the new dredger fill is inevitable.

Nevertheless, due to the limited permeability of soft marine clay, the foundation treatment only results in the formation of a solid soil layer on top, while the underlying layer of soft marine clay remains incompletely consolidated, referred to as an underconsolidated condition [1,6–13]. During the consolidation of a compressible foundation layer under consolidation, residual excess pore water pressures are present, which result from previous loads or self-weight. This means the primary consolidation of the compressible foundation has not been completed, and the residual excess pore water pressures equal the reduction of the initial overburden pressures. As the average overconsolidation ratio (OCR) in the underconsolidated case is less than unity, the preconsolidation effective stress, σ'_p (less than the effective overburden pressure, σ'_0), becomes the new initial overburden pressure in the middle, as shown in Figure 2 [9]. The dredger fill foundation is characterized by high water content, high compressibility, and low strength [3]. A mass of civil engineering projects, such as buildings, roads, bridges, ports, and docks projects, have been built or are under construction on the underconsolidated soft soil foundation in the reclamation area. How to accurately obtain the geotechnical design parameters of the underconsolidated soft soil foundation in the reclamation area where these structures are located during the engineering survey process has become a key issue in engineering design and construction. Typically, the new manual fills sites, making sampling and laboratory evaluation difficult. Geotechnical in situ tests have become an effective method for determining geotechnical design parameters [14,15].

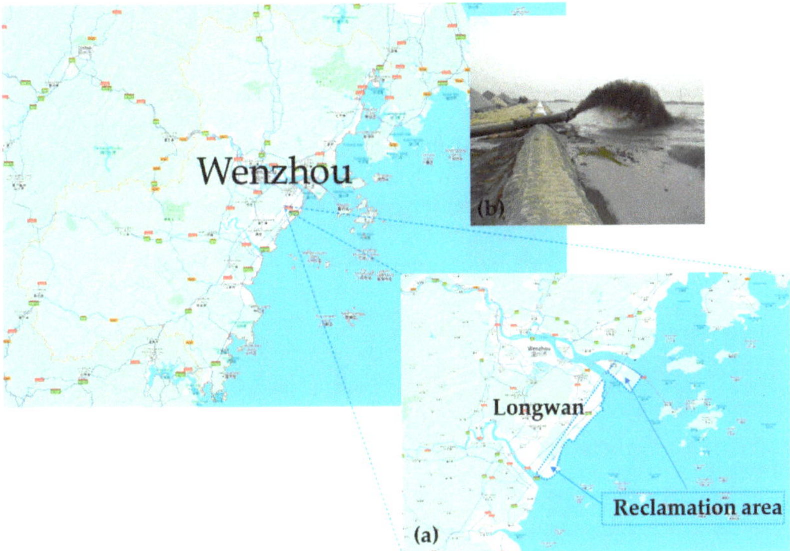

Figure 1. The Oufei project reclamation site, Longwan, in Wenzhou, China. (**a**) The position of the view; (**b**) The photo of the dredger fill in the field.

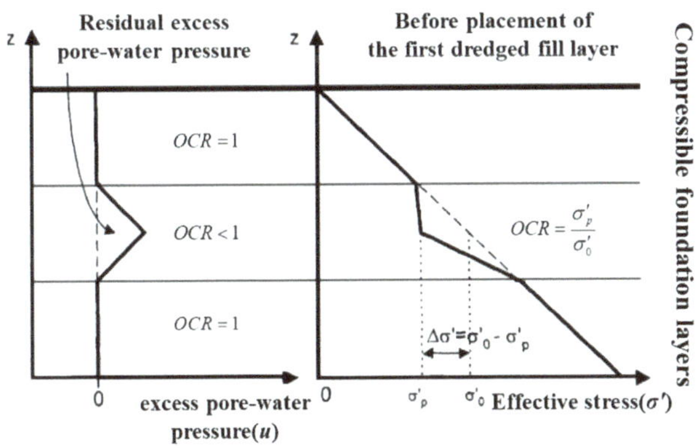

Figure 2. Distribution of residual excess pore -water pressure and effective overburden pressure to reflect underconsolidated compressible foundation material [9].

Recently, the full-flow penetration testing technique has gained popularity as a novel in situ testing method for onshore and offshore engineering projects. It is commonly employed to measure undrained soft clay sediments, both onshore and offshore, using instruments like Ball and T-bar penetrometers [16]. In comparison to conventional piezocone penetration testing, full-flow penetration testing offers several benefits. The penetrometer compels the soil to flow around it, excluding the small drill string area, resulting in equal pressure above and below the probe, which reduces the importance of overburden correction. In soft sediments, the full-flow penetrometers have a larger projected area compared to traditional CPTu, enabling them to test larger volumes of soil and provide more precise measurement outcomes. A plastic solution and numerical model have been created for the flow of soil around the penetrometer [17]. Furthermore, the ball overcomes the susceptibility to the potential eccentric loading of the T-bar, increasingly becoming prevalent owing to its inherent merits over the T-bar penetrometer. Full-flow penetration testing is a more efficient method than vane shear test (VST) as it enables continuous analysis of undrained strength instead of analyzing at discrete depths. Both T-bar and Ball penetrometers possess equal projected areas. The projected areas of both the full-scale T-bar and Ball penetrometers are equal (100 cm^2). There is no conclusive comparison between the T-bar and Ball penetrometers because both are still evolving, and there are limited test data on the full-flow penetrometer. It is recommended that one of the full-flow penetrometers (T-bar or Ball) is used for a site investigation based on the project requirements, the soil conditions likely to be encountered, and the geotechnical problems to be addressed. Table 1 summarizes the relative merits of both T-bar and Ball penetrometers.

Table 1. Advantages and disadvantages of Ball and T-bar penetrometers.

Full-Flow Penetrometers (Full-Scale, Projected Area of 100 cm^2)	Ball	T-Bar
Advantages	• particularly suited for offshore use as the axisymmetry avoids load cell bending • the symmetrical geometry is suited to downhole testing • assesses consolidation characteristics by adding pore water pressure sensors	• viewed as a model pipeline element, thus providing direct information for pipeline and riser design • reduces the effect of buoyancy and sediment density on the surficial penetration resistance and bearing capacity factor • standardized by NORSOK G-001, Marine Soil Investigations (2004)
Disadvantages	• presently not standardized	• more susceptible to bending moments being induced in the load cell

Prior research has indicated that the tip resistance acquired through CPTu in loosely packed soil differs considerably from the composition of the normally consolidated clay stratum. According to Tanaka et al. [18], the mid-depths of the clay stratum in an underconsolidated clay exhibit the lowest levels of both penetration resistance and pore water pressure, as observed through CPTu. In the case of normally consolidated clay, the tip resistance and pore water pressure both exhibit a linear increase as depth increases. According to Karakouzian et al. [19], the N_{kt} value obtained from VST data in underconsolidated clay is greater than that of normally consolidated clays at a specific location when estimating the in situ shear strength using CPTu values. On the contrary, according to the study conducted by Lunne et al. [20], it was found that the N_{kt} value remains consistent for both underconsolidated clay and normally consolidated clay.

The result of CPTu and laboratory tests have effectively showcased the correlation between penetration rate and the recorded penetration resistance and undrained shear strength in clay [20]. The penetration resistance of full-flow penetrometers is also influenced by the rate of penetration. Nevertheless, prior research on the influence of penetration rate on penetration resistance in full-flow penetration testing has primarily concentrated on normally consolidated and overconsolidated clay [21–23]. This article introduces data on underconsolidated clay (new dredged fill) in laboratory modeling tests. The authors examined and deliberated on the outcomes of penetration tests conducted at different rates and using both twitch penetration tests and full-flow penetrometers. The objective was to explore how the rate of penetration impacts the resistance encountered in underconsolidated marine clay.

2. Model Box Testing System

The apparatus used for the test primarily included a model box, penetration equipment, depth encoder, penetrometer, and adaptable support frame, as Figure 3 shows.

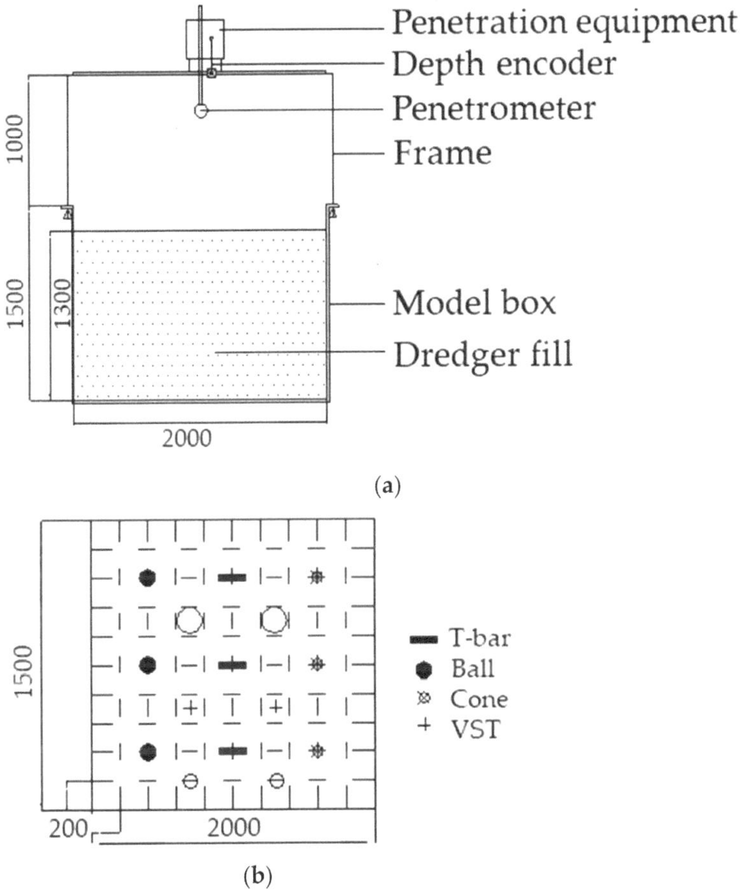

Figure 3. Schematic diagram testing apparatus. (**a**) Penetration actuator assembly; (**b**) Testing layout (unit in mm).

2.1. Model Box and Penetration Equipment

The model box is made of stainless steel, and it has 200 cm of length and width, 150 cm of height, and 1.25 cm of thickness. Figure 4 displays the system designed for the penetrometer to penetrate the soil at a predetermined velocity. This adjustable support frame system consisted of two channeled metal frames 100 cm high, with a width of 250 cm, braced with crossbars. Two bushes were located in the middle of the crossbars to facilitate the linear movement of the penetrometer. The penetrometer is driven by a

TCH-1 electrical machinery drive (Nanjing Soil Instrument Factory Co., Ltd., Nanjing, China). The movement of the penetrometer was recorded using a displacement sensor that had a travel distance of 50 cm. To stop the penetrometer from penetrating beyond the base of the clay bed, limit switches were positioned on the supporting frame, ensuring a 10 cm gap. In this test, the penetration was conducted using equipment that allowed for adjustable speed. The penetration rate range designed for the test was 0 to 50 mm/s, which determined the five required penetration rates for the variable speed test: 2 mm/s, 5 mm/s, 10 mm/s, 20 mm/s, and 40 mm/s. The penetration equipment is capable of causing different drainage conditions of penetrometers in the clay ranging from partial drainage to undrained conditions by adjusting the rate of penetration. The encoder of depth connects the penetrometer to the steel frame. A wireline potentiometer was used to record measurements of the depth meter.

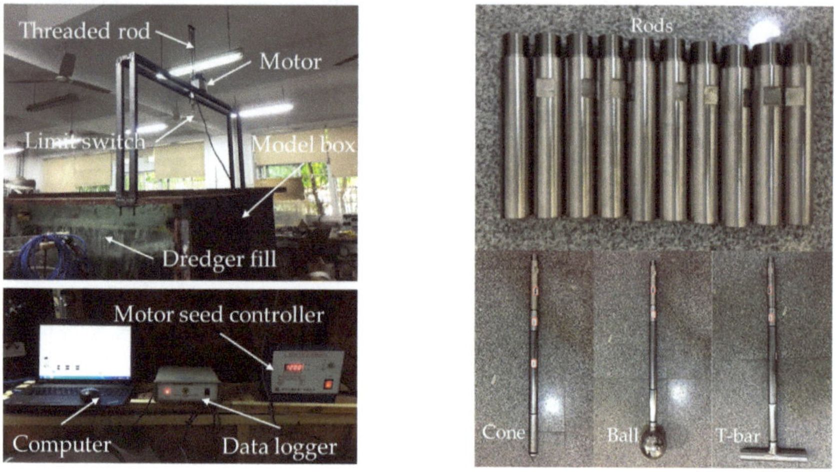

(a) (b)

Figure 4. Testing equipment. (**a**) Penetration and data acquisition equipment; (**b**) Penetrometers and rods.

2.2. Penetrometers and Rods

These tests utilized a piezocone, a Ball, and T-bar full-flow penetrometers. Piezocone penetration tests were conducted by using a standard cone with a 60° tip angle and 35.7 mm diameter. The cone is depicted in Figure 4b, which installed a high-density polypropylene filter element, measuring u_2, with a thickness of 1.5 mm, positioned behind the cone shoulder that is 5 mm in height. In this study, the T-bar and Ball penetrometers with the projected areas of 100 cm^2 and the area ratios, AR, of 10:1 ($AR = A_p/A_s$) were used. The T-bar penetrometer has a length of 250 mm and a diameter of 40 mm, while the Ball has a diameter of 113 mm. To guarantee a coarse interface, the penetrometers' surfaces underwent a gentle sandblasting process. The load cell is positioned directly behind the penetrometer, aligned with the rods, and was specifically engineered to account for flexing and variations in temperature. Table 2 presents the detailed parameters of the cone and full-flow penetrometers. A set of rods consists of 10 stainless steel short rods with a 180 mm length, including a 30 mm thread length, which means the effective length after tightening is 150 mm.

Table 2. Summary of parameters for cone and full-flow penetrometers.

System Design		Cone	Ball	T-Bar
Specification	Dimensions	35.7 mm diameter	113 mm diameter	250 mm × 40 mm ($L \times D$)
	Area ratio	1:1	10:1	10:1
	Pore water pressure measurement	Located at a connection between full-flow probe and push rods (conventional u_2 filter measurement if CPTu mandrel is used); CPTu pore fluid and saturation procedures apply		
	Load cell calibration pore	Calibrated following CPT specifications		
	Load cell design	Calibrated for tension and compression following CPT specifications; report load cell sensitivity to temperature shift		
	Maximum sampling interval	≤5 mm (2 Hz sampling frequency @ 2 cm/s penetration rate)		

2.3. Vane

The high-precision portable laboratory vane shear test apparatus (PS-VST-P, PENESON, Guangzhou, China) was used to conduct tests on all VSTs. This apparatus, as depicted in Figure 5, offered a vane with a shear strength resolution of 0.8125 kPa.

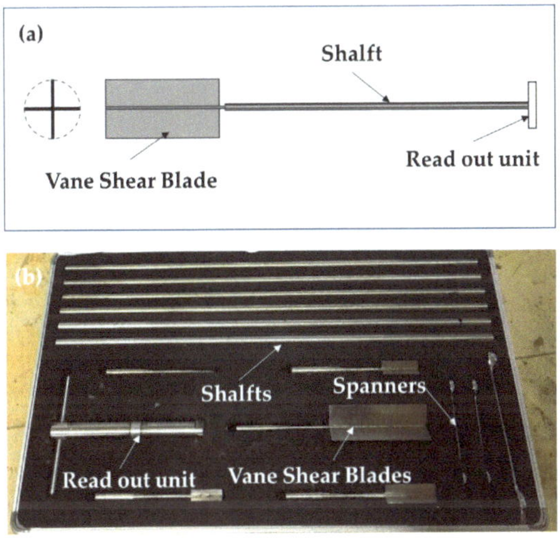

Figure 5. Portable laboratory vane. (**a**) Vane construction view; (**b**) Actual photo of the vane.

2.4. Sample Preparation

Tests were performed on clay samples obtained from the Oufei project site, Longwan, in Wenzhou, China, and the reclamation site is shown in Figure 1b. The clay samples were acquired through hydraulic filling after dredging of the sediment. Fill material for land reclamation projects is generally dredged from the nearby seabed. Dredged soil can reach a depth of five meters. Figure 6 plots the geological profile of soil in a typical reclamation site of this project.

The X-ray diffraction (XRD) technique was served as the analytical method for determining the mineral composition of the clay in the dredged silt samples collected from the Wenzhou Oufei Project. To ascertain the material composition of the blown silt with precision, XRD was employed to generate the mineral composition diffractograms (Figure 7). These diffractograms were analyzed against the standard mineral diffraction patterns of

XRD, revealing that halite predominates the mineral composition of the clay, succeeded by illite, with lesser amounts of quartz, an illite–smectite mixed layer, and chlorite. The analytical results, presented in Table 3, indicate a significant presence of clay minerals in the blown silt, with halite comprising 32% of the clay minerals' weight, illite 17%, quartz 16%, an illite–smectite mixed layer 11%, and chlorite 9%.

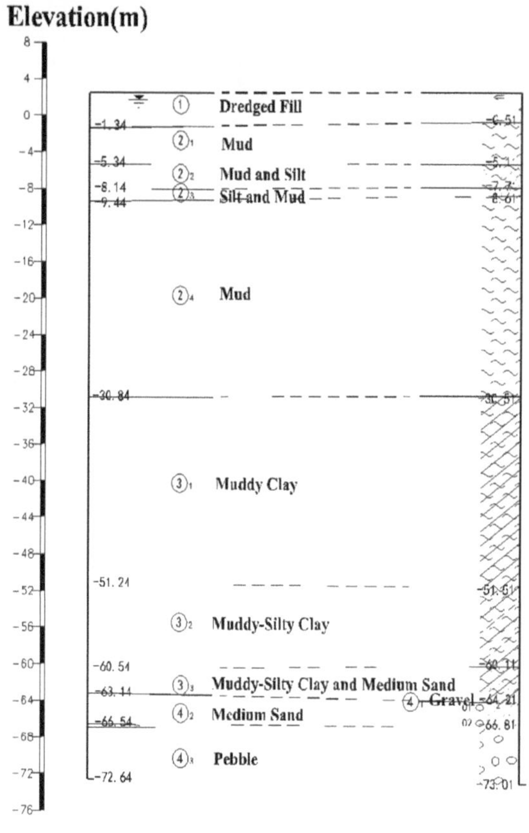

Figure 6. Geological profile of soil in the Wenzhou Oufei project site.

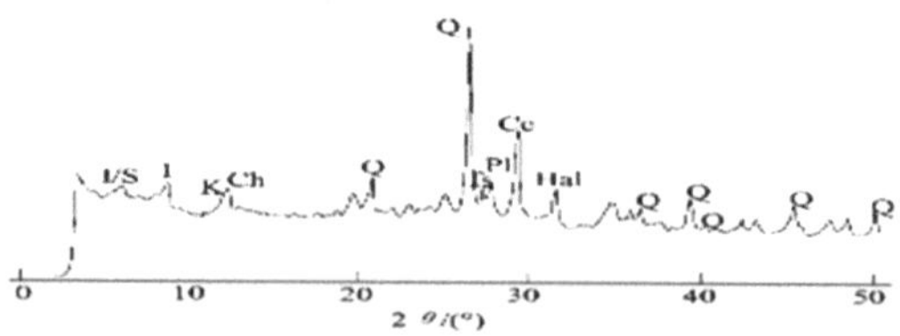

Note: Q: Halite; Fs: Alkaline feldspar;
Pl: Plagioclase feldspar; S: Montmorillonite;
I: Illite; K: Quartz; Ch: Chlorite; Cc: Calcite; Hal: Kaolin

Figure 7. The XRD patterns of dredged silts from Wenzhou Oufei Project.

Table 3. Classification and statistics of mineral composition in dredged silts.

Halite	Illite	Chlorite	Illite–Smectite Mixed Layer	Quartz	Feldspar	Calcite	Kaolinite
32%	17%	9%	11%	16%	5%	4%	6%

Laboratory tests were conducted to determine the soil characteristics, as shown in Table 4. The clay had a high-water content of 109.4%, with a liquid limit of 56%, a plastic limit of 27%, and a specific gravity of 2.73. A Rowe cell oedometer test was conducted to determine the consolidation properties of the soil. As the vertical effective stress level ($\sigma_v{}'$) rises from 12.5 kPa to 100 kPa, the coefficient of consolidation (c_v) also rises, ranging from 3.09 to 7.37 $\times$ 10^{-3} cm^2/s. Other properties are provided in Table 4. At the time of the test, it had been nearly two years since the soil sample was obtained, and the soil sample had been naturally deposited in the model box, which has certain structural characteristics. Self-weight consolidation of newly placed fill has not yet been completed. The clay samples were ensured to be submerged completely under water during tests. The presence of free water on the surface of clay samples had a profound impact on the penetration resistance [24].

Table 4. Summary of dredged soil properties.

Water content (%)	109.4
Relative density of particle	2.73
Unit weight (kN/m^3)	14.43
Liquid limit (%)	56
Plastic limit (%)	27
Plasticity index (%)	29
Coefficient of consolidation, c_v (cm^2/s)	3.09–7.37 $\times$ 10^{-3}
Void ratio	2.37
Saturation (%)	99.6

3. Test Methodology

Newly placed fill was used to conduct penetration tests on the underconsolidated clay specimens. To examine the impact of two significant factors, namely, shape (for determining resistance factor N) and penetration rate on penetration resistance in underconsolidated clay, each location underwent multiple penetrations using the three penetrometers (cone, Ball, and T-bar). To investigate the impact of the tip shape, the researchers penetrated the cone with a diameter of 35.7 mm, the Ball with a diameter of 113 m, and the T-bar with a diameter of 40 mm and a length of 250 mm ($D \times L$).To examine the impact of the penetration rate, the penetrometers were penetrated at various speeds, namely, 2 mm/s, 20 mm/s, and 40 mm/s. Lastly, the twitch penetrations were performed with three differently shaped penetrometers to examine the drainage conditions' effect. Following the completion of all penetration tests, vane shear tests were conducted at ten varying depths ranging from 10 cm to 100 cm with increments of 10 cm. VSTs were conducted at two distinct positions within the plane. The vane was spun at a rate of approximately 2.0° per second until the clay's strength was completely activated to determine the undrained shear strength of both intact and remolded samples. Various penetration tests were conducted on the underconsolidated clay samples in the model container specified in Table 5. The locations of the penetration test are indicated in Figure 3b. To monitor the soil's pore water pressure, electric miniature piezometers were placed at depths of 20 cm, 40 cm, 60 cm, 80 cm, and 100 cm. At each location of the laboratory model, piezometer data were collected simultaneously from the pore water pressure acquisition instrument.

Table 5. Summary of vane shear tests and penetrometer tests.

Test Method	Test Type	Test Sample Number	Penetrometer Type	Number of Tests
Vane shear test	Intact	VST1	Vane	4
	Remolded	VST2		
Penetrometer test	Variable rate monotonic penetration testing	CPT1	Cone	3
		CPT2		
		CPT3		
		BPT1	Ball	3
		BPT2		
		BPT3		
		TPT1	T-bar	3
		TPT2		
		TPT3		
	Twist penetration test	TCPT1	Cone	3
		TBPT1	Ball	
		TTPT1	T-bar	

3.1. Vane Shear Test (VST)

In order to ascertain the undrained strength characteristics, a total of four laboratory vane shear experiments were conducted on samples of dredger fill in the model box. These tests encompassed two undisturbed and remolded undrained shear tests, which were carried out at four distinct positions within the plane. The VSTs were conducted at 10 depths within every 100 mm. To ensure undrained conditions during the shearing process, the rotation rate recommended in ASTM D2573 (2002) was used for all the VSTs [25,26]. The VSTs were carried out inside the model box, where the available test space is more than five times the diameter of the vane shear blade. It can be expected to avoid boundary effects and ensure enough distance from the container's edges and bottom, as well as between each measurement [27]. To prevent the buildup of excessive pore water pressure, the vane was carefully inserted into the soil sample at a rate of less than 1 mm per second [28].

3.2. Piezocone Penetration Test (CPTu)

1. Correction of measured cone tip resistance

The calculated value of cone tip resistance, q_t, is determined using the following relationship [20]:

$$q_t = q_c + u_2(1 - a) \tag{1}$$

where q_c is the cone tip resistance, u_2 is the pore water pressure measured behind the shoulder of the cone, and a is the area ratio of the cone. The ratio of the cone area is approximately equal to the ratio between the load cell or shaft's cross-sectional area and the cone's projected area. The calculation of $u_2(1-a)$ determines the pore water pressure that adds to cone resistance, taking into account the tip shape. According to the research by Lunne et al. [20], the range is valued at approximately 0.55 to nearly 0.9. In the study, a cone penetrometer with a net area ratio of 0.75 was adopted. The pore water pressure behind the shoulder of the cone, u_2, is determined by utilizing the measured pore water pressure values on the cone, u_1, according to the relationship:

$$u_2 = K(u_1 - u_0) + u_0 \tag{2}$$

where u_0 is the in situ pore water pressure at time 0 (before inserting the cone), u_1 is the pore water pressure measured on the piezocone, and K is the adjustment factor. Karakouzian et al. [19] indicated that the recommended values of the adjustment factor, K, for underconsolidated clay in CPTu were dependent on the stress history of the soil deposit and soil type. In this study, the adjustment value, K, of 0.75, was adopted to calculate u_2. For all these corrections, the in situ pore water pressure has been taken as hydrostatic with a water table about 1 cm above the ground surface, and the average unit weight for the soil was 14.43 kN/m^3 below the water table. The net tip resistance of the piezocone is calculated as below:

$$q_{net,c} = q_t - \sigma_{v0} = q_t - \gamma h \tag{3}$$

where q_t is the corrected tip resistance by the pore water pressure, σ_{v0} is the total overburden stress, γ is the unit weight, and h is the soil thickness.

2. Calculation of undrained shear strength from CPTu

In clays, an estimate of undrained shear strength, S_u is calculated according to the equation [29]:

$$S_u = \frac{q_{net,c}}{N_{kt}} = \frac{q_t - \sigma_{vo}}{N_{kt}} \tag{4}$$

where q_t is the corrected tip resistance by pore water pressure, σ_{vo} is the total overburden stress, and N_{kt} is cone resistance factor. For practical use, errors in determining q_t from the measured tip resistance, q_c, and in estimating σ_{vo} and N_{kt} lead to quite large inaccuracies, particularly for soft clay [29].

3.3. Full-Flow Penetrometer Tests

1. Correction of measured penetration resistance

For the full-flow penetrometer, correction is required for the net penetration resistance observed by the load cell due to the effects of measured pore water pressure and overburden stress. The net resistance of the full-flow penetrometer, a relationship suggested by Chung et al. [29], is as follows:

$$q_{net} = q_m - \frac{[\sigma_v - u_2(1-\alpha)]A_s}{A_p} \tag{5}$$

where q_m is the measured penetration resistance, σ_v is the total vertical overburden stress, u_2 is the measured pore water pressure, and α is the load cell area ratio. A_s is the projected areas of the push rod and A_p is the projected area of the penetrometer.

In practice, σ_{v0} is often uncharted, and the measurement of u_2 in real-time is challenging and rarely conducted during experiments. Chung et al. [29] made a few minor alterations to the aforementioned equation, which did not lead to substantial changes in the calculations and removed the need for precise measurement of u_2 in the full-flow penetrometer. As Nanda et al. [30] reported, there is no need to correct for pore water pressure when using the improved full-flow penetrometer. Only the overburden correction is necessary, resulting in the following modified equation:

$$q_{net} = q_m - (\sigma_{v0})\frac{A_s}{A_p} \tag{6}$$

where q_m is the measured penetration resistance, σ_{v0} is the in situ total vertical overburden stress, A_s is the projected area of the push rod, and A_p is the projected area of the penetrometer.

2. Evaluation of undrained shear strength by the full-flow penetrometer test

The estimation of undrained shear strength, s_u, can be obtained by dividing the net penetration resistance, q_{net}, by the resistance factor, N [29]:

$$s_u = \frac{q_{net}}{N} = \frac{q_{net,\ T\text{-}bar}}{N_{T\text{-}bar}} = \frac{q_{net,\ Ball}}{N_{Ball}} \tag{7}$$

where $N_{\text{T-bar}}$ and N_{Ball} are the resistance factors for the T-bar and Ball, respectively.

3. Influence of penetration rate

In general, the shear strength increases with increasing strain rate, which is dependent on factors such as stress history and soil type. The consolidated undrained triaxial compression (CK_0UC) experiments in resedimented Boston Blue Clay indicated an increased effect at lower *OCR* values, which is pertinent to the soils examined [31]. The literature has previously reported studies on the strain-rate effect using full-flow penetrometers [21,24,32–38]. Strain-rate effects can be studied under two conditions based on the rate of penetration of full-flow penetrometers: partial drainage and undrained conditions [39]. The dissipation of partial pore water at very slow penetration rates increases the soil strength and penetration resistance as well as the subsequent consolidation of the soil. From a very slow penetration rate, penetration resistance gradually decreases until fully undrained conditions are reached. As the rate of penetration increases from fully undrained conditions, the penetration resistance increases largely as a result of penetration rate effects. Generally, when the value of the normalized speed parameter, V, is greater than 20 for the full-flow penetrometers, fully undrained conditions will be predominant [40]. In terms of the dimensionless speed parameter, V, it is possible to define it as follows:

$$V = \frac{vd}{c_v} \tag{8}$$

where v is the rate of the penetrometer, d is the diameter of the penetrometer, and c_v is the coefficient of consolidation.

For both normally consolidated and overconsolidated soft clays with intact strengths, various relationships have been proposed: the hyperbolic sine equation [30,34], the semi-logarithmic equation [21], and the power law. Based on the reference penetration rate, all of these relationships are fitted with parameters. Most previous works assume a penetration rate of 2 cm/s for full-flow penetrometer testing. Table 6 summarizes the reported relationships between rate and strength for normally consolidated and overconsolidated soft clays.

Table 6. Reported relationships of the rate–strength for soft clays.

Relationship	Type of Function	Drainage Condition	Consolidation State	Reference
$\frac{q}{q_{ref}} = \frac{1+\frac{\lambda}{\ln(10)}\cdot\sinh^{-1}(v/d\dot{\gamma}_o)}{1+\frac{\lambda}{\ln(10)}\cdot\sinh^{-1}(v_{ref}/d\dot{\gamma}_o)}$	Hyperbolic sine function	Undrained	Normally consolidated	Chung et al., 2006 [34]
$\frac{q_{in}}{q_{\text{in }(ref)}} = a + \mu\log\left(\frac{v/d}{(v/d)_{ref}}\right)$	Semi-logarithmic function	Undrained	Normally consolidated and lightly overconsolidated	Yafrate et al., 2007 [21]
$\left.\frac{q}{\sigma'_v}\right\vert_{\text{undrained}} = a\left[\frac{v/d}{(v/d)_{\text{ref}}}\right]^m$	Power law	Undrained	Normally consolidated and overconsolidated	Lehane et al., 2009 [33]

Note: q: penetration resistance at any rate; q_{ref}: reference penetration resistance; λ: rate parameter; v: penetration rate; $\dot{\gamma}_o$: strain rate for which the viscous effects start to decay; v_{ref}: reference penetration rate; $\dot{\gamma}_{ref}$: reference strain rate corresponding to q_{ref}; q_{in}: intact penetration resistance; $q_{in(ref)}$: reference intact penetration resistance; a: intercept factor; μ: rate factor; v: sensitivity; v/d: normalized penetrometer velocity; $(v/d)_{ref}$: reference normalized penetrometer velocity; q: penetration resistance; σ'_v: initial vertical effective stress; a: curve-fitting parameter; and m: rate exponent.

These types of relationships are generally interpreted using hyperbolic sine, semilogarithmic, and power functions. It is usually necessary to fit the testing results with nondimensional parameters to interpret the results. In general, there are no clear guidelines for determining which function to use, but it largely depends on the type of soil and penetrometers used [30]. Schlue et al. [28] indicated that both a semilogarithmic and a power law can be used to describe the relationship between peripheral velocity in the VSTs and measured shear strength of dredged harbor mud, and the semi-logarithmic function produces a slightly better curve. In this article, the soil sample was also obtained by hydraulic fill after dredging. Accordingly, the following semi-logarithmic equation was adopted to fit the penetration resistance, q_{in}, with the penetration rates:

$$\frac{q_{in}}{q_{in(ref)}} = a + \mu \log\left(\frac{v/d}{(v/d)_{ref}}\right) \quad (9)$$

where a is the intercept factor, μ is the rate factor, v/d is the normalized penetrometer velocity, and $(v/d)_{ref}$ is the reference normalized penetrometer velocity.

4. Results and Discussion

4.1. Vane Shear Test (VST)

The uniform underconsolidated marine clay in the model box has a rising undrained shear strength of 4.9 kPa near the surface to 22.5 kPa at the bottom of the clay bed. The sensitivity range is from 1.5 to 2.5. Table 7 provides the result of the undrained shear strength (intact and remolded strength, s_u and s_{ur}) and sensitivity values of soil at ten various depths, which is based on the average of the two vane shear tests.

Table 7. Undrained shear strength (intact and remolded strength, s_u and s_{ur}) over depth by the VST.

Depth (cm)	Undrained Shear Strength (kPa)		
	s_u	s_{ur}	s_t
0~10	4.9	3.2	1.53
10~20	12.6	5.1	2.47
20~30	9.2	4.8	1.92
30~40	13.5	5.5	2.45
40~50	11.3	6.9	1.64
50~60	16.4	7.4	2.22
60~70	17	8	2.13
70~80	19.3	10.1	1.91
80~90	20.1	10.2	1.97
90~100	22.5	12.1	1.86

4.2. Piezocone Penetration Test (CPTu) in Underconsolidated Clay

Test profiles of CPTu including cone tip resistance (q_t), sleeve friction (f_s), pore water pressure (u_2), and intact and remolded undrained shear strength (s_u and s_{ur}) from VST are presented in Figure 8. Also shown in Figure 8c are the pore water pressure readings at five different depths from the electric piezometers in the soil.

Figure 8c displays that the pore water pressure in the soil is higher than the corresponding hydrostatic pressure. The calculation of hydrostatic pressure is equal to the unit weight of pore water multiplied by the depth below the water table as the dashed straight line in Figure 8c shows. This indicates that the remaining excess pore water pressure resulting from the consolidation of dredger fill due to its self-weight has not completely dissipated, indicating that the clay is still in an underconsolidated condition [19]. The mineral analysis of XRD also indicates that the material composition of the dredged silt is characterized by a high proportion of viscous mineral particles, adversely affecting the self-settling behavior of the silt.

Figure 9 shows the measured cone tip resistance (q_c), the corrected cone tip resistance (q_t), and net cone tip resistance (q_n) profiles by piezocone penetration tests. The corrected cone tip resistance (q_t) is used in Equation (1) and the net cone tip resistance (q_n) is calculated by Equation (3). The test results of CPTu show that cone tip resistance (q_t) is less than 10kPa and sleeve friction (f_s) is less than 5kPa. Accordingly, corrections need to be applied to the raw measured cone tip resistance. As may be seen in Figure 9, the correction of tip resistance for the cone is significant. This is mainly because the measured tip resistance is relatively small, and hence, increases the weight of the pore water pressure (u_2) and overburden stress. The errors in determining q_t from the measured resistance, q_c, lead to quite large inaccuracies, particularly for soft clay of low strength. Obviously, the strength of underconsolidated clay in the model box is too low to be measured accurately by conventional piezocone. In addition, as seen in Figure 8d, intact and remolded undrained shear strength of soil shows a consistent rise with depth. This trend was similar to self-weight consolidation clays.

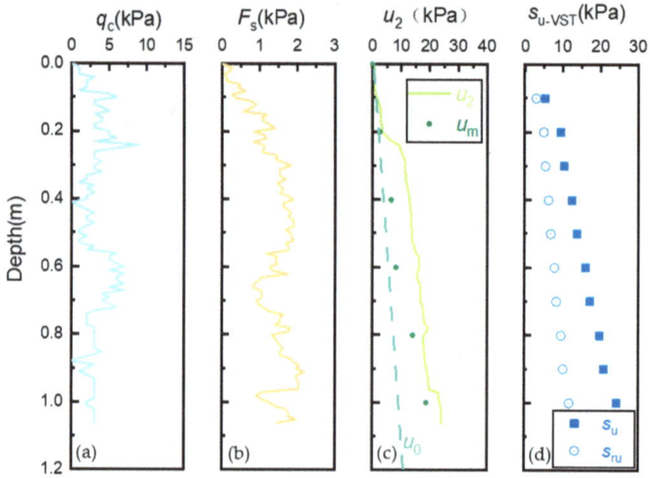

Figure 8. Piezocone penetration test (CPTu) profiles of underconsolidated clay in laboratory model from Wenzhou Bay (Oufei project site). (**a**) Cone tip resistance profiles; (**b**) Sleeve friction profiles; (**c**) Pore water pressure profiles (u_2 measured by piezocone and u_m measured by electric piezometers); (**d**) Vane shear test data.

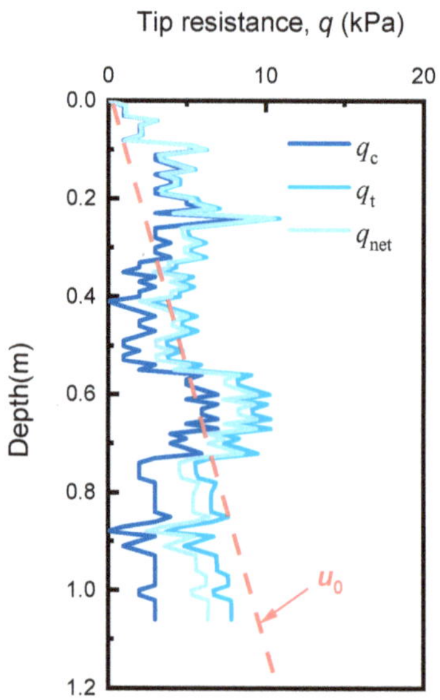

Figure 9. Correction of cone tip resistance data (cone tip resistance, q_c, corrected cone tip resistance, q_t, and net tip resistance, q_{net}).

4.3. Full-Flow Penetrometer Tests in Underconsolidated Clay

As may be seen in Figure 10, the profiles of penetration resistances for the full-flow penetrometers (T-bar and Ball) were corrected by Equation (5), as the penetrometers have a significantly larger projected area than the shaft, rendering the correction insignificant [29]. Both the T-bar and Ball penetrometers exhibit comparable resistance profiles, with the T-bar indicating lesser penetration resistance compared to the Ball. Low et al. [41] and DeJong et al. [42] also observed similar results in field tests. The penetration resistance of the Ball exhibits roughly 20% greater penetration resistance compared to the T-bar at the bottom of the clay bed as Figure 10 shows. It is necessary to mention that two sharp drops exhibited in the Ball resistance profile are not seen in the T-bar resistance profile. This occurred due to their immediate suspension, which is an unintentional incident that happened during continuous penetration tests.

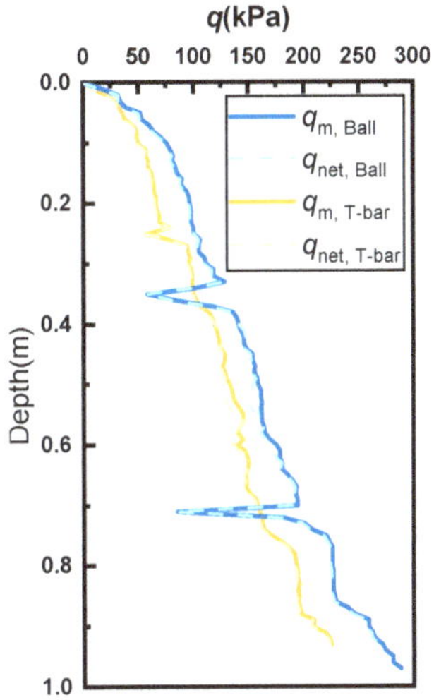

Figure 10. Correction of Ball and T-bar net penetration resistance profile.

1. Determination of resistance factors N of full-flow penetrometers in underconsolidated clay using strength data of VST

To determine the values of penetration factors N for underconsolidated clay from Wenzhou Bay, China, the VSTs were conducted 1 m around the penetration point. The resistance factor N for a specific site can be approximated by determining the slope of undrained shear strength, s_u (reference strength from VST data) versus net penetration resistance, q_{net}, as Figure 11 shows. This estimation is based on the fact that the reference strengths increase uniformly with depth. The averages calculated for the resistance factor N_{Ball} and $N_{T\text{-}bar}$ of the spot, as determined by the VST data, are 10.9 and 12.0, correspondingly; both are within the recommended value range of normally consolidated clay.

DeJong et al. [42] reported that four highly characterized soft sediment sites (from No. 1 to 4) were all overconsolidated soil with overconsolidation ratio, OCR, from 1.5 to 2.7 [42]. The mean resistance factors (N) with different OCR from four sites for each profile based on reference strength from VST are shown in Figure 12 and Table 8. The resistance factors N_{kt} for the CPTu range from 8.1 to 15.0. The resistance factors N_{Ball} for Ball range from 5.8 to 15.9. The resistance factors $N_{T\text{-}bar}$ for T-bar range from 5.7 to 14.9. The No. 5 site is a reclamation site using dredged material. Consolidation of the dredged marine clay at Craney Island has been occurring since 1955. For underconsolidated marine clays, the N_{kt} value of 15 is adopted to estimate in situ undrained shear strength using the results of CPTu and VST [19,43]. Therefore, it is critically important to ascertain the suitable N_{kt} value for a specific site if CPT values will be used to evaluate the undrained shear strength in the field. The most desirable approach involves directly calibrating the N using the empirical relationship between reference strength from laboratory or VST data and net tip resistance from CPT data of specific sites [19,43]. When using VST data as the reference strength, the recommended values of the penetration factors N for full-flow penetrometer in underconsolidated clay are average values of range in overconsolidated clay reported by the previous in situ full-flow testing data.

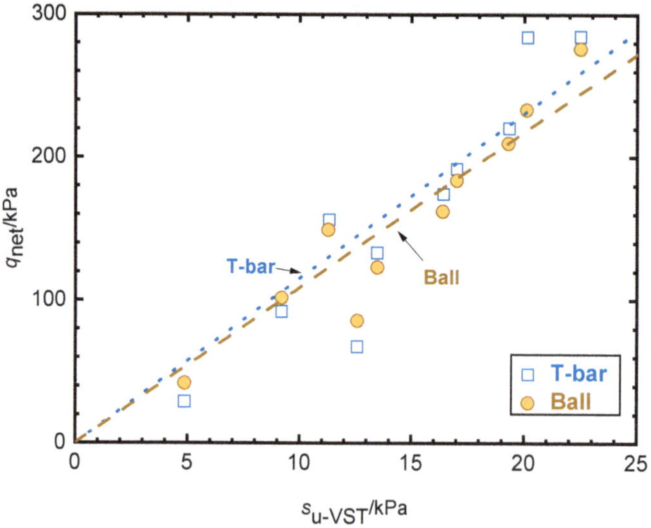

Figure 11. Relationship between net penetration resistance, q_{in}, and undrained shear strength, s_u, for underconsolidated clay at Wenzhou.

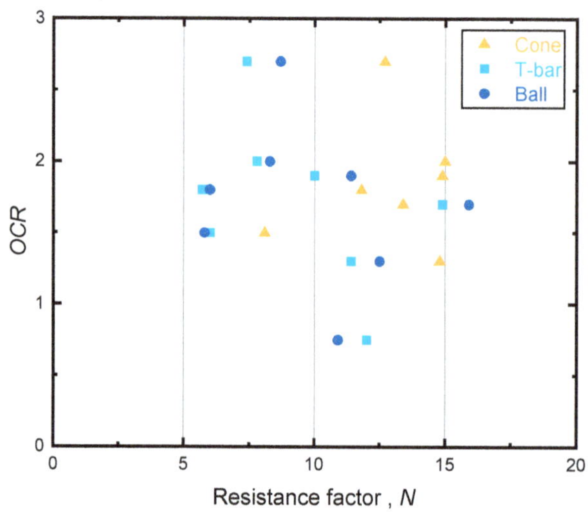

Figure 12. Range of measured resistance factors (N) with overconsolidation ratio (OCR) for cone, Ball, and T-bar [19,21,42].

Table 8. Values of resistance factor N for normally consolidated and underconsolidated clay [19,21,42].

Test Sites, Country	Property				
	Depth (m)	OCR	$N_{k,\,VST}$	$N_{T\text{-}bar,\,VST}$	$N_{Ball,\,VST}$
No. site 1: Burswood, Australia	8.8	1.7	13.7	13.3	13.6
	13.7	1.7	13.4	14.9	15.9
No. site 2: Onsoy, Norway	5.4	1.9	14.9	10.0	11.4
	15.2	1.3	14.8	11.4	12.5
No. site 3: Louiseville, Canada	7.8	2.7	12.7	7.4	8.7
	11.6	2	15.0	7.8	8.3
No. site 4: Gloucester, Canada	4.4	1.8	11.8	5.7	6.0
	8.3	1.5	8.1	6.0	5.8
No. site 5: Craney Island, USA	10–30	<1	15	-	-
No. site 6: Wenzhou, China	0.4	<1	-	12.0	10.9
(This study)	0.8	<1	-	12.0	10.9

Note: VST is vane shear test.

2. Variable rate monotonic penetration testing

For both the Ball and T-bar penetrometer, the penetration resistance profiles at three different velocities of 2 mm/s, 20 mm/s, and 40 mm/s are drawn in Figure 13. The test result shows that the q_m at the rates of 20 mm/s and 40 mm/s was significantly greater than that at 2 mm/s. The profile of the penetration resistance remains relatively unchanged and is similar at both shallow and deep embedment in the monotonic penetration test, despite the change in penetration rate from 20 mm/s to 40 mm/s, which is under an order of magnitude.

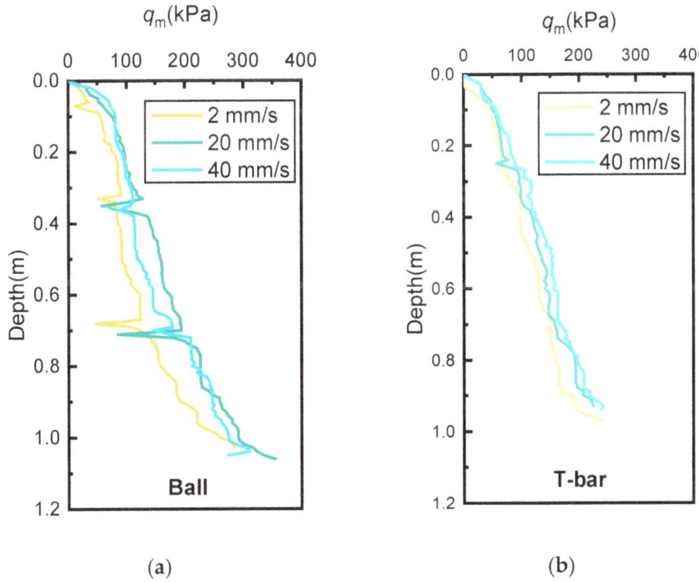

Figure 13. Penetration resistance profiles of different constant velocities for the full-flow penetrometers. (**a**) Ball; (**b**) T-bar.

This study extends the field research of Yafrate et al. [21] by examining the rate dependence with the full-flow penetrometers (Ball and T-bar) in underconsolidated soil, considering the effect of the consolidation condition. Table 9 shows that the testing results presented by Yafrate et al. [21] and DeJong et al. [42] were four high-quality in situ test data with different overconsolidation ratios, *OCR*, and sensitivities, s_t. The author compared and analyzed the field data of four sites with the monotonic penetration test result at three different penetration rates. As Figure 14 shows, the penetration resistance in the underconsolidated clay increases with the increase in the penetration rate at rates from 2 mm/s to 40 mm/s. The relationship between the penetration resistance and the velocity is presented as Figure 14 shows. The plots demonstrate a consistent pattern of higher penetration resistance as the penetration velocity increases for the underconsolidated clay, resembling the findings for the normally and overconsolidated clay presented by Yafrate et al. [21] except for one test site in Louiseville, Norway. That site is likely an accidental event.

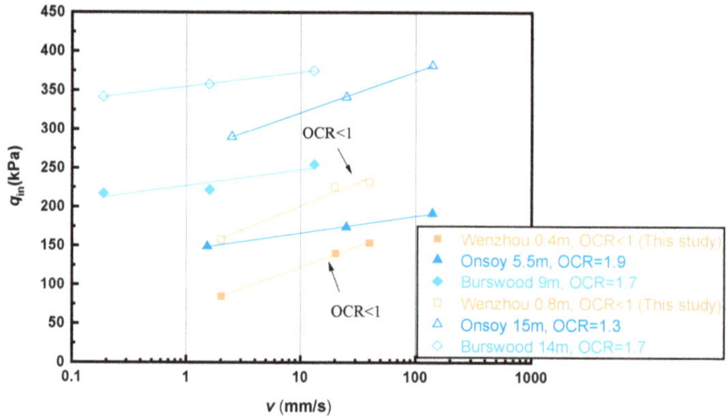

(a)

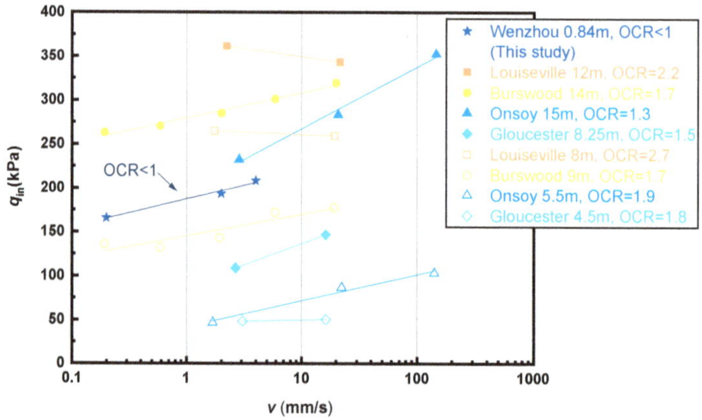

(b)

Figure 14. Penetration resistance (q_{in}) versus penetration velocity (v). (**a**) Ball; (**b**) T-bar.

In order to combine the Ball and T-bar penetrometer testing data, the penetration velocity, v, was normalized by dividing it by the diameter of the penetrometer, d, which was 40 mm and 113 mm, respectively. Comparing the resistance values at different depths is possible by normalizing the intact penetration resistance qin with a reference standard penetration resistance $q_{in(ref)}$ at a velocity of 20 mm/s. Figure 15 illustrates a consistent pattern of higher penetration resistance as the velocity increases.

Table 9. Summary of normally consolidated and overconsolidated clay test site data (Adapted from Yafrate et al. [21] and DeJong et al. [42]).

Penetrometer	Site, Country	Depth (m)	$s_{u,\,VST}$ (kPa)	S_t	OCR
T-bar	Burswood,	9.0	21	3	1.7
	Australia	14.0	28	4	1.7
	Gloucester,	4.5	20	67	1.8
	Canada	8.5	34	85	1.5
	Louiseville,	8.0	40	18	2.7
	Canada	12.0	50	22	2.0
	Onsoy,	5.5	13	7	1.9
	Norway	15.5	27	6	1.3
Ball	Burswood,	9.0	21	3	1.7
	Australia	14.0	28	4	1.7
	Onsoy,	5.5	13	7	1.9
	Norway	15.5	27	6	1.3

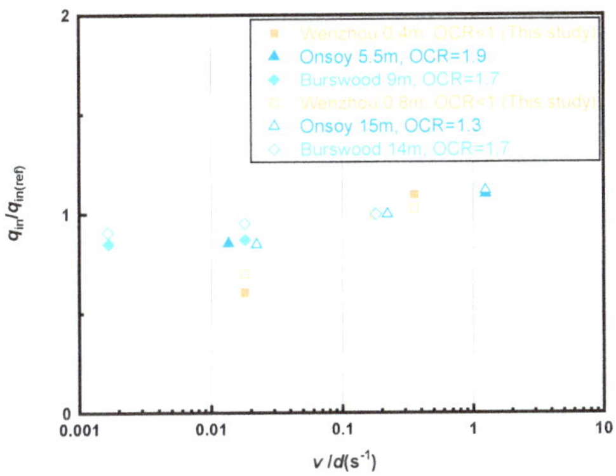

(**a**)

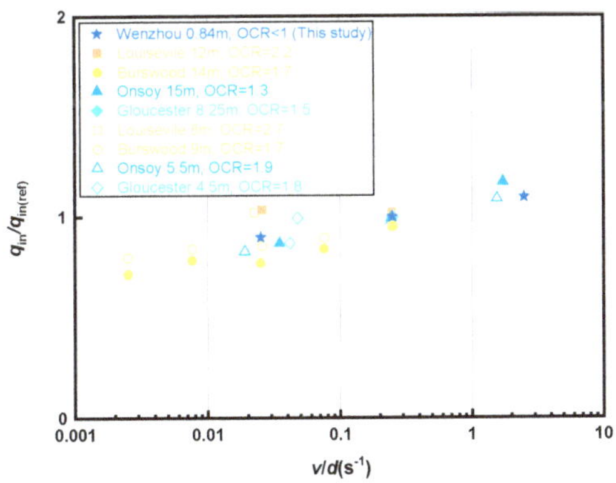

(**b**)

Figure 15. Normalized penetration resistance ($q_{in}/q_{in(ref)}$) versus normalized penetration velocity (v/d). (**a**) Ball; (**b**) T-bar.

Yafrate et al. [21] presented the values of the velocity parameters μ for Ball and T-bar as 0.14 and 0.12, respectively (Table 10), which is based on Equation (9). The testing results of the full-flow penetrometers in underconsolidated clay are subsequently compared to the stress conditions at the site and the characteristics of the soft clay to deduce the following semi-logarithmic relationships for interpretation over the underconsolidated, normally consolidated, and overconsolidated clay:

$$\frac{q_{\text{in,Ball}}}{q_{\text{in }(ref),\text{Ball}}} = 0.989 + 0.113\log\left(\frac{v/d}{(v/d)_{ref}}\right) \tag{10}$$

$$\frac{q_{\text{in,T-bar}}}{q_{\text{in }(ref),\text{T-bar}}} = 0.999 + 0.117\log\left(\frac{v/d}{(v/d)_{ref}}\right) \tag{11}$$

where $q_{\text{in,Ball}}$ is the intact penetration resistance of Ball, $q_{\text{in}(ref),\text{Ball}}$ is the reference intact penetration resistance of Ball, v/d is the normalized penetrometer velocity, $(v/d)_{ref}$ is reference normalized penetrometer velocity, $q_{\text{in,T-bar}}$ is the intact penetration resistance of T-bar, and $q_{\text{in}(ref),\text{T-bar}}$ is the reference intact penetration resistance of T-bar.

Table 10. Reported rate effect parameters for Equation 9 (Adapted from Yafrate et al. [21]).

Parameter	Normally Consolidated and Overconsolidated Clay		Underconsolidated Clay (This Study)	
	Ball	T-Bar	Ball	T-Bar
μ	0.14	0.12	0.113	0.117
a	0.98	1.02	0.989	0.999

The rate factors for penetration resistance shown in Figure 16 are slightly lower than the rate factor presented by Yafrate et al. [21] for both Ball and T-bar penetrometer variable penetration rate testing for the field data. The difference could be that they reported the rate parameter for a rate of different consolidation conditions, whereas a rate factor of 0.113 and 0.117, respectively, seems to be more appropriate for the underconsolidated clay from the Gulf of Wenzhou with weak structure behavior.

Meanwhile, the rate factors for the intact penetration resistance of Ball, $q_{\text{in,Ball}}$, are lower than those for the intact penetration resistance of T-bar, $q_{\text{in,T-bar}}$, which is consistent with the results presented by Low et al. [44]. A lower rate parameter implies Wenzhou Bay clay is slightly less rate-dependent than both normally consolidated and overconsolidated clay, which has implications for the penetration of the full-flow penetrometer in newly dredger fill.

3. Twitch penetration testing

Twitch penetration testing for the full-flow penetrometers, namely, the penetration rate is gradient-varying during penetration with the order of magnitude changes, which improved the efficiency of investigating the rate effect on penetration resistance [34]. The accelerated twitch penetration testing methodology was applied in this paper, where the velocity of penetration is gradually increased throughout a single penetration test. The penetrometers were driven into the underconsolidated clay at a velocity that varied from 2 mm/s, 5 mm/s, 10 mm/s, 20 mm/s to 40 mm/s. The results of the twitch penetration tests for the differently shaped penetrometers (cone, Ball, and T-bar) are as Figure 17 shows.

Figure 17a shows the testing profiles of tip resistance (q_t), sleeve friction (f_s), and pore water pressure (u_2), from the twitch penetration tests of the piezocone. Figure 15(a3) shows the total pore water pressure measurement during the twitch penetration test. Pore water pressure measurements at the cone's shoulder have recorded the development of pore water pressure during twitch penetration. When penetrating at a velocity of 2 mm/s, the excess pore water pressure escalates uniformly as the depth increases. Following the elevation of the penetration velocity to 20 mm/s, the pore water pressure persists in its ascent, albeit at a diminished pace. After a further increase in the penetration velocity to 40 mm/s, the measured pore water pressure u_2 remains relatively stable, indicating that the dissipation rate of pore water pressure is equivalent to the generation rate [23].

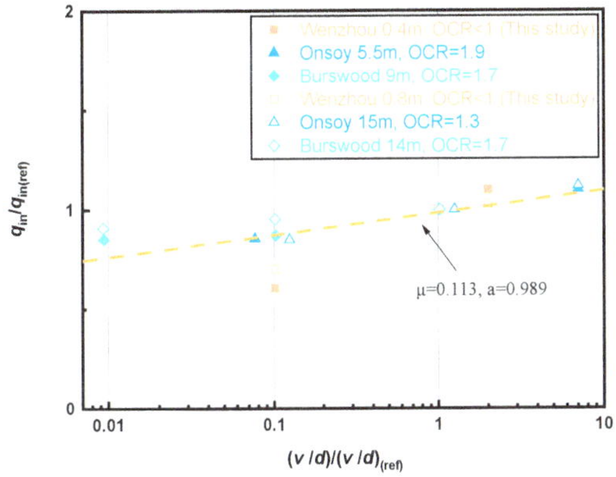

(a)

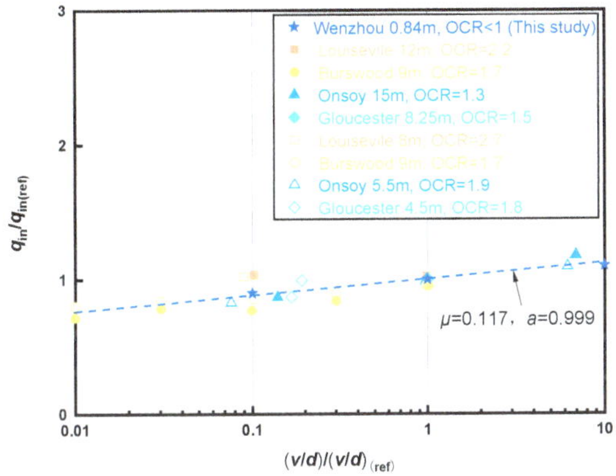

(b)

Figure 16. Relationship normalized penetration resistance versus normalized penetration velocity. (**a**) Ball; (**b**) T-bar.

The net tip resistance for the piezocone in twitch penetration testing has been corrected by Equation (3), as Figure 17b shows. As the velocity is increased during penetration, there was a transformation from the partially drained condition to the undrained condition, implying that the given normalized penetration rate range experienced critical normalized velocity (transition point), which leads to a decrease firstly and an increase subsequently of penetration resistance [45].

According to Figure 17c, the penetration resistance of the T-bar penetrometer is slightly lower at 15% than that of the Ball penetrometer. The causes can be ascribed to the geometry and flow behavior of the penetrometers, and the fact that consolidation under conditions of axial symmetry happens at a faster rate compared to plane strain [46].

Figure 17d compares the penetration resistance of the T-bar penetrometer measured in twitch penetration testing with the penetration resistance measured in constant penetration rate testing performed at a velocity of 2 mm/s. It may be seen that the T-bar twitch penetration testing initially exhibits a similar trend of penetration resistance to the monotonic penetration testing at a velocity of 2 mm/s up to a velocity of 20 mm/s from where the twitch penetration testing begins to exhibit resistance, raising more greatly with depth than the monotonic penetration testing. In the final step, with the rate of 40 mm/s, the twitch test shows approximately 25% greater resistance than the constant test.

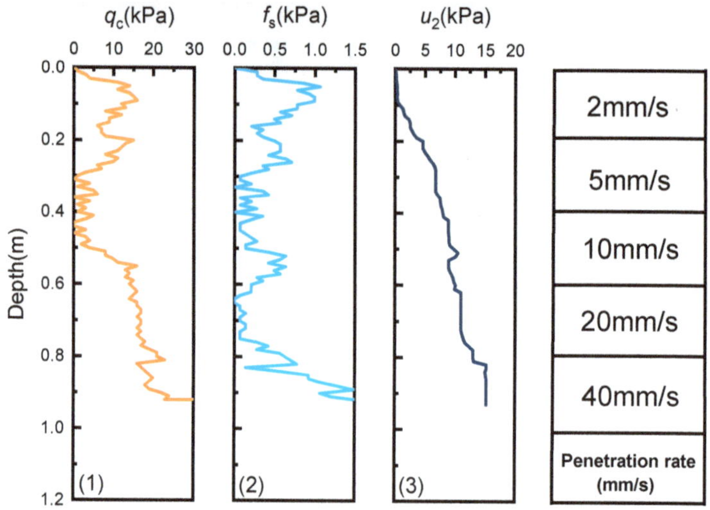

(a)

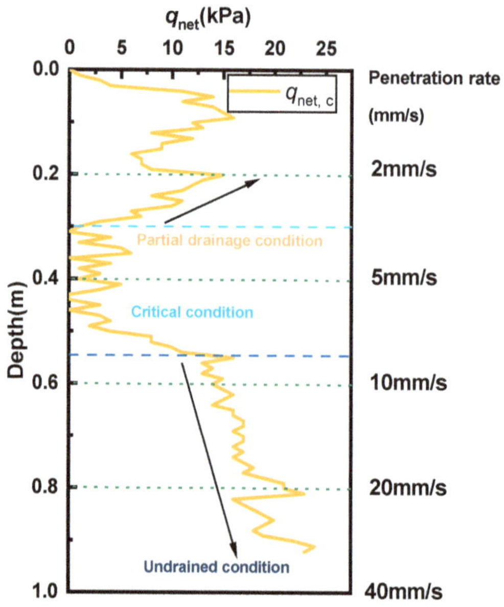

(b)

Figure 17. *Cont.*

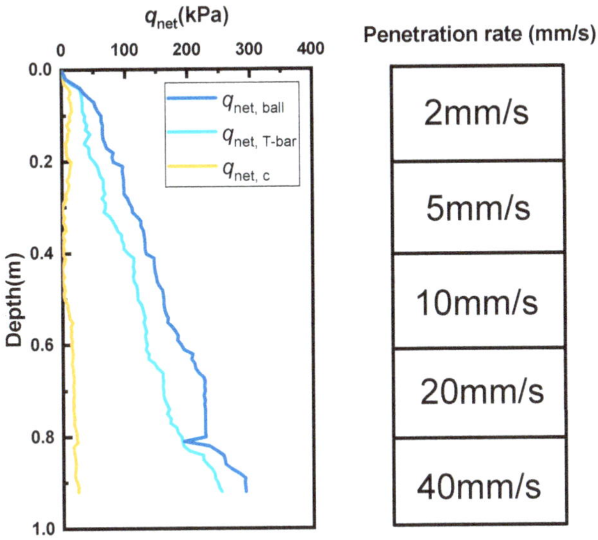

(c)

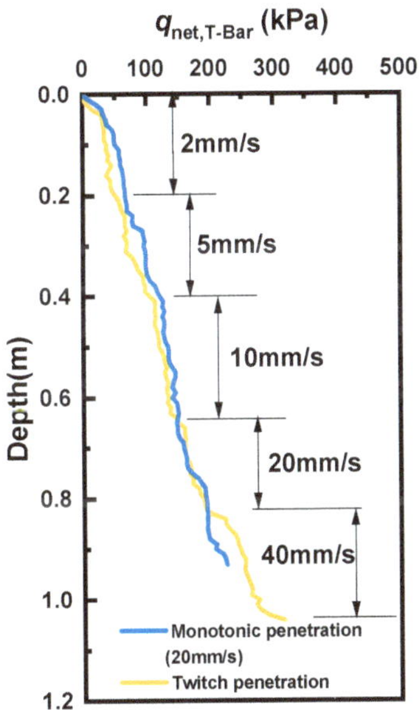

(d)

Figure 17. Twitch penetration tests: (**a**) Profiles of twitch tests for piezocone; (**b**) Tip resistance profiles for piezocone; (**c**) Penetration resistance profiles of for cone, Ball, and T-bar; (**d**) Comparison between twitch test and constant rate penetration test (20 mm/s) for T-bar.

5. Conclusions

This paper has presented the results of the variable rate monotonic penetration testing and twitch penetration testing conducted in the laboratory model box for differently shaped penetrometers (cone, Ball, and T-bar). The effects of penetration rate on the penetration resistance in underconsolidated clay have been discussed, considering the effect of the consolidation condition of clay:

1. The strength of underconsolidated clay is too low to be measured accurately by conventional piezocone. The Ball penetrometer produces slightly greater penetration resistance (10–20%) than that of the T-bar penetrometer in the underconsolidated clay, and the correction for the Ball and T-bar both are far less significant than that for the cone, which is similar to that in normally consolidated clay.
2. Using VST data as the reference strength, the values of the penetration resistance factors, N, for full-flow penetrometer in underconsolidated clay, are recommended as values of 10.9, and 12.0, respectively, for Ball and T-bar penetrometer, which both are within the recommended value range of normally consolidated clay;
3. Under undrained conditions, like in normally consolidated and overconsolidated clay, there was a general trend of increasing penetration resistance with increasing penetration rate in underconsolidated clay. Considering the influence of the consolidation state of the soil, one modified semi-logarithmic equation relating the penetration rate to the resistance penetration is applied, and the rate factor, μ, for Ball and T-bar penetrometer with this test condition is 0.113 and 0.117, respectively, with the relatively minor influence of rate-dependency behavior;
4. The rate effect on penetration resistance for full-flow penetrometer in underconsolidated clay is lower than that in overconsolidated clay, which has implications for the penetration of the full-flow penetrometer in newly dredger fill. Therefore, caution should be exercised when extrapolating the penetration resistance of the full-flow penetrometer in underconsolidated clays for direct application in design.

In this paper, we have solely evaluated the rate effect on the penetration resistance of the full-flow penetration test in underconsolidated clay. However, the impact of the rate effect on penetration resistance of the full-flow penetrometers in underconsolidated clays with different degrees of consolidation is unclear. Moreover, due to limitations in speed control of the penetration equipment, the full-flow penetration test in this study was primarily conducted under undrained conditions. In future work, we will carry out the parameter sensitivity analysis of the penetration resistance rate effect in underconsolidated soils with varying degrees of consolidation. We will also involve extending the magnitudes of penetration rates and further study the characteristics of the rate effect on penetration resistance under both drained and partially drained conditions with the full-flow penetration test.

Author Contributions: Writing—original draft preparation, H.Q. and P.P.; conceptualization, P.P.; supervision and writing—review and editing, H.H.; investigation, X.L.; data curation and providing language, R.Z.; investigation—performing the experiments and data collection, Z.J. All authors have read and agreed to the published version of the manuscript.

Funding: This research was funded by the Project of Jiangsu Province Transportation Engineering Construction Bureau, grant number CX-2019GC02, National Natural Science Foundation of China for Distinguished Young Scholars (Grant No. 42225206), and National Natural Science Foundation of China (Grant No. 42072299).

Institutional Review Board Statement: Not applicable.

Informed Consent Statement: Not applicable.

Data Availability Statement: Data are contained within the article.

Conflicts of Interest: The authors declare no conflicts of interest.

References

1. Liang, F.; Zhao, M.; Qin, C.; Jia, Y.; Wang, Z.; Yue, G. Centrifugal test of a road embankment built after new dredger fill on thick marine clay. *Mar. Georesources Geotechnol.* **2020**, *38*, 114–121. [CrossRef]
2. Wang, W.; Liu, H.; Li, Y.; Su, J. Development and management of land reclamation in China. *Ocean Coast. Manag.* **2014**, *102*, 415–425. [CrossRef]
3. Cai, Y.; Qiao, H.; Wang, J.; Geng, X.; Wang, P. Experimental tests on effect of deformed prefabricated vertical drains in dredged soil on consolidation via vacuum preloading. *Eng. Geol.* **2017**, *222*, 10–19. [CrossRef]
4. Wang, J.; Fang, Z.; Cai, Y.; Chai, J.; Wang, P.; Geng, X. Preloading using fill surcharge and prefabricated vertical drains for an airport. *Geotext. Geomembr.* **2018**, *46*, 575–585. [CrossRef]
5. Wang, J.; Gao, Z.; Fu, H.; Ding, G.; Cai, Y.; Geng, X.; Shi, C. Effect of surcharge loading rate and mobilized load ratio on the performance of vacuum–surcharge preloading with PVDs. *Geotext. Geomembr.* **2019**, *47*, 121–127. [CrossRef]
6. Zheng, J.; Hu, X.; Gao, S.; Wu, L.; Yao, S.; Dai, M.; Liu, Z.; Wang, J. Undrained cyclic behavior of under-consolidated soft marine clay with different degrees of consolidation. *Mar. Georesources Geotechnol.* **2022**, *42*, 176–183. [CrossRef]
7. Zhu, W.; Wang, J.; Zhuang, H. Influence of cyclic deviator stress and consolidation degree on permanent strain of "under-consolidated" marine sediment. *Mar. Georesources Geotechnol.* **2023**, *41*, 764–773. [CrossRef]
8. Park, H.; Lee, S.-R.; Jee, S.-H. Cover Placement on Dredged Marine Clay Reclaimed Deposit. *Mar. Georesources Geotechnol.* **2014**, *32*, 38–61. [CrossRef]
9. Stark, T.D.; Choi, H.; Schroeder, P.R. Settlement of dredged and contaminated material placement areas. I: Theory and use of primary consolidation, secondary compression, and desiccation of dredged fill. *J. Waterw. Port Coast. Ocean Eng.* **2005**, *131*, 43–51. [CrossRef]
10. Stark, T.D.; Choi, H.; Schroeder, P.R. Settlement of dredged and contaminated material placement areas. II: Primary consolidation, secondary compression, and desiccation of dredged fill input parameters. *J. Waterw. Port Coast. Ocean Eng.* **2005**, *131*, 52–61. [CrossRef]
11. Gu, C.; Wang, J.; Cai, Y.; Sun, L.; Wang, P.; Dong, Q.Y. Deformation characteristics of overconsolidated clay sheared under constant and variable confining pressure. *Soils Found.* **2016**, *56*, 427–439. [CrossRef]
12. Rotte, V.M.; Sutar, A.A.; Patel, A.; Patel, A. Effect of various parameters on electrokinetic dewatering of saturated clay. In Proceedings of the Indian Geotechnical Conference 2019: IGC-2019, Surat, Gujarat, 19–21 December 2019; Volume III, pp. 183–194.
13. Flora, A.; Gargano, S.; Lirer, S.; Mele, L. Effect of electro-kinetic consolidation on fine grained dredged sediments. *Procedia Eng.* **2016**, *158*, 3–8. [CrossRef]
14. Cavallaro, A. The use of Cone Penetration Tests (CPT) for the study of the dynamic characteristics of the soils. In Proceedings of the IMEKO TC4 International Conference on Metrology for Archaeology a Cultural Heritage, Trento, Italy, 22–24 October 2020; pp. 242–247.
15. Cavallaro, A. Dynamic characteristics of the soils by Cone Penetration Tests (CPT). In Proceedings of the 5th International Symposium on Cone Penetration Testing, CPT22, Bologna, Italy, 8–10 June 2022; pp. 311–316.
16. Qiao, H.; Liu, L.; He, H.; Liu, X.; Liu, X.; Peng, P. The Practice and Development of T-Bar Penetrometer Tests in Offshore Engineering Investigation: A Comprehensive Review. *J. Mar. Sci. Eng. Geol.* **2023**, *11*, 1160. [CrossRef]
17. Einav, I.; Randolph, M.F. Combining upper bound and strain path methods for evaluating penetration resistance. *Int. J. Numer. Methods Eng.* **2005**, *63*, 1991–2016. [CrossRef]
18. Tanaka, Y.; Sakagami, T. Piezocone testing in underconsolidated clay. *Can. Geotech. J.* **1989**, *26*, 563–567. [CrossRef]
19. Karakouzian, M.; Avar, B.B.; Hudyma, N.; Moss, J.A. Field measurements of shear strength of an underconsolidated marine clay. *Eng. Geol.* **2003**, *67*, 233–242. [CrossRef]
20. Lunne, T.; Powell, J.J.; Robertson, P.K. *Cone Penetration Testing in Geotechnical Practice*; CRC Press: Boca Raton, FL, USA, 2002.
21. Yafrate, N.J.; DeJong, J.T.; DeGroot, D.J. The influence of full-flow penetrometer area ratio on penetration resistance and undrained and remoulded shear strength. Proceedings of Offshore Site Investigation and Geotechnics: Confronting New Challenges and Sharing Knowledge, London, UK, 11–13 September 2007.
22. Taukoor, V.; Rutherford, C. Displacement rate effects during T-bar cycling in remoulded Gulf of Mexico clay. *Géotechnique* **2017**, *67*, 553–557. [CrossRef]
23. Mahmoodzadeh, H.; Boylan, N.; Randolph, M.; Cassidy, M. The Effect of Partial Drainage on Measurements by a Piezoball Penetrometer. Proceedings of 30th International Conference on Ocean, Offshore and Arctic Engineering, Rotterdam, The Netherlands, 19–24 June 2011; pp. 1007–1016.
24. Ganesan, S.A.; Bolton, M.D. Characterisation of a high plasticity marine clay using a T-bar penetrometer. *Underw. Technol.* **2013**, *31*, 179–185. [CrossRef]
25. Blight, G. A note on field vane testing of silty soils. *Can. Geotech. J.* **1968**, *5*, 142–149. [CrossRef]
26. Matsui, T.; Abe, N. Shear mechanisms of vane test in soft clays. *Soils Found.* **1981**, *21*, 69–80. [CrossRef]
27. Flaate, K. Factors influencing the results of vane tests. *Can. Geotech. J.* **1966**, *3*, 18–31. [CrossRef]
28. Schlue, B.F.; Moerz, T.; Kreiter, S. Influence of shear rate on undrained vane shear strength of organic harbor mud. *J. Geotech. Geoenviron. Eng.* **2010**, *136*, 1437–1447. [CrossRef]
29. Chung, S.; Randolph, M. Penetration resistance in soft clay for different shaped penetrometers. In Proceedings of the 2nd International Conference on Site Characterization, Porto, Portugal, 19–22 September 2004; pp. 671–677.

30. Nanda, S.; Sivakumar, V.; Hoyer, P.; Bradshaw, A.; Gavin, K.; Gerkus, H.; Jalilvand, S.; Gilbert, R.; Doherty, P.; Fanning, J. Effects of strain rates on the undrained shear strength of kaolin. *Geotech. Test. J.* **2017**, *40*, 951–962. [CrossRef]
31. Sheahan, T.C.; Ladd, C.C.; Germaine, J.T. Rate-dependent undrained shear behavior of saturated clay. *J. Geotech. Eng.* **1996**, *122*, 99–108. [CrossRef]
32. Lunne, T.; Andersen, K.H.; Low, H.E.; Randolph, M.F.; Sjursen, M. Guidelines for offshore in situ testing and interpretation in deepwater soft clays. *Can. Geotech. J.* **2011**, *48*, 543–556. [CrossRef]
33. Lehane, B.; O'loughlin, C.; Gaudin, C.; Randolph, M. Rate effects on penetrometer resistance in kaolin. *Géotechnique* **2009**, *59*, 41–52. [CrossRef]
34. Chung, S.F.; Randolph, M.F.; Schneider, J.A. Effect of penetration rate on penetrometer resistance in clay. *J. Geotech. Geoenviron. Eng.* **2006**, *132*, 1188–1196. [CrossRef]
35. Guo, X.; Liu, X.; Zhang, H.; Shan, Z.; Sun, M. Improved predictive model for the strength of fluidized seabed sediments with rate effect characteristics by full-scale spherical penetrometer tests. *Comput. Geotech.* **2023**, *161*, 105535. [CrossRef]
36. Sespeñe, S.M.; Choo, Y.W. Empirical factors for miniature cone and T-bar penetrometers for kaolin clay. *KSCE J. Civ. Eng.* **2019**, *23*, 4675–4686. [CrossRef]
37. Mahmoodzadeh, H.; Randolph, M.F. Penetrometer testing: Effect of partial consolidation on subsequent dissipation response. *J. Geotech. Geoenviron. Eng.* **2014**, *140*, 04014022. [CrossRef]
38. Zhu, B.; Dai, J.; Kong, D. Assess the effects of loading rate and interface roughness on T-bar penetration resistance. *Soils Found.* **2020**, *60*, 266–273. [CrossRef]
39. Randolph, M.; Hope, S. Effect of cone velocity on cone resistance and excess pore pressures. In Proceedings of the Engineering Practice and Performance of Soft Deposits, Osaka, Japan, 1 January 2004; pp. 147–152.
40. House, A.; Oliveira, J.; Randolph, M. Evaluating the coefficient of consolidation using penetration tests. *Int. J. Phys. Model. Geotech.* **2001**, *1*, 17–26. [CrossRef]
41. Low, H.; Lunne, T.; Andersen, K.; Sjursen, M.; Li, X.; Randolph, M. Estimation of intact and remoulded undrained shear strengths from penetration tests in soft clays. *Géotechnique* **2010**, *60*, 843–859. [CrossRef]
42. DeJong, J.T.; Yafrate, N.J.; DeGroot, D.J. Evaluation of undrained shear strength using full-flow penetrometers. *J. Geotech. Geoenviron. Eng.* **2011**, *137*, 14–26. [CrossRef]
43. Cai, G.; Liu, S.; Tong, L.; Du, G. Field evaluation of undrained shear strength from piezocone penetration tests in soft marine clay. *Mar. Georesources Geotechnol.* **2010**, *28*, 143–153. [CrossRef]
44. Low, H.; Randolph, M.; DeJong, J.; Yafrate, N. Variable rate full-flow penetration tests intact and remoulded soil. In Proceedings of the Geotechnical and Geophysical Site Characterization, London, UK, 1 January 2009; pp. 1087–1092.
45. Bemben, S.; Myers, H. The influence of rate of penetration on static cone resistance in Connecticut river valley varved clay. In Proceedings of the European Symposium on Penetration Testing, Stockholm, Sweden, 5–7 June 1974; pp. 33–43.
46. Davis, E.H.; Poulos, H. Rate of settlement under two-and three dimensional conditions. *Geotechnique* **1972**, *22*, 95–114. [CrossRef]

Journal of
Marine Science and Engineering

MDPI

Article

Installation Disturbance of Helical Anchor in Dense Sand and the Effect on Uplift Capacity Based on Discrete Element Method

Rong Chen [1], Hu Liu [1], Dongxue Hao [1,*], Zhaoguo Liu [2] and Chi Yuan [3]

1 School of Civil Engineering and Architecture, Northeast Electric Power University, Jilin 132012, China; 20112384@neepu.edu.cn (R.C.); lhu0221@163.com (H.L.)
2 China Energy Engineering Group Heilongjiang Electric Design Co., Ltd., Harbin 150090, China; zgliu0019@ceec.net.cn
3 College of Architecture and Civil Engineering, Beijing University of Technology, Beijing 110124, China; yuanc@emails.bjut.edu.cn
* Correspondence: 20102291@neepu.edu.cn; Tel.:+86-432-6480-6481

Abstract: Helical anchors have been extensively employed as foundation systems for carrying tension loads due to their installation efficiency and large uplift capacity. However, the installation influences of helical anchors are still not well understood, especially for multi-helical anchors. The matrix discrete element method was used to model the process of helical anchor penetration and pull-out in dense sand to investigate the effects of the anchor geometry and advancement ratio (AR, the relative vertical movement per rotation) on soil disturbance, the particle flow mechanism, and the uplift capacity. For shallow helical anchors, the overall disturbance zone is the shape of an inverted cone after installation, while for deep helical anchors, it is funnel-shaped. The advancement ratio has significant effects on the soil particle movement and uplift capacity of helical anchors. The soil particle flow mechanism around helical plates has been identified for single-helix anchors at various advancement ratios, and for double-helix anchors, the influence of the top plate on particle movement during installation was investigated. The uplift capacities of both single- and double-helix anchors increase with the decrease in the AR (AR = 0.5~1), and the influence decreases with the anchor embedment ratio. The efficiency of double-helix anchors induced by installation is close to 1 at pitch-matched installation (AR = 1), indicating that the impact of the top plate during installation is minimal in this case.

Keywords: DEM; helical anchor; installation effect; particle movement; uplift capacity

Citation: Chen, R.; Liu, H.; Hao, D.; Liu, Z.; Yuan, C. Installation Disturbance of Helical Anchor in Dense Sand and the Effect on Uplift Capacity Based on Discrete Element Method. *J. Mar. Sci. Eng.* **2024**, *12*, 422. https://doi.org/10.3390/jmse12030422

Academic Editors: Dong-Sheng Jeng and Pan Hu

Received: 1 February 2024
Revised: 20 February 2024
Accepted: 26 February 2024
Published: 27 February 2024

1. Introduction

A helical anchor, consisting of one or more helical plates welded to a steel shaft, is a deep foundation system used to support or resist any load or application. It has been extensively employed in many engineering systems, such as offshore platforms, pipelines, transmission towers, slopes and embankments, due to its rapid installation and immediate service, relatively large load-bearing capacity, and no associated environmental damage [1–4]. It is a form of displacement pile or anchor, screwed into the ground by applying a torque at the top of the anchor together with a crowd force [5–7]. There exists a coupling effect of compression and shear on the surrounding soil during the installation process. The installation effects cannot be ignored in practical engineering.

In recent years, research on the installation effect and required load of helical anchors has been increasing, including studies on the disturbed zone, density change, soil parameters after disturbance [8–10], lateral earth pressure and uplift or compressive performance after disturbance [11–14], installation requirements, particle motion during screwing and the torque correlation factor [15–18]. Some studies have investigated the impact of the installation advancement ratio (AR) on the load-bearing performance of single-helix anchors, where the AR is defined as the ratio of vertical displacement per one rotation Δz to helix pitch p_h [16,19,20]. Cerfontine et al. [16,20] found that a lower vertical displacement per rotation can significantly reduce the necessary crowd force during installation or even generate some pull-in, and the uplift stiffness and capacity of the pile were enhanced by this installation process, but the torque remained relatively unchanged. In addition, the pile capacity in tension generally increases as the AR is reduced and reaches a maximum for AR = 0.5, while the compressive capacity reduces.

Extensive experiments [8,13,21–30], numerical analyses [12,31–37], and theoretical analyses [38–40] have been conducted on the load-bearing performance of helix anchors. Tsuha et al. [28] performed centrifuge uplift loading tests on 12 different types of piles installed in two sands with different densities to examine the effect of the number of helices on the performance of helical anchors in sand. This investigation indicated that in double- and triple-helix anchors, the contributions of the second and third plate to the total anchor uplift capacity decreased with the increase in the sand relative density and plate diameter because the upper soil layers were penetrated more times. Lutenegger et al. [13] conducted field load tests on the compression and tension of helical anchors and measured the degree of installation disturbance using the ratio of uplift and compression-bearing capacity. Wang et al. [29] conducted a series of model tests in dense saturated sand and found the installation speed and helix diameter have a significant influence on the uplift capacity of the helical piles. Nagai et al. [30] studied the effect of the installation method on the performance of helix piles through calibration chamber tests and found that the load–displacement relationship strongly depended on the installation method, but that the second-limit uplift resistance was almost unaffected. Cerfontine et al. [41] used the finite element method to predict the full tensile load–displacement response of shallow helix anchors installed in sand for practical use, incorporating the effects of a pitch-matched installation, and found the compression (crowd) load applied during the anchor installation phase modified the stress field around the anchor, which in turn affected the anchor uplift stiffness. The DEM has also been used to investigate the installation effects of helical anchors [31–33]. Cerfontine et al. [31] found that when a single-helix pile is installed at AR < 1, the soil particles below the helical plate show an upward movement trend. Sharif et al. [32,33] investigated the effects of the base geometry, shaft diameter, and helix pitch of single-helix anchors by simulating the full installation process prior to conducting axial compression and tension tests and proposed the possible optimization of the geometry of the screw pile to reduce installation requirements.

Among these methods, the DEM can visually observe the movement of particles during installation and its impact on bearing capacity. The discrete element method (DEM) is based on molecular dynamics and was first developed in the fields of physics and fluid dynamics. In this method, the granular assemblies are made up of a series of soft particles, which obey Newton's equations of motion. The behavior of granular material can be simulated by investigating the movement of the discrete assemblies [41–44]. However, existing research on pile installation effects based on the DEM has focused on complex screw piles (augers) [18] and single-helical piles at a certain depth [31–33]. The installation effects of multi-helix anchors and single-helix anchors at different embedment depths on particle displacement and uplift capacity are still unclear.

Therefore, this investigation used the discrete element method (MatDEM software version 3.4) to simulate the installation and pull-out process of single- and double-helix anchors to analyze the installation disturbance caused by different advancement ratios and the influence on the subsequent uplift performance. Firstly, the discrete element method and procedures, and the particle contact parameter calibration and simulation scheme are described (Section 2). Then, the disturbance zones, the soil particle motion, around single- and double-helix anchors during anchor installation were analyzed and the particle flow mechanism at different advancement ratios were discerned (Section 3). Finally, the effects of installation on the uplift capacity were investigated by comparing the results of wish-in-place and installed single-helix anchors, and the results of anchors with different advancement ratios are presented. Moreover, the efficiency of double-helix anchors at different advancement ratios is compared (Section 4).

2. Method and Models

2.1. Introduction to MatDEM and Contact Model

The discrete element method (DEM) starts with analyzing the inter-block contact of discrete elements to identify their constitutive relationship, establishing a physical and mechanical model of the contact, and simulating discontinuous and discrete elements according to Newton's second law. This study employed the MatDEM software, developed by Nanjing University. It runs in the Matlab environment, with data stored in matrix form in objects. All operations are converted into pure matrix operations, which can be analyzed in real time. It supports mixed CPU and GPU computing, and can switch between the two in real time during program operation, greatly improving the computational speed of particle simulation [45].

In the most basic linear elastic model, it is assumed that particles interact with each other and generate forces through springs. The normal force F_n and normal deformation X_n between particles can be simulated through the normal spring between particles, as shown in Figure 1a. The particles are initially connected to each other and interact with each other through tension (positive tension) or compression. When the relative displacement between the two particles exceeds the fracture displacement X_b, springs are broken, the tension disappears and only compression exists.

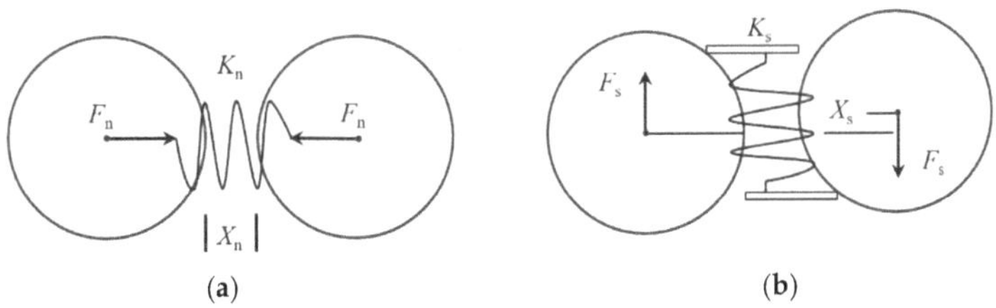

Figure 1. Particle contact: (**a**) normal spring contact, (**b**) tangential spring contact.

Similarly, the simulation of the shear force and shear deformation between particles through tangential springs is shown in Figure 1b. When two particle surfaces contact with each other and slide in the shear direction, the tangential force generated is opposite to the relative movement direction. This tangential force can be considered the action of tangential springs between particles. The interaction force between particles is determined by Equations (1)–(4) [41].

$$F_n = \begin{cases} K_n X_n, \ X_n < X_b \ \text{(Complete connection)} \\ 0, \qquad\qquad\quad \text{(Connection broken)} \end{cases} \tag{1}$$

$$F_s = K_s X_s \tag{2}$$

For a complete connection of particles, the maximum tangential force allowed by the Coulomb criterion is

$$F_{smax} = F_{s0} - \mu_p F_n \ \text{(Complete connection)} \tag{3}$$

$$F_{smax} = -\mu_p F_n \ \text{(Connection broken)} \tag{4}$$

In the formulas, F_n is the normal force; K_n is the normal stiffness; X_n is the normal relative displacement; X_b is the fracture displacement; F_s is the tangential force; K_s is the tangential stiffness; X_s is the tangential relative displacement; F_{s0} is the inter-granular shear resistance; μ_p is the inter-particle friction coefficient. F_{smax} is the maximum shear force; F_{s0} is the shear resistance between particles.

When the tangential force exceeds the maximum allowable tangential force, the tangential connection between the two particles is broken, the shear resistance (F_{s0}) is zero, and only sliding friction force ($-\mu_p F_n$) exists [41].

In the case of equal particle size, the microscopic mechanical parameters K_n, K_s, X_b, F_{s0}, μ_p between particles can be calculated through five macroscopic mechanical parameters, including Young's modulus E, Poisson's ratio v, compressive strength C_u, tensile strength T_u, and internal friction coefficient μ_i, as well as particle diameter d. Equations (5)–(9) represent analytical solutions between macroscopic and microscopic mechanical parameters under the same particle size.

$$X_b = \frac{(3K_n + K_s)T_u d^2}{6\sqrt{2}K_n(K_n + K_s)} \tag{5}$$

$$K_n = \frac{\sqrt{2}}{4(1-2v)}Ed \tag{6}$$

$$K_s = \frac{\sqrt{2}(1-5v)}{4(1+v)(1-2v)}Ed \tag{7}$$

$$F_{s0} = \frac{(1-2\mu_p)}{6}C_u d^2 \tag{8}$$

$$\mu_p = \frac{-2\sqrt{2}+\sqrt{2}I}{2+2I},\ I = \left[\left(1+\mu_i^2\right)^{1/2} + \mu_i\right]^2 \tag{9}$$

2.2. Model Establishment Steps and Model Parameter Calibration

To calibrate the micro-mechanical parameters in the DEM numerical calculation model, a model test of a single-helix anchor in reference [30] (test numbers De-A, Lo-A) was first simulated, where the anchor was installed continuously in homogeneous dense sand and then pulled out. The size of the calibration chamber is 1 m × 1 m × 1.2 m, in which a uniformly distributed load of 100 kPa was applied on the soil surface. The single-helix anchor rod diameter d_0 and plate diameter d_w are 48.6 mm and 97.2 mm, respectively, and the embedment ratio H/d_w equals 5.

The installation and pull-out process of a helical anchor in MatDEM can be achieved in four steps by establishing four sub-files for analysis: (1) Establish an initial accumulation model, namely the soil model; (2) establish an anchor model and assemble it with the soil, then assign material parameters; (3) set the installation load for iterative calculation of helical anchor installation. The installation is controlled by vertical displacement and angular displacement, both of which satisfy the relationship of penetrating into the soil at one pitch after one rotation ($1p_h/r$), that is, the advancement ratio AR = 1; (4) set up the uplift displacement for iterative calculation of helical anchor pull-out.

When establishing the initial soil layer, to reduce the run time of the simulation, a particle distribution scaling value of 56 was used in the whole sand layer, which was selected based on the minimum recommended ratio of the diameter of the pile core to the median particle size d_{50} of 2.69 [46]. Then, the median particle size d_{50} was set as 18 mm, 1/2.7 times the rod diameter d_0 of 48.6 mm. The dispersion coefficient was 0.25, and the ratio of maximum to minimum particle size (d_{100}/d_0) is $(1 + 0.25)^2$. And to ensure the soil particles passing through the helix pitch opening do so smoothly, helix pitch p_h was set as 50 mm. In order to balance the boundary effect and calculation time, the sand layer width is set to 10 times the helix diameter (d_w). The computational domain size (10 d_w) can cause negligible effects because the sand displacement at the boundary during the installation and pulling out of the anchor is less than 1 mm (0.056 d_{50}), which is close to the results (13 times pile diameter) for pile installation [32]. The length of the simulated sand layer is set to 1.2 m, that is, anchor embedment depth H plus 7 d_w, which is large enough to avoid boundary effects. It is judged from this simulation of the calibration model that the distance of 1 d_w from anchor bottom to bottom boundary will not produce boundary effects. Therefore, in subsequent simulation models in Section 3 and 4, the length of the sand layer is set to 1.33 m (more than embedment depth plus 1 d_w). The number of soil particles in calculation domain is 150,000~160,000. Pressure plates above the soil was installed to achieve gravity accumulation. Due to the setting of an overburden pressure of 100 kPa above the soil in the calibration chamber, an overburden pressure of 100 kPa was also set in the validation model simulation, but it was not set in the subsequent analysis of the influence of the installation advancement ratio. The helical anchor size was set according to the actual size of the model test. The simulation model is shown in Figure 2.

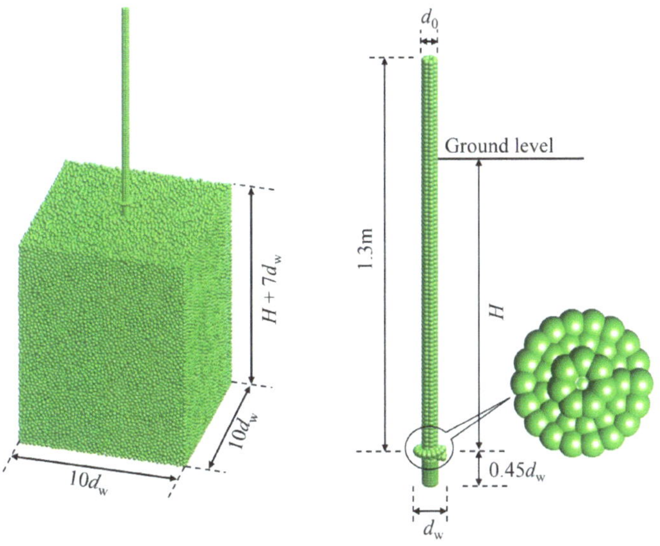

Figure 2. Soil and anchor model diagram.

When simulating the installation, the displacement loading conditions for the anchor are set as follows: vertical displacement is the embedment depth H, and angular displacement is $2\pi H/p_h$. After installation, apply an upward displacement of 0.01 m (10% d_w) to the top of the anchor to pull the anchor. In order to control the calculation time, 50 incremental steps were set during the installation and pulling out phases, with each incremental step corresponding to 20 sub-incremental steps, totally including 2000 iterative calculations with 400,000 iterations.

2.3. Particle Contact Parameter Calibration

The macro-parameters input in MatDEM need to be determined first based on the original parameter values, multiplying the "rate" matrix that was trained by the "material training" module. The default rate matrix for five macro-mechanical parameters E, v, T_u, C_u, μ_i, and ρ is [2.7, 0.8, 6.0, 6.5, 1, 1.19]. And then, the micro-mechanical parameters were determined in terms of macro-mechanical parameters, as the analytical Formula (5)~(9). The original macro-parameters μ_i and ρ originate from the literature [30], and other parameters are determined by several attempts based on empirical parameters of sand. The final particle contact parameters were calibrated by comparing the results obtained from the pull-out model test with the numerical simulation results of variable parameters.

The numerical results used to determine the particle contact parameters were plotted against model test results, as shown in Figure 3, in which u is upward displacement, d_w is helix diameter, δ is the ratio of displacement to helix diameter, q_p represents the average pressure on the entire cross-section of the anchor plate. And the calculated torque value in terms of the torque–uplift capacity relationship formula [7] was compared to the value obtained in the test, as shown in Figure 4.

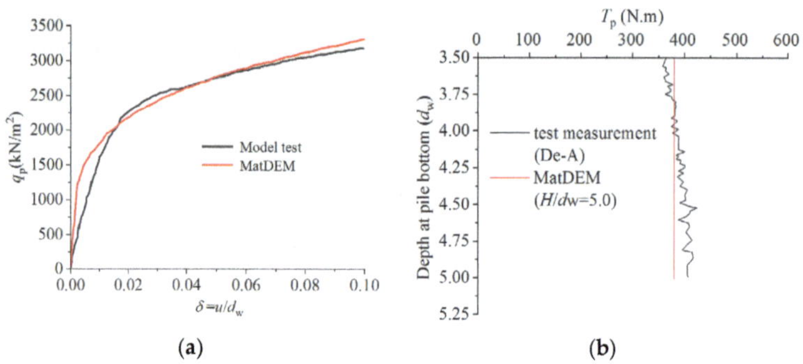

Figure 3. Comparisons of numerical and test results: (**a**) uplift resistance–displacement relationship, (**b**) installation torque.

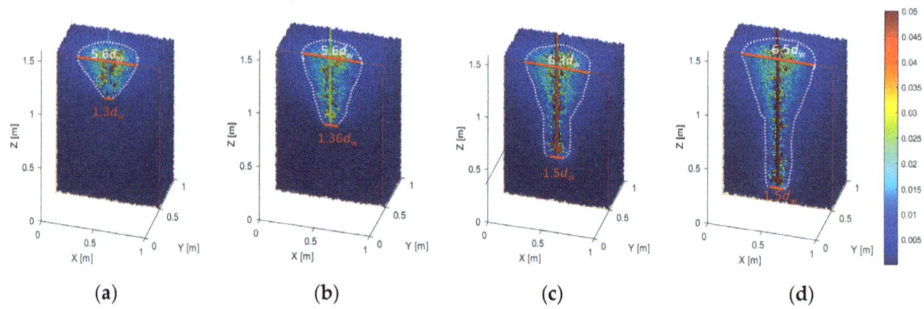

Figure 4. Displacement nephogram of single-helix anchor with H/d_w = 12 during installation (surcharge p = 100 kPa) (**a**) L = 3 d_w (**b**) L = 6 d_w (**c**) L = 9 d_w (**d**) L = 12 d_w (unit: m).

As the upward displacement increases, the anchor plate pressure increases, and it does not reach a stable stage within the displacement loading. The curve exhibits three stages of straight, curved, and straight characteristics. The initial slope of the straight line segment in the numerical result is significantly higher than the experimental result, but the final straight line stage of the two curves basically coincides and the starting displacement point of the straight line is similar. This study takes the uplift capacity corresponding to 10% d_w displacement as the ultimate uplift capacity [47]. Then, the ultimate uplift capacity only differs by 4.2%, and the torque at a position of 5 dw differs by only 7.2%. Overall, the ultimate capacity and trend of the load–displacement curve obtained from the numerical simulation are in good agreement with the model test results, and these input parameters are used as subsequent calculation parameters, as shown in Table 1.

Table 1. Calibration values of macroscopic and microscopic mechanical parameters.

Macro-Parameters			Micro-Parameters		
	Dense Sand	**Anchor**		**Dense Sand**	**Anchor**
Modulus, (E/GPa)	0.162	5.4	Normal stiffness K_n/(MN/m)	3.0322	112.31
Poisson's ratio, v	0.16	0.24	Shear stiffness Ks (MN/m)	0.5228	19.363
Tensile strength (T_u/kPa)	0.06	1.2×10^4	Breaking displacement X_b (m)	2.04×10^{-9}	1.363×10^{-5}
Compression strength (C_u/kPa)	1.3	1.3×10^4	Friction coefficient, μ_p	0.2828	0
Internal friction coefficient, μ_i	0.75	0.3	Shear resistance T_f (N)	0.0421	866.67
Density ρ/(kg·m^{-3})	2011	9341			

2.4. Simulation Scheme

Based on the calibration of micro-mechanical parameters, the numerical simulations varying the anchor geometry and installing advancement ratio were conducted to investigate the installation disturbance and its effects on uplift capacity. The simulation program is listed in Table 2. The first group is used to examine the effects of the embedment ratio and advancement ratio on installation effects for single-helix anchors, including the comparisons of pre-embedment (wish-in-place) and installation at AR = 1. The second group is used to examine the effects of helix number and their relative spacing for double-helix anchors. The helix pitch and shaft diameter d_0 are 50 mm and 0.5 d_w.

Table 2. Modeling program.

Group	Influence Factor	Installation Mode	Helix Diameter, d_w(mm)	Embedment Ratio, H/d_w	Relative Spacing S/d_w	Surcharge Pressure (kPa)
Single-helix anchor	Embedment depth, installation mode	Wish-in-place AR = 1	100	6, 8, 9, 10, 12	-	100
		AR = 0.5, 0.8, 1	100	3, 6, 9, 12	-	-
Double-helix anchor	Helix number, relative spacing	AR = 0.5, 0.8, 1	100	3, 6, 9, 12	3, 4.5	-

AR = $\Delta z/p_h$, Δz—the vertical displacement per rotation, p_h—the geometric pitch of helical plate.

3. Particle Motion Induced by Installation

3.1. Single-Helix Anchor Installed at AR = 1 (Surcharge p = 100 kPa)

An arbitrary face through vertical axis z can be cut to examine the particle motion characteristics due to the axial symmetry of the geometry and the installation. The Y-axis middle section was selected to exhibit the displacement nephogram of the soil particles during the drilling process, as shown as Figure 4.

Figure 4 shows the results of a single-helix anchor with H/d_w =12 (ground surcharge p = 100 kPa) during installation at AR = 1. It can be seen that the displacement of particles around the anchor is significantly greater than that of the outer surrounding particles, and the particles near the ground surface around the shaft are uplifted. The white dashed line is the contour with a particle displacement of half a particle's radius (approximate 5 mm). It is shown that at the drilling depth L, which is shallow and less than 6 d_w, the particles within the zone of 1.4 d_w around the plate move. The disturbed zone from the helical plate to the ground surface presents as an inverted cone shape. The radius of the disturbed zone on the ground surface is about 5.6 times the radius of the helical plate.

As the anchor drilling proceeds, the overlying pressure on the anchor plate increases, and the lateral movement of particles around the anchor plate is gradually limited. The particle movement area below 6 d_w ground level does not expand; the radial compression around the plate is more obvious. The range of soil particle movement around the anchor gradually transforms from an approximately inverted cone (L = 3~6 d_w) to a funnel shape (white line) (L = 9~12 d_w), approaching a shape that is a combination of an inverted cone and a cylinder. The diameter of the cylinder below remains, basically, non-varied, at 1.5 d_w, and the influence radius on the ground surface slowly increases with the drilling depth. For the single-helix anchor, the overall disturbance zone is funnel-shaped after installation.

The horizontal displacement of the soil particles u_x were extracted at the axial radius $r \approx d_w/2$ along the ground surface to the bottom of the soil layer at different drilling depths, and the scatter points are plotted in Figure 5, with solid lines representing the envelopes of u_x. As the drilling depth increases, the particle motion becomes more intense and the maximum horizontal displacement value gradually increases. However, when the drilling depth L = 9 d_w and 12 d_w, the maximum horizontal displacement is basically the same, with a value of about the radius of the helical plate, and occurs about 2 d_w above the plate. From the maximum displacement position to the depth of the anchor plate, the horizontal displacement gradually decreases until its value approaches zero; the soil particles mainly move vertically within this range.

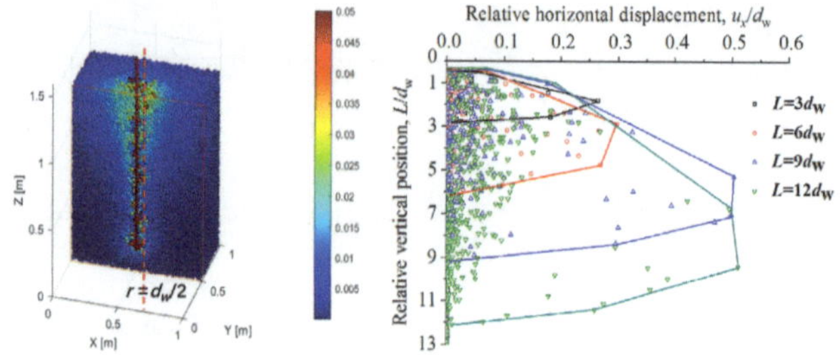

Figure 5. Horizontal displacement of particles at the vertical line of $r \approx d_w/2$ at various penetration depths.

The upper portion of the envelopes at different drilling depths are basically overlapped, which indicates that the displacement of the upper soil particles has inheritance and is no longer affected by the lower helix continuous penetration.

3.2. Single-Helix Anchors Installed at Different Advancement Ratios (ARs)

Figure 6 shows the comparisons of the particle displacement during the installation for AR = 0.5 and 1. The particle displacements are greater in the case of the smaller advancement ratio. When the anchor drills into the soil for less than 6 d_w, the influence zone caused by the installation are basically the same. When the anchor drills greater than 9 d_w, the difference in the particle motion zone for the different AR is obvious; the influence zone increases as the AR decreases.

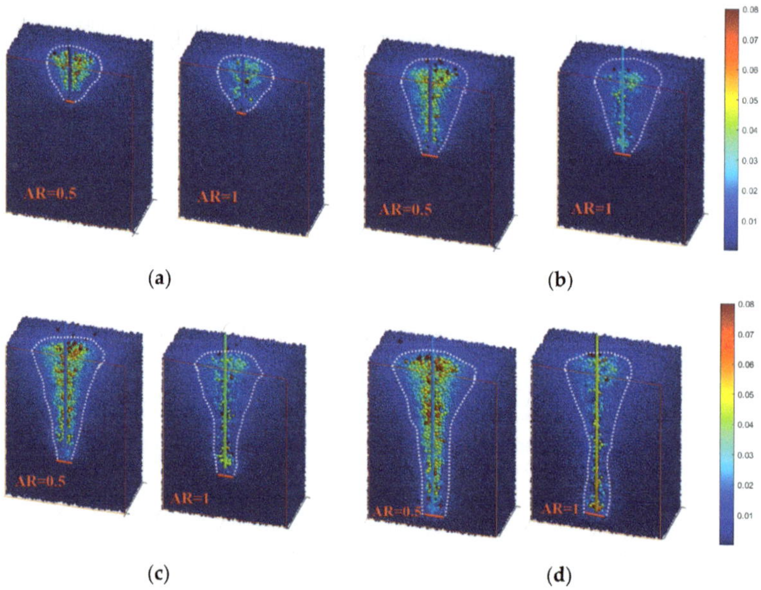

Figure 6. Displacement nephogram during installation for different AR: (**a**) $L = 3\ d_w$, (**b**) $L = 6\ d_w$, (**c**) $L = 9\ d_w$, (**d**) $L = 12\ d_w$ (unit: m).

In order to observe the particle displacement more clearly, the vertical and horizontal particle displacements after the installation of the anchor with an embedment ratio of 12 at different ARs are shown in Figure 7. The white dashed boxes is the periphery of helix diameter and the red boxes represent the area of significant particle movement.

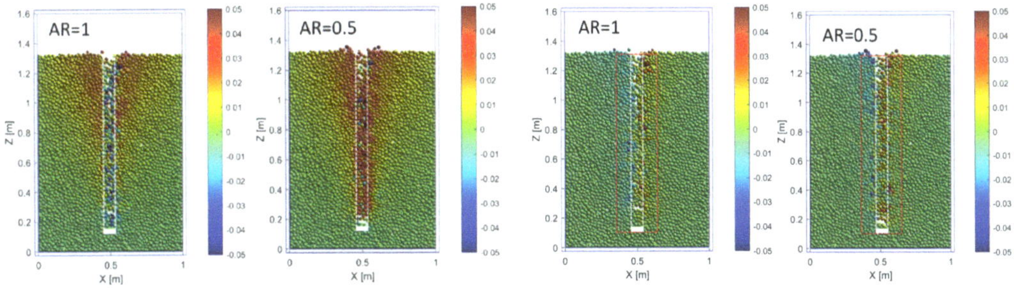

(a) (b)

Figure 7. Comparison of particle displacement after installation for single-helix anchor with H/d_w = 12 at AR = 0.5 and 1.0. (**a**) Vertical displacement, (**b**) lateral displacement (unit: m).

It can be observed from the displacement nephogram for AR = 1 that the particles around the anchor shaft are primarily displaced downwards and laterally and soil particles outside the diameter of the anchor plate display an upward displacement to different extents. The downwards flow-around mechanism may cause the sand around the shaft and above the plate to loosen, which is the reason for the decrease in the uplift capacity [18,26].

It can be seen from the nephogram of AR = 0.5 that the particles around the anchor shaft are primarily displaced upwards and laterally. Within the white dashed area, the cylindrical zone of the plate diameter, the particle lateral displacements for AR = 0.5 are larger than those for AR = 1. But the range value of the particle lateral displacements in the outer zone, represented by the red dashed area, are basically the same. The radial motion of particles within the cylindrical zone along the helical plate will not change the sand density. This indicates that the installation with AR = 0.5 (over-flighted) makes particles denser than the installation with AR = 1.0 (pitch-matched).

3.3. Double-Helix Anchors Installed at Different Advancement Ratios

Figure 8 shows the soil particle displacement nephogram of the double-helix anchors with different relative helix spacing at AR = 0.5. Blue and red circles roughly indicate the particle motion zones above the top helix and between the two helices in Figure 8a. And the boxes in Figure 8b roughly represent the area where particles exhibit lateral motion. For the double-helix anchors, the soil above the upper helical plate is disturbed more obviously than for the single-helix anchor because it has been penetrated twice. The particles above the upper helical plate move upwards and laterally more intensely during drilling. The obvious heave of the ground surface occurs. The displacement of soil particles between the two plates also changes significantly compared to that of a single-helix anchor, although the soil between the plates is only penetrated once, which indicates that the installation of the top plate will affect the movement of soil particles between the plates.

When the relative helix spacing is from 4.5 to 3, the vertical displacement nephograms show the red color of the particles between the two plates gradually decreases compared with the color of the single-helix anchor at the same depth, which indicates the amount of particles below the top plate rolling to the above of the top plate probably decreases. Compared to the single-helix anchor, the lateral displacement of the soil particles on the top helix increases significantly, which probably decreases the friction resistance around the shaft.

To further explore the impact of the top helix on the movement of soil particles between two helical plates, the particle displacement with an absolute vertical displacement and lateral displacement greater than 9 mm (0.5 d_{50}) are extracted and plotted in Figures 9 and 10, with dashed lines representing envelope lines. Black, red and blue dashed lines corresponds to the cases of single-helix, double-helix with S/d_w = 3 and double-helix with S/d_w = 4.5 respectively.

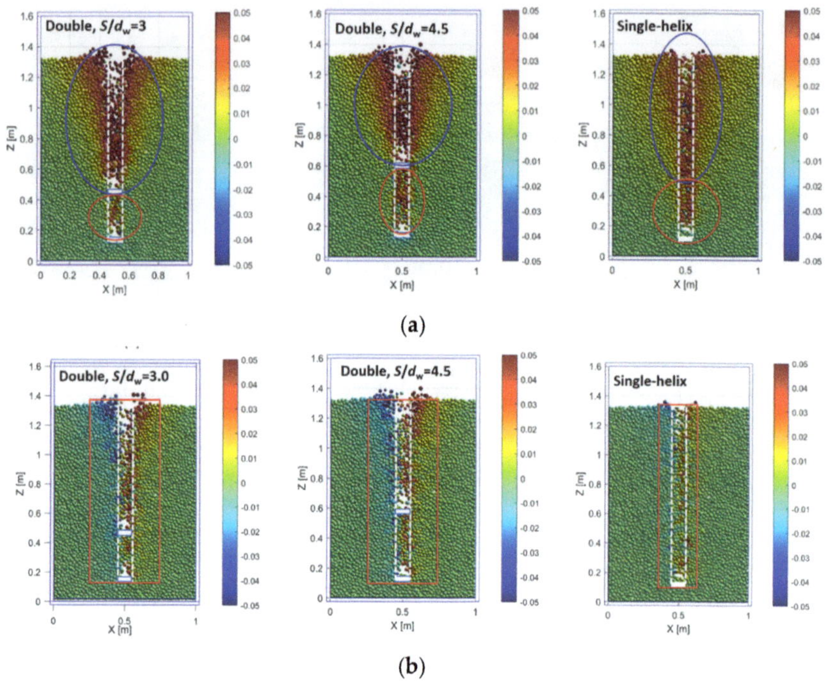

Figure 8. Soil particle displacement nephogram of double-helix anchors with different relative helix spacing at AR = 0.5. (a) Vertical displacement, (b) lateral displacement (unit: m).

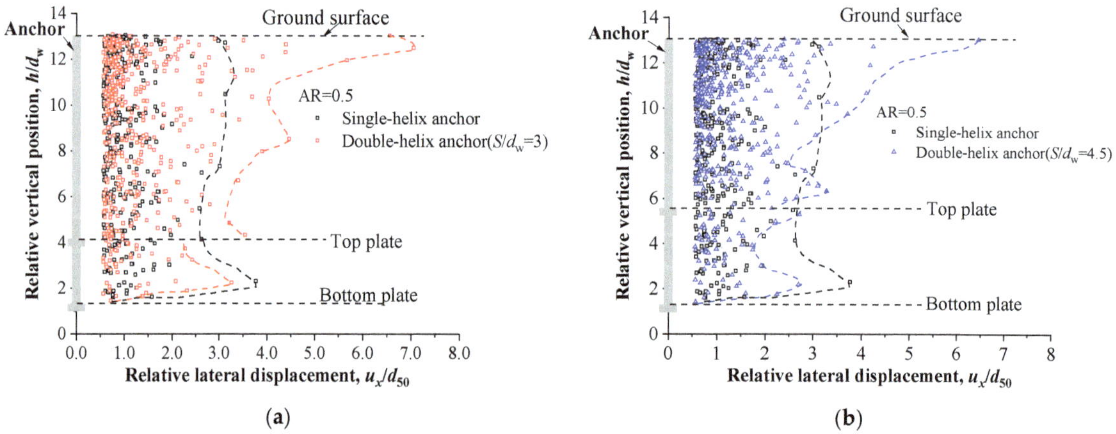

Figure 9. Comparisons of particle lateral displacement of double-helix anchor with single-helix anchor at AR = 0.5. (a) $S/d_w = 3$, (b) $S/d_w = 4.5$.

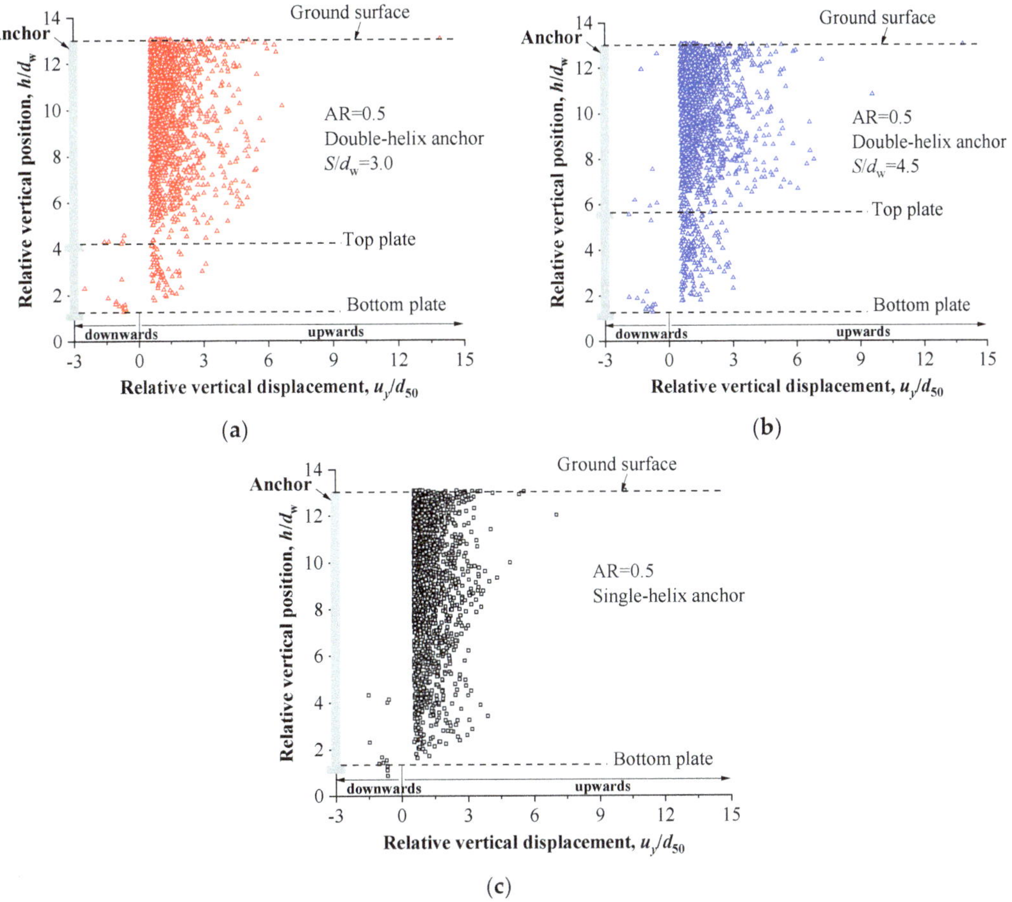

Figure 10. Comparisons of particle vertical displacement of double-helix anchor with single-helix anchor at AR = 0.5. (**a**) Double-helix anchor with S/d_W =3. (**b**) Double-helix anchor with S/d_W =4.5. (**c**) Single-helix anchor.

The lateral displacement of soil particles between the two helical plates is significantly reduced by the constraint of the two plates, especially for the anchor with a relative spacing of 3. There is a significant mutation in the lateral displacement of soil particles above and below the top helix for S/d_W = 3.

In the case of AR = 0.5, vertical displacements of soil particles along the depth of the single-helix and double-helix anchors are basically upwards. The vertical displacement of the particles above the top plate are greater than that of the single-helix anchor. Combined with the lateral displacement rule mentioned before, this will make the sand above the top plate looser than that of the single-helix anchor. The vertical displacement of the soil particles between the two plates is less than that of the singe-helix anchor, which indicates that the top plate also has a limitation for the upwards movement of particles. And the limitation of the top plate is greater in the case of smaller helix spacing. Then, the soil particles between the two plates for a relative spacing of 3 is probably denser than those for a relative spacing of 4.5.

Figure 11 shows the particle displacement nephograms of double-helix anchors at AR = 1. The downward movement of soil particles in the columnar area (white dashed line) of the double-helix anchor is more severe than that of the single-helix anchor. This phenomenon above the top plate is because the upper soil zone is penetrated twice. And this phenomenon between the two plates is due to the effects of the downwards flow-around mechanism. The bottom plate is a lead plate, and the downward flow of particles above the bottom plate might cause the soil to become loose, which can make the particles above the top plate flow downwards easier. At the periphery of the white dashed columnar area with a depth of less than 6 times the helix diameter (red ellipse), there is a significant upward displacement of soil particles. And the lateral displacement of the double-helix anchor in the peripheral area is more pronounced than that of a single-helix anchor.

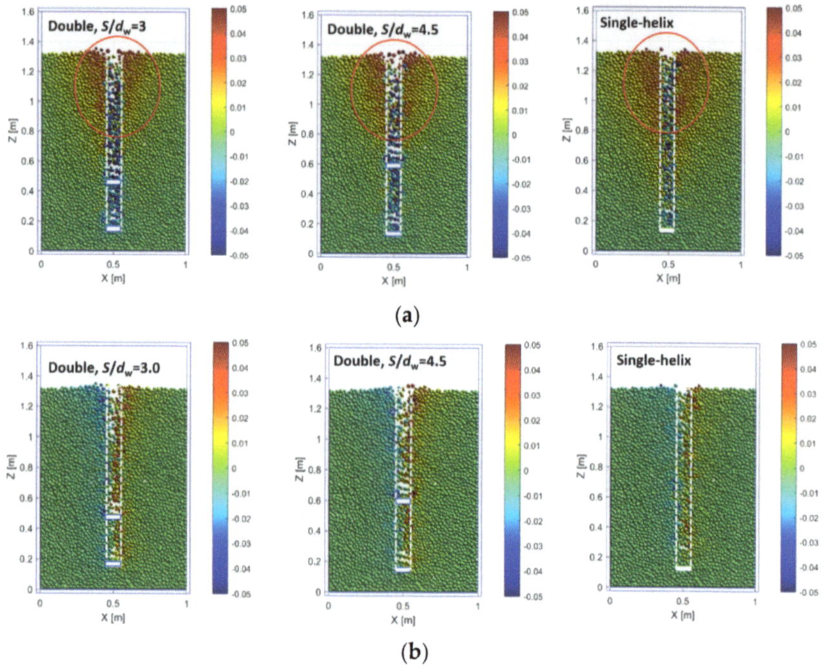

Figure 11. Soil particle displacement nephogram of double-helix anchors with different relative helix spacing at AR = 1. (**a**) Vertical displacement, (**b**) lateral displacement.

Similarly, particle displacement with absolute vertical displacement and lateral displacement greater than $0.5d_{50}$ were extracted and are plotted in Figures 12 and 13. The red and blue dashed lines in Figure 12 indicate envelope lines of lateral displacement between two helices of double-helix anchors with S/d_W = 3.0 and 4.5 respectively, and black one is the envelope line of single-helix anchor corresponding to the same position. Same as with the case of AR = 0.5, the lateral displacement of the soil particles between the plates is significantly reduced by the constraint of the two plates compared to that for the single-helix anchor, but the difference in the lateral displacements for the different relative spacing is not obvious. Unlike AR = 0.5, there is no obvious difference in the lateral displacement of the soil particles above the top plate for the case of the double-helix and single-helix anchors when AR = 1. This might not result in an obvious change in the shaft friction.

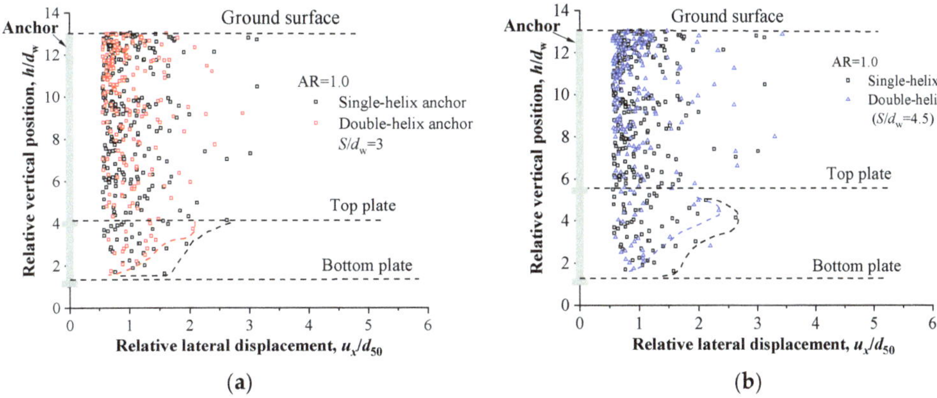

Figure 12. Soil particle lateral displacement of double-helix anchors with different relative helix spacing at AR = 1. (**a**) S/d_w = 3, (**b**) S/d_w = 4.5.

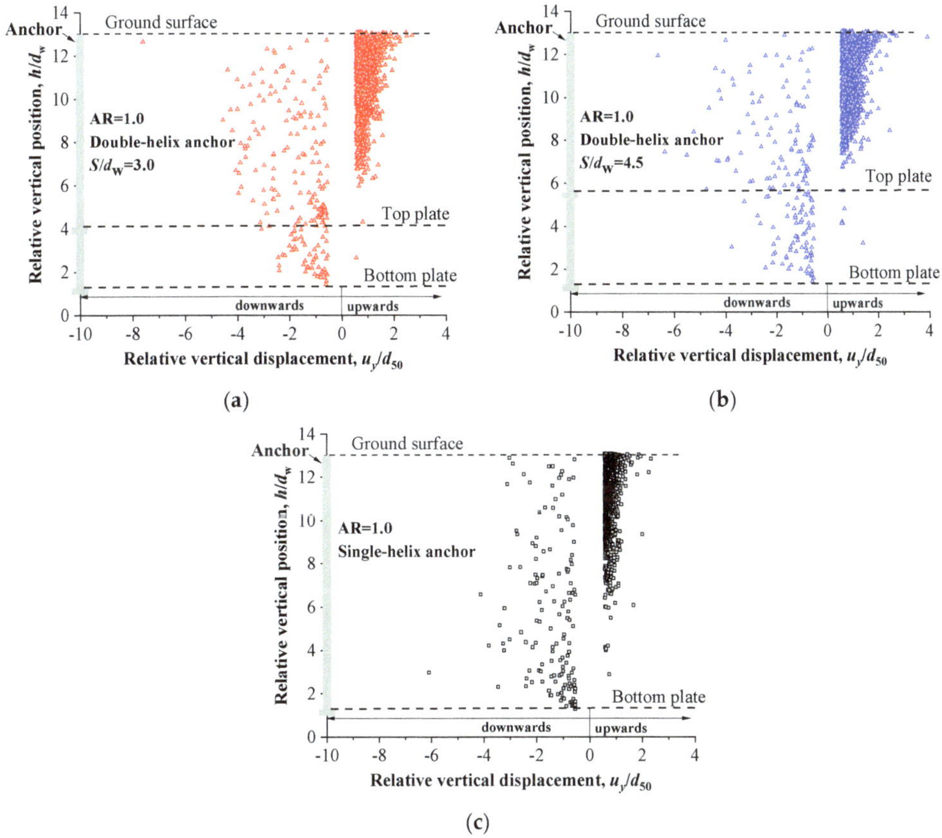

Figure 13. Soil particle vertical displacement diagram of double-helix anchors with different relative helix spacing at AR = 1. (**a**) Double-helix anchor with S/d_w = 3. (**b**) Double-helix anchor with S/d_w = 4.5. (**c**) Single-helix anchor.

It can be seen from Figure 13 that there is a difference in the vertical displacement of the soil particles between the two plates compared to a single-helix anchor at the same depth, although the change in the particle displacement between the plates is not as significant as in the case of AR = 0.5. Combined with the lateral and vertical displacement of particles between the two plates, it can be inferred that the soil between the two plates may become loose or dense. And more soil particles are encouraged to move downwards around the top plate compared to the case of a single-helix anchor.

In order to summarize the particle motion mechanisms during the installation of the single-helix and double-helix anchors, a schematic diagram of the particle motion for different ARs is plotted in Figure 14.

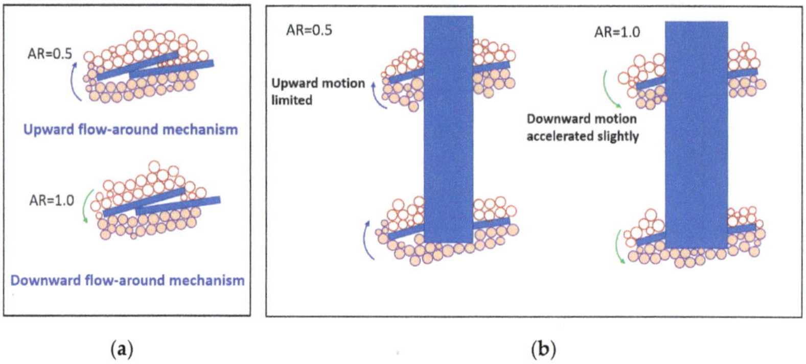

Figure 14. Schematic diagram of particle motion for different ARs. (**a**) Single-helix anchor. (**b**) Double-helix anchor.

For the single-helix anchor, when AR = 0.5, the particles above the plate move upwards, which is known as the upward flow-around mechanism, and when AR = 1, the particles above the plate move downwards, which is known as the downward flow-around mechanism. For the double-helix anchor, the lateral displacements of the particles between the two plates are smaller than those for the single-helix anchor regardless of the AR value. When the AR = 0.5, the upward motions of the particles below the top plate are limited and the upward displacement is smaller than that of the single-helix anchor, while when the AR = 1, the downward motion of the particles above the top plate is accelerated slightly and the downwards displacement is slightly greater than that of the single-helix anchor. Therefore, it is concluded that the density change in the sand between the two plates will be also affected by the top plate although the sand in this area is only penetrated once by the bottom plate. And the effects are greater with the decrease in the relative helix spacing. As a result, it will affect the uplift capacities.

4. Effects of Installation on Uplift Capacity

4.1. Single-Helix Anchors Pre-Embedded and Installed at AR = 1 (Surcharge p = 100 kPa)

Figure 15 shows the load–displacement curves of the single-helix anchor for installation at AR = 1 and wish-in-place conditions with a surcharge p of 100 kPa. The trends of these curves are similar, and the slope of the initial straight section of the wish-in-place curve is slightly higher than that of the installation curve, which indicates that the soil has been disturbed and has become loose at AR = 1. As the uplift displacement increases, the difference in the uplift resistance between the two placement methods becomes more obvious. Taking a displacement of 10% d_w as a control criterion of the ultimate uplift capacity, the ultimate uplift capacity for the torsional installation, $Q_{u,1.0}$, is 12~18% lower than the value for the pre-embedded condition Q_{up}.

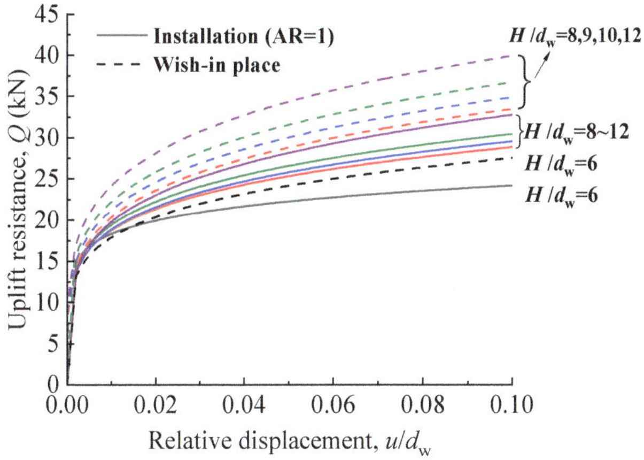

Figure 15. Q-u/d_w curves for single-helix anchors with various embedment ratios under different installation methods (surcharge p = 100 kPa).

To better compare the ultimate uplift capacities of the two placement methods, Q_{up}, $Q_{u,1.0}$, and $Q_{u,1.0}/Q_{up}$ are plotted in Figure 16. When embedment ratio is no more than 10, the uplift capacity loss caused by installation disturbance gradually increases as the embedment ratio increases. However, when embedment ratio is more than 10, the impact of installation disturbance on the uplift capacity becomes less with the embedment depth. This may be related to the failure mode. For small and medium-sized helical anchors in dense sand, when $H/d_w > 10$, it may be in the deep mode. The failure area for a deep anchor does not increase with the increase in the embedment depth, so the reduction in the uplift capacity caused by installation slows down with the increase in the embedment depth.

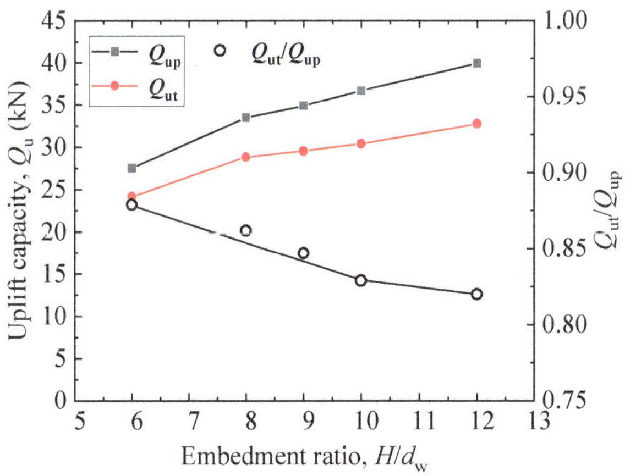

Figure 16. Comparisons of Q_u between the wish-in-place and installation with various embedment ratios (surcharge p = 100 kPa).

4.2. Single-Helix Anchors Installed at Different Advancement Ratios

Figure 17 shows the relationships of the uplift resistance and displacement for various ARs and embedment ratios. To analyze the impact of the installation advancement ratio on the uplift capacity at different embedment ratios, the uplift capacity for post-installation at AR = 0.5 and 0.8 were compared with the uplift capacity at AR = 1, as shown in Figure 18.

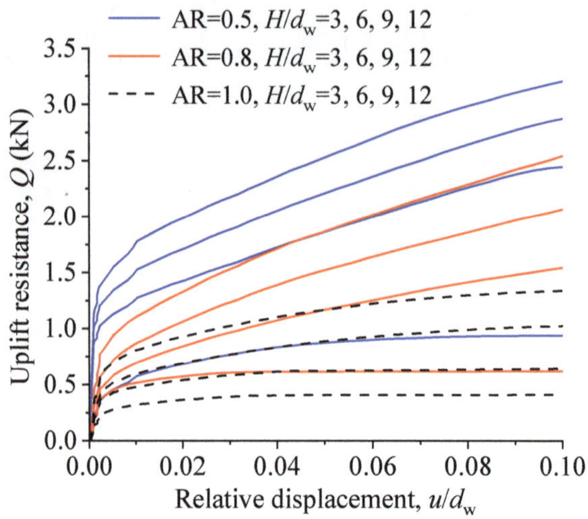

Figure 17. Q-u/d_w curves for various ARs and embedment ratios.

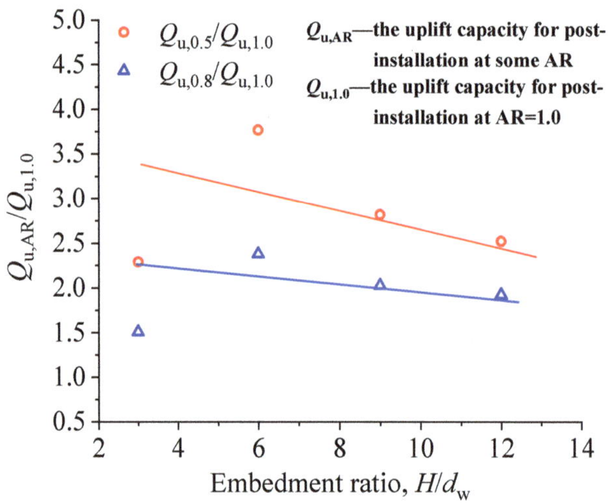

Figure 18. The comparisons of uplift capacity for different ARs.

The advancement ratio has significant effects on the uplift capacity of helical anchors. The uplift resistance of helix anchors increases with the decrease in the AR, which is closely related to the soil particle motion caused by installation. As mentioned before, the soil particles below the plate move upwards, and the lateral displacements of the soil in severely disturbed areas around anchors decrease when anchors are installed at a low AR, which makes the soil around the anchor more compact than that at a high AR. And when the AR reaches 1, the pitch-matched installation will make the sand above the plate loose.

As shown in Figure 18, the overall trend is that for the cases of over-flight installations (AR < 1), the increment in the uplift capacity compared with the cases of pitch-matched installations (AR = 1) decreases as the embedment ratio increases. This change is related to the pressure acted on the moving particle during the installation process. As the embedment ratio increases, the confinement pressure around the helical plate increases, and the downward or upward movement mechanisms of the particles are restricted, resulting in a decrease in the change in the uplift capacity caused by installation at various advancement ratios.

The comparisons between the results of this study and other results were made, as shown in Figure 19. The solid symbols are the results of the centrifuge tests from Cerfontaine et al. [16], the hollow symbols are the results of the DEM from Sharif et al. [32], the star and multiplication symbols are the results of this study. Figure 19 includes the centrifuge test ID 1~11 of a flat-base pile P1 (prototype embedment depth $H \approx 8$ m, shaft and helix diameter ratio $D_s/D_h = 0.52$) in dense sand and medium dense sand and the ID 12~17 of an asymmetric-base pile P2 ($D_s/D_h = 0.52$, $H = 8.65$ m), P5 ($D_s/D_h = 0.66$, $H = 9.05$ m), P4 ($D_s/D_h = 0.38$, $H = 8.45$ m) in dense sand. All the piles have the same prototype helix diameter D_h of 1.06 m and a helix pitch p_h of 0.35 m.

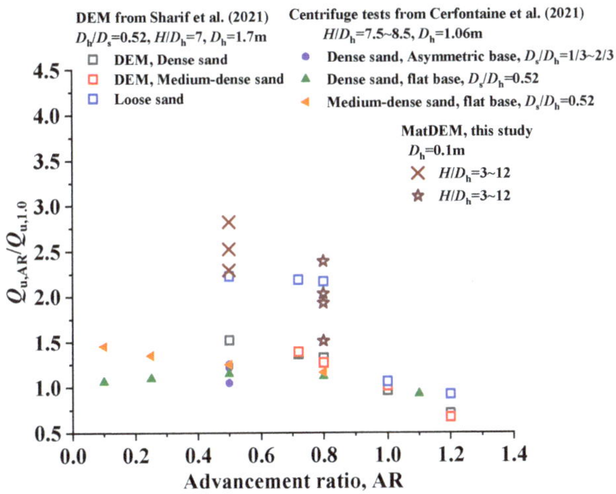

Figure 19. Comparisons of the effects of AR on uplift capacity [16,32].

It can be seen that all of these studies present the same trend of the impact of the AR on the uplift capacity. The uplift capacity increases with the decrease in the AR, and at AR = 0.8, the calculation results of this study are similar to the results of loose sand in Sharif et al. [26], while the remaining results are higher than the other research results. This may be due to the smaller diameter of the helical plate in this numerical model, where the constraint pressure around the plate is small, and the upward flow-around mechanism for the particles is formed easily, which makes the soil above the plate become denser.

4.3. Double-Helix Anchors Installed at Different Advancement Ratio

The uplift capacities of the double-helix anchors were plotted in Figure 20. Meanwhile, the efficiency of the uplift capacity of the double-helix anchor was defined as Formula (10) to reflect the interaction of the two plates [3,21]. For wish-in-place anchors, the efficiency is only affected by helix spacing [21]. In comparison with pre-embedded conditions, the efficiency varies with both the helix spacing and installation disturbance.

$$\eta = Q_u / \sum Q_{ui} \quad (10)$$

where η is the ratio of the uplift capacity of the double-helix anchor to the sum of the uplift capacities of each single-helix anchor.

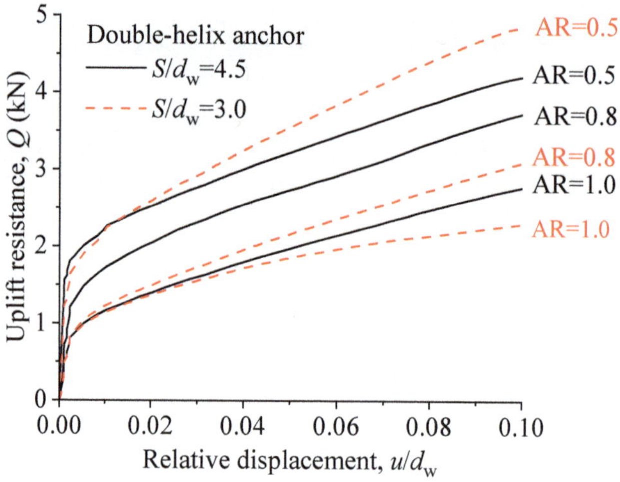

Figure 20. Relationships of uplift resistance and relative displacement for various ARs and embedment ratios.

In order to analyze the impact of the advancement ratio on the uplift capacity and efficiency of double-helix anchors, the uplift capacities of double-helix anchors with different spacing and efficiency of double-helix anchors with a relative spacing of 3 were plotted in Figure 21. It can be seen from Figure 21 that the uplift capacities of the double-helix anchors also increase with the decrease in the AR. When the AR = 1.0 and 0.8, the uplift capacity of the double-helix anchors with a relative spacing of 4.5 is higher than that of the anchor with a relative spacing of 3, while the relationship is opposite for the case of AR = 0.5. This change is related to particle movements induced by installation. When AR = 0.5, the top plate has restrictions both on the vertical upward and lateral displacement of the soil between the plates during installation. The smaller spacing between the plates will produce more obvious restrictions, and thus make the soil between the plates denser. Therefore, at AR = 0.5, the double-helical anchor with a small spacing has a higher uplift capacity.

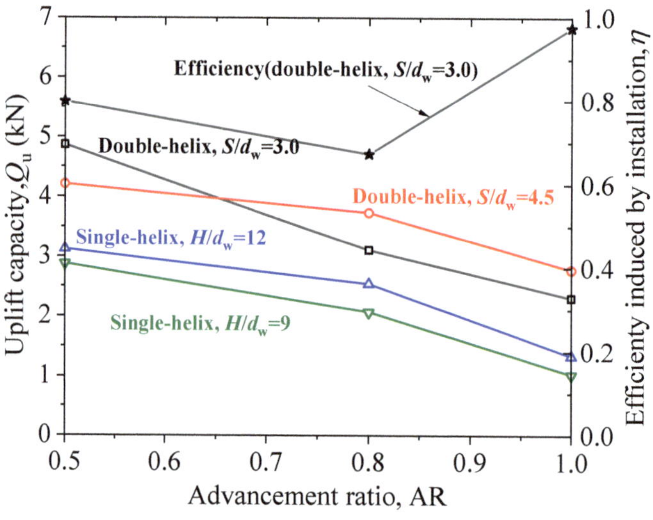

Figure 21. The uplift capacities and efficiency of double-helix anchors for different ARs.

Figure 22 shows the displacement nephogram at the uplift failure of a double-helix anchor with a relative spacing of 4.5. The yellow dotted lines indicate the particle motion zone at uplift failure. It can be seen that the height of the particle motion area above the bottom plate does not exceed 3 d_w at failure regardless of AR = 1 or 0.5, which indicates for installation situations, a relative spacing of 3 can ensure the two plates do not interact each other during the pulling process. Then, the efficiency change is only caused by the installation. It can be seen from the relationship of the AR and efficiency (in Figure 21) that although the uplift capacity of the double-helix anchor increases for the cases of AR <1.0 compared to that of AR = 1, the efficiency coefficient is close to 1 only for the case of AR = 1, indicating that the impact of the top plate during installation is minimal when AR = 1. This is consistent with the particle movement rule during installation. The twice disturbance of the top plate in the case of AR = 1 are not obvious, but are significant for AR = 0.5.

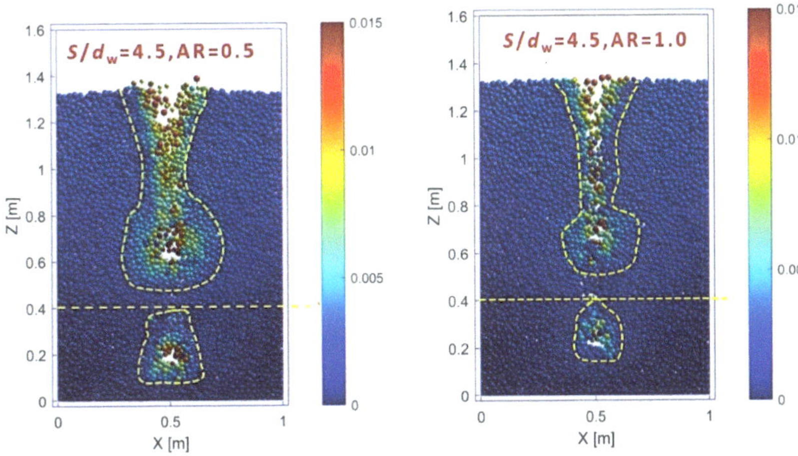

(**a**) (**b**)

Figure 22. Displacement nephogram of double-helix anchor with S/d_w = 4.5, (**a**) AR = 0.5, (**b**) AR = 1.

5. Conclusions

This study investigated the effects of the helical anchor geometry and installation advancement ratio on the soil disturbance zone, particle movement mechanism, and the uplift capacity by simulating the installation and pull-out process of single- and double-helix anchors in dense sand based on matrix discrete element software. The main investigations and conclusions are as follows:

(1) When the helical anchor is drilled into a shallow depth, the particle motion zone is the shape of an inverted cone, and when drilled into a deep depth ($L > 9\ d_w$), the overall disturbed zone is funnel-shaped. For the double-helix anchor, the soil above the upper helical plate is disturbed more obviously than with the single-helix anchor because it has been penetrated twice.

(2) When a single-helix anchor is installed at an AR of less than 1, the particles around the plate are encouraged to move upwards, and the back flow-around mechanism occurs. And when installed at AR = 1, the flow-around mechanism occurs.

(3) For a double-helix anchor, the lateral displacements of the particles between the two plates are smaller than those for a single-helix anchor regardless of the AR value. When AR = 0.5, the upward motions of the particles below the top plate are limited and the upward displacement is smaller than that of a single-helix anchor, while when AR = 1, the downward motion of the particles above the top plate is accelerated slightly and the downward displacement is slightly greater than that of a single-helix anchor.

(4) The pitch-matched installation (AR = 1) of a single-helix anchor will make the uplift capacity decrease compared with the cases of wish-in-place conditions (undisturbed conditions). As the embedment ratio increases, the uplift capacity loss caused by installation disturbance (AR = 1) gradually increases. However, when the embedment depth can make deep failure mode occur, the impact rate of the installation disturbance on the uplift capacity becomes slow with the increase of embedment depth.

(5) The uplift capacities of both single-helix and double-helix anchors increase with the decrease in the AR (AR = 0.5~1). The efficiency induced by installation is close to 1 for the case of AR = 1, indicating that the impact of the top plate during installation is minimal when AR = 1.

The investigation of the observed disturbance zone and the comparison of the uplift capacity under pitch-matched installation and wish-in-place conditions can provide a beneficial reference for analyzing the considered installation effects. The influence of the installation advancement ratio (AR) on the uplift capacity of double-helical anchors is the same as that of single-helix anchors. But it is also affected by helix spacing and diameter. Further experimental evidence is still required to demonstrate the impact of installation on the uplift capacity of multi-helical anchors with various embedment ratios and anchor geometries in sands with different levels of compactness.

Author Contributions: Conceptualization, formal analysis, and writing—review and editing: R.C. and D.H.; investigation and writing—original draft preparation: H.L. and Z.L.; and investigation and data curation: C.Y. All authors have read and agreed to the published version of the manuscript.

Funding: This research was funded by the National Natural Science Foundation of China (grant number 52078108); Jilin Province Youth Science and Technology Innovation Leader, Team Project of Provincial Department of Science and Technology (grant number 20210509058RQ); and the Scientific Research Project of Jilin Provincial Department of Education (grant number JJKH20210103KJ).

Institutional Review Board Statement: Not applicable.

Informed Consent Statement: Not applicable.

Data Availability Statement: Data are contained within the article.

Conflicts of Interest: Zhaoguo Liu was employed by China Energy Engineering Group Heilongjiang Electric Design Co., Ltd. The remaining authors declare that the research was conducted in the absence of any commercial or financial relationships that could be construed as a potential conflict of interest.

References

1. Hao, D.; Che, J.; Chen, R.; Zhang, X.; Yuan, C.; Chen, X. Experimental investigation on behavior of single-helix anchor in sand subjected to uplift cyclic loading. *J. Mar. Sci. Eng.* **2022**, *10*, 1338. [CrossRef]
2. Lin, Y.; Xiao, J.; Le, C.; Zhang, P.; Chen, Q.; Ding, H. Bearing characteristics of helical pile foundations for offshore wind turbines in sandy Soil. *J. Mar. Sci. Eng.* **2022**, *10*, 889. [CrossRef]
3. Lutenegger, A.J. Historical development of iron screw-pile foundations: 1836–1900. *Int. J. Hist. Eng. Technol.* **2011**, *81*, 108–128. [CrossRef]
4. Meng, Z.; Chen, J.; Zhang, L.; Wang, J.; Yao, J. Field tests to investigate the installation effects of drilled displacement piles with screw-shaped shaft in clay. *J. Geotech. Geoenviron. Eng.* **2015**, *141*, 06015010. [CrossRef]
5. Huang, Z.M. Construction and development of EHV transmission line in China. *China Electr. Power* **1996**, *29*, 9–12. (In Chinese)
6. Cerato, A.B.; Victor, R. Effects of helical anchor geometry on long-term performance of small wind tower foundations subject to dynamic loads. *J. Deep Found. Inst.* **2008**, *2*, 30–41. [CrossRef]
7. Tsuha, C.H.C.; Aoki, N. Relationship between installation torque and uplift capacity of deep helical piles in sand. *Can. Geotech. J.* **2010**, *47*, 635–647. [CrossRef]
8. Kulhawy, F.H. Uplift behavior of shallow soil anchors—An overview: Uplift behavior of anchor foundations in soil. In *Uplift Behavior of Anchor Foundations in Soil*; ASCE: Reston, VA, USA, 1985; pp. 1–25.
9. Dijkstra, J.; Broere, W.; Bezuijen, A.; Bezuijen, A.; Van Tol, A.F. Density changes near an advancing displacement pile in sand. In Proceedings of the Second BGA International Conference on Foundations, Dundee, UK, 24–27 June 2008; pp. 545–554.
10. Perez, Z.A.; Schiavon, J.A.; Tsuha, C.H.C.; Dias, D.; Thorel, L. Numerical and experimental study on influence of installation effects on behaviour of helical anchors in very dense sand. *Can. Geotech. J.* **2018**, *55*, 1067–1080. [CrossRef]

11. Clemence, S.P.; Pepe, F.D. Measurement of lateral stress around multihelix anchors in sand. *Geotech. Test. J.* **1984**, *7*, 145–152. [CrossRef]
12. George, B.E.; Banerjee, S.; Gandhi, S.R. Helical piles installed in cohesionless soil by displacement method. *Int. J. Geomech.* **2019**, *19*, 0401907. [CrossRef]
13. Lutenegger, A.J.; Tsuha, C.H.C. Evaluating installation disturbance from helical piles and anchors using compression and tension tests. In Proceedings of the Pan-American Conference on Soil Mechanics and Geotechnical Engineering, Buenos Aires, Argentina, 15–19 November 2015; pp. 373–381.
14. Saleem, M.A.; Malik, A.A.; Kuwano, J. Model study of screw pile installation impact on ground disturbance and vertical bearing behaviour in dense sand. *Earth Environ. Sci.* **2021**, *710*, 012056. [CrossRef]
15. Davidson, C.; Brown, M.J.; Cerfontaine, B.; Al-Baghdadi, T.; Knappett, J.; Brennan, A.; Augarde, C.; Coombs, W.; Wang, L.; Blake, A.; et al. Physical modelling to demonstrate the feasibility of screw piles for offshore jacket-supported wind energy structures. *Géotechnique* **2020**, *72*, 108–126. [CrossRef]
16. Cerfontaine, B.; Brown, M.J.; Knappett, J.A.; Davidson, C.; Sharif, Y.; Huisman, M.; Ottolini, M.; Ball, J.D. Control of screw pile installation to optimise performance for offshore energy applications. *Géotechnique* **2023**, *73*, 234–249. [CrossRef]
17. Sakr, M. Relationship between installation torque and axial capacities of helical piles in cohesionless soils. *Can. Geotech. J.* **2015**, *52*, 747–759. [CrossRef]
18. Shi, D.; Yang, Y.; Deng, Y.; Xue, J. DEM modelling of screw pile penetration in loose granular assemblies considering the effect of drilling velocity ratio. *Granul. Matter* **2019**, *21*, 74. [CrossRef]
19. Bradshaw, A.S.; Zuelke, R.; Hildebrandt, L.; Robertson, T.; Mandujano, R. Physical modelling of a helical pile installed in sand under constant crowd. In Proceedings of the 1st International Symposium on Screw Piles for Energy Applications (ISSPEA), Dundee, UK, 27–28 May 2019; pp. 109–115.
20. Cerfontaine, B.; Brown, M.J.; Davidson, C.; Sharif, Y.U.; Huisman, M.; Ottolini, M. Optimised screw pile design for offshore jacket foundations in medium–dense sand. *Geotech. Lett.* **2022**, *12*, 114–119. [CrossRef]
21. Hao, D.X.; Chen, R.; Fu, S.N. Experimental study on uplift capacity of multi-helix anchors in sand. *Chin. J. Geotech. Eng.* **2015**, *37*, 126–132. (In Chinese)
22. Jeffrey, J.; Brown, M.J.; Knappett, J.A.; Ball, J.D.; Caucis, K. CHD pile performance: Part I-physical modelling. *Proc. ICE Geotech. Eng.* **2016**, *169*, 43–54. [CrossRef]
23. Komatsu, A. Development on battered pile with screw pile method (NS-ECO pile). In Proceedings of the International Workshop on Recent Advances of Deep Foundations, Yokosuka, Japan, 1–2 February 2007; pp. 253–257.
24. Clemence, S.P.; Smithling, A.P. Dynamic uplift capacity of helical anchors in sand. In Proceedings of the 4th Australia-New Zealand Conference on Geomechanics, Perth, Australia, 14–18 May 1984; Volume 1, pp. 88–93.
25. Ilamparuthi, K.; Dickin, E.A.; Muthukrisnaiah, K. Experimental investigation of the uplift behaviour of circular plate anchors embedded in sand. *Can. Geotech. J.* **2002**, *39*, 648–664. [CrossRef]
26. Cerato, A.B.; Victor, R. Effects of long-term dynamic loading and fluctuating water table on helical anchor performance for small wind tower foundations. *J. Perform. Constr. Facil. ASCE* **2009**, *23*, 251–261. [CrossRef]
27. Hao, D.; Wang, D.; O'Loughlin, C.D.; Gaudin, C. Tensile monotonic capacity of helical anchors in sand: Interaction between helices. *Can. Geotech. J.* **2019**, *56*, 1534–1543. [CrossRef]
28. Tsuha, C.H.C.; Aoki, N.; Rault, G.; Thorel, L.; Garnier, J. Evaluation of the efficiencies of helical anchor plates in sand by centrifuge model tests. *Can. Geotech. J.* **2012**, *49*, 1102–1114. [CrossRef]
29. Wang, L.; Zhang, P.; Ding, H.; Tian, Y.; Qi, X. The uplift capacity of single-plate helical pile in shallow dense sand including the influence of installation. *Mar. Struct.* **2020**, *71*, 102697. [CrossRef]
30. Nagai, H.; Tsuchiya, T.; Shimada, M. Influence of installation method on performance of screwed pile and evaluation of pulling resistance. *Soils Found.* **2018**, *58*, 355–369. [CrossRef]
31. Cerfontaine, B.; Ciantia, M.; Brown, M.J.; Sharif, Y.U. DEM study of particle scale and penetration rate on the installation mechanisms of screw piles in sand. *Comput. Geotech.* **2021**, *139*, 104380. [CrossRef]
32. Sharif, Y.U.; Brown, M.J.; Cerfontaine, B.; Davidson, C.; Ciantia, M.O.; Knappett, J.A.; Brennan, A.; Ball, J.D.; Augarde, C.; Coombs, W.M.; et al. Effects of screw pile installation on installation requirements and in-service performance using the discrete element method. *Can. Geotech. J.* **2021**, *58*, 1334–1350. [CrossRef]
33. Sharif, Y.U.; Brown, M.J.; Ciantia, M.O.; Cerfontaine, B.; Davidson, C.; Knappett, J.A.; Ball, J.D. Assessing single-helix screw pile geometry on offshore installation and axial capacity. *Proc. Inst. Civ. Eng. Geotech. Eng.* **2021**, *174*, 512–529. [CrossRef]
34. Kurian, N.P.; Shah, S.J. Studies on the behavior of screw piles by the finite element method. *Can. Geotech. J.* **2009**, *46*, 627–638. [CrossRef]
35. Rawat, S.; Gupta, A.K. Numerical modelling of pullout of helical soil nail. *J. Rock Mech. Geotech. Eng.* **2017**, *9*, 648–658. [CrossRef]
36. Chen, R.; Fu, S.N.; Hao, D.X.; Shi, D.D. Scale effects of uplift capacity of circular anchors in dense sand. *Chin. J. Geotech. Eng.* **2019**, *41*, 78–85. (In Chinese)
37. Cerfontaine, B.; Knappett, J.A.; Brown, M.J.; Davidson, C.S.; Al-Baghdadi, T.; Sharif, Y.U.; Brennan, A.; Augarde, C.; Coombs, W.M.; Wang, L. A finite element approach for determining the full load–displacement relationship of axially loaded shallow screw anchors, incorporating installation effects. *Can. Geotech. J.* **2021**, *58*, 565–582. [CrossRef]

38. Shao, K.; Su, Q.; Liu, J.; Liu, K.; Xiong, Z.; Wang, T. Optimization of inter-helix spacing for helical piles in sand. *J. Rock Mech. Geotech. Eng.* **2022**, *14*, 936–952. [CrossRef]
39. Yuan, C.; Hao, D.; Chen, R.; Zhang, N. Numerical investigation of uplift failure mode and capacity estimation for deep helical anchors in sand. *J. Mar. Sci. Eng.* **2023**, *11*, 1547. [CrossRef]
40. Hao, D.X.; Fu, S.N.; Chen, R.; Zhang, Y.J.; Hou, L.Q. Experimental investigation of uplift behavior of anchors and estimation of uplift capacity in sands. *Chin. J. Geotech. Eng.* **2015**, *37*, 2101–2106. (In Chinese)
41. Liu, C.; Pollard, D.D.; Shi, B. Analytical solutions and numerical tests of elastic and failure behaviors of close-packed lattice for brittle rocks and crystals. *J. Geophys. Res. Solid Earth* **2013**, *118*, 71–82. [CrossRef]
42. Morgan, J.K.; Boettcher, M.S. Numerical simulations of granular shear zones using the distinct element method. Shear zone kinematics and the micromechanics of localization. *J. Geophys. Res. Solid Earth* **1999**, *104*, 2703–2719. [CrossRef]
43. Kuhn, M.R.; Bagi, K. Contact rolling and deformation in granular media. *Int. J. Solids Struct.* **2004**, *41*, 5793–5820. [CrossRef]
44. Ergenzinger, C.; Seifried, R.; Eberhard, P. A discrete element model to describe failure of strong rock in uniaxial compression. *Granul. Matter* **2011**, *13*, 341–364. [CrossRef]
45. Liu, C.; Shi, B.; Gu, K.; Sun, Y.J. Development and application of large-scale discrete element simulation system for rock and soil. *J. Eng. Geol.* **2014**, *22*, 551–557. (In Chinese)
46. Arroyo, M.; Butlanska, J.; Gens, A.; Calvetti, F.; Jamiolkowski, M. Cone penetration tests in a virtual calibration chamber. *Géotechnique* **2011**, *61*, 525–531. [CrossRef]
47. Bradshaw, A.S.; Cullen, L.; Miller, Z. Field study of group effects on the pullout capacity of "deep" helical piles in sand. *Can. Geotech. J.* **2022**, *59*, 538–545. [CrossRef]

Journal of
Marine Science
and Engineering

MDPI

Article

Data-Driven Prediction of Maximum Settlement in Pipe Piles under Seismic Loads

Sajjad E. Rasheed [1], Duaa Al-Jeznawi [2], Musab Aied Qissab Al-Janabi [2] and Luís Filipe Almeida Bernardo [3,*]

1 Department of Civil Engineering, College of Engineering, University of Kerbala, Kerbala 56001, Iraq; sajjad.e@uokerbala.edu.iq
2 Department of Civil Engineering, College of Engineering, Al-Nahrain University, Jadriya, Baghdad 10881, Iraq; duaa.a.al-jeznawi@nahrainuniv.edu.iq (D.A.-J.); musab.a.jindeel@nahrainuniv.edu.iq (M.A.Q.A.-J.)
3 Department of Civil Engineering and Architecture, University of Beira Interior, GeoBioTec-UBI, 6201-001 Covilhã, Portugal
* Correspondence: lfb@ubi.pt

Abstract: The structural stability of pipe pile foundations under seismic loading stands as a critical concern, demanding an accurate assessment of the maximum settlement. Traditionally, this task has been addressed through complex numerical modeling, accounting for the complicated interaction between soil and pile structures. Although significant progress has been made in machine learning, there remains a critical demand for data-driven models that can predict these parameters without depending on numerical simulations. This study aims to bridge the disparity between conventional analytical approaches and modern data-driven methodologies, with the objective of improving the precision and efficiency of settlement predictions. The results carry substantial implications for the marine engineering field, providing valuable perspectives to optimize the design and performance of pipe pile foundations in marine environments. This approach notably reduces the dependence on numerical simulations, enhancing the efficiency and accuracy of the prediction process. Thus, this study integrates Random Forest (RF) models to estimate the maximum pile settlement under seismic loading conditions, significantly supporting the reliability of the previously proposed methodology. The models presented in this research are established using seven key input variables, including the corrected SPT test blow count $(N_1)_{60}$, pile length (L), soil Young's modulus (E), soil relative density (Dr), friction angle (ϕ), soil unit weight (γ), and peak ground acceleration (PGA). The findings of this study confirm the high precision and generalizability of the developed data-driven RF approach for seismic settlement prediction compared to traditional simulation methods, establishing it as an efficient and viable alternative.

Keywords: pipe piles; settlement; data-driven prediction; random forest; seismic loads

Citation: Rasheed, S.E.; Al-Jeznawi, D.; Al-Janabi, M.A.Q.; Bernardo, L.F.A. Data-Driven Prediction of Maximum Settlement in Pipe Piles under Seismic Loads. *J. Mar. Sci. Eng.* **2024**, *12*, 274. https://doi.org/10.3390/jmse12020274

Academic Editors: Pan Hu and Dong-Sheng Jeng

Received: 30 December 2023
Revised: 29 January 2024
Accepted: 31 January 2024
Published: 2 February 2024

1. Introduction

The phenomenon of seismic-induced pile settlement is a significant concern in structural engineering and foundation design due to its potential impact on the stability and performance of buildings and infrastructure during and after seismic events [1]. Pile foundations are extensively employed in various infrastructure projects, such as ports, offshore bridges, and offshore wind power generation [2]. Among these, pipe piles have gained considerable interest due to their handling, simplification, and quality at low costs. In the extreme marine environment, a foundation not only faces the operational load transmitted by the structure but also the cyclic loading induced by waves and wind. Assessing the stability and deformation of the foundation under such cyclic loading is crucial, and employing the appropriate methods for this evaluation holds significant importance [3]. When subjected to seismic forces, the ground undergoes dynamic movements, which can result in the settlement of the piles [4]. This settlement, in turn, affects the stability of the entire structure, leading to structural damage or failure. Consequently, the study of seismic-induced pile settlement is essential for ensuring the seismic resilience of structures. After a moderate-to-severe earthquake in liquefiable zones, it has been noted that piled foundations often experience both tilting and settlement. Bhattacharya, in 2003 [5], conducted research that proposed an explanation, acknowledging the common occurrence of significant axial loads in pile foundations during earthquakes. When the soil surrounding the piles undergoes liquefaction, it experiences a substantial reduction in its stiffness and strength. Consequently, the piles essentially transform into unsupported, long, slender columns, and they buckle under the influence of these axial loads. Thus, the behavior of pipe pile foundations is a significant concern within the field of geotechnical engineering, particularly in the areas prone to earthquakes. Accurately anticipating how pipe piles will react horizontally is essential for creating strong foundations for various structures, such as buildings, bridges, and offshore platforms [6]. Recently, there has been substantial interest in investigating how piles respond to seismic actions. Many researchers have explored the characteristics of ground motion inputs and the mechanisms involved in the interaction between the soil and piles [7–11].

Based on empirical evidence, the simultaneous development of methods involves establishing the foundation of the pile predominantly in a stratum beneath the soil, succeeded by a layer with lower compressibility [12–16]. Consequently, the layers of compression underneath the piles have been widely acknowledged as a critical design concern and a potential risk, given their potential to significantly increase pile settlement [12]. A particular study from Poulos in 2017 [17] suggested an additional subsidence rate due to the underlying layers, which can be influenced by the geometry of the piles and the physical properties of the soil, depending on the limited analysis available. It is worth noting that research on this essential issue is limited, and manual calculations and analytical methods often do not apply well to the unique properties of individual soil layers. Therefore, in the current study, the collected data considered the impact of multilayered soil in combination with a single homogeneous soil layer. This consideration allows for a comprehensive analysis of the soil structure, acknowledging the presence of multiple layers and their potential influence on the outcomes. Moreover, innovative solutions, such as artificial neural networks (ANNs) and advanced machine learning techniques, have emerged as a result of the extensive research conducted by several authors [18–25]. Recently, ANNs have found various applications in geotechnical engineering, showing promising results. ANNs are a type of artificial intelligence that first aimed to replicate the biological design of the human brain and nervous system through their architecture. While the idea of artificial neurons was initially introduced in 1943, the research into ANNs gained significant drive with the introduction of the backpropagation training algorithm for feedforward ANNs in 1986, as demonstrated by Rumelhart et al. in 1986 [26].

The prediction technique has been applied to estimate damage progression, mixed-mode fracture, and fatigue durability (as indicated in [27–30]). This predictive approach facilitates future engineering judgments by selectively sampling from the available data set in a wide range of phenomena, including engineering science. Furthermore, prediction aids in reducing the complexity of engineering analytical processes and the time required for product design. Qian et al., in 2019 [31], applied a statistical technique to determine the material strength and the possibility of failure based on the fracture strength of irregularly shaped particles. Similarly, Lei et al., in 2019 [32], utilized statistical methods to assess the stress distribution in rock and cohesive soils when dealing with diagonal cross-sectional specimens, and they also evaluated the interactions during loading.

In the construction field, historical methods for determining pile settlement, such as static and dynamic load tests, have been proven to be reliable but are criticized for being time-consuming and uneconomical [3,10]. To address this issue, some researchers propose semi-empirical formulas using in situ test results [15,16,33], while others employ finite element simulations with software tools like MIDAS GTS (version 2019) [10]. Recognizing the limitations, recent efforts explore the application of artificial intelligence, with this specific study focused on the efficiency of the Random Forest model to predict pile settlement under seismic excitation based on shaking table tests and intensive numerical studies. Unlike traditional models, Random Forest models demonstrate a faster training speed and resistance to overfitting, offering a promising avenue for optimizing machine learning solutions in construction design [19]. Random Forest, a machine learning algorithm, operates by constructing multiple decision trees, each trained on a randomly sampled subset of the data (training data), and outputs an aggregated result, either the mode of predictions for classification or the mean for regression. This approach, known as ensemble learning, significantly reduces the risk of overfitting, making Random Forest particularly effective for complex datasets. While powerful and versatile in handling various data types, including in soil engineering for predictive modeling, its limitations include its reduced interpretability and potentially high computational demands compared to simpler models.

Raman et al. in 2008 [34] documented that in previous seismic events, pile foundations in liquefiable soil were highly susceptible to damage or failure, often resulting in the significant tilting and settling of structures, while lateral ground spreading is a typical explanation for these failures. A closer analysis of specific cases indicates that pile foundation settling can also contribute to structural tilting.

Pile behavior may vary due to several factors, including pile geometry, construction materials, applied load, and soil type. Accounting for these factors may not provide precise predictions of pile seismic responses, and applying seismic loads in the analytical process can be time-consuming. To address this challenge, a statistical model can be applied to analyze the settlement of piles in seismic conditions. While Random Forest (RF) is often considered one of the most effective and widely used machine learning algorithms, a thorough review of the existing literature reveals that this method has not been employed to predict the pile settlement under seismic excitation [19].

In this study, the feasibility of rapid pile settlement estimation using the Random Forest (RF) algorithm is being investigated. To achieve this, 542 data points from previous research and laboratory experiments have been gathered. The dataset comprised 271 data points for pipe piles embedded in dry soil conditions and another 271 data points for pipe piles embedded in saturated soil conditions. The models presented in this research are established using seven key input variables, including the corrected SPT test blow count $(N_1)_{60}$, pile length (L), soil Young's modulus (E), soil relative density (Dr), friction angle (ϕ), soil unit weight (γ), and peak ground acceleration (PGA). The model's performance was assessed by using three evaluation criteria: the Mean Absolute Error (MAE), Mean Square Error (MSE), and the coefficient of determination (R^2).

2. Methodology

2.1. Data Collecting

The current database was initially introduced by Al-Jeznawi et al. in 2023a [10], and it was subsequently expanded by Al-Jeznawi et al. in 2023b [11] to incorporate additional valuable insights. The dataset covers 271 data points specifically related to the seismic responses of pipe piles. The dataset covers various attributes, including the corrected standard penetration test (SPT) blow count $(N_1)_{60}$, the peak ground acceleration (PGA), and the pile slenderness ratio (L/D), where 'L' denotes the length of the pile, and 'D' denotes the diameter of the pile. Table 1 provides a summary of the current soil properties, with both soils undergoing drying and sieving using a No. #10 sieve before testing. The primary data derived from these tests underwent numerical analysis to facilitate a more in-depth exploration, addressing the difficulties associated with directly calculating pile settlement through experimental means, considering a wide range of soil–pile parameters.

Table 1. Main ground properties [10].

Parameter	Soil #1	Soil #2	Soil #3
	Dr = 30%	Dr = 70%	Dr = 65%
Poisson's ratio (ν)	0.33	0.33	0.3
K_o	0.47	0.426	0.45
E (kPa)	11,000	28,000	25,000
Secant elastic modulus in shear hardening (kPa)	5639	15,037	15,400
Tangential stiffness primary oedometer test loading ($Eoed_{ref}$) (kPa)	5639	15,038	15,400
Elastic modulus at unloading (Eur_{ref}) (kPa)	22,225	59,265	46,200
K_G^e (unitless)	902	1093	1019
K_G^P (unitless)	320	940	617
$Ø_p$ (°)	34	36	35
Failure ratio (R_f) (%)	0.9	0.9	0.9
Porosity (%)	0.8	0.6	0.77
$Ø_{cv}$ (°)	32	35	34
Dilatancy angle (ψ) (°)	2	5	4
Cohesion (c) (kPa)	0.1	0.1	0.1

The laboratory box, measuring 60 × 60 × 70 cm, housed a pile with a 26 mm diameter, a 1.5 mm wall thickness, and a 400 mm embedded length. The numerical soil box design followed a non-elastic concept (17 times outer diameter) while adhering to the influence area limitations [34,35]. Despite a minimal impact on seismic lateral behavior, the lower boundary exceeded four times the pile's outer diameter, as validated through sensitivity assessments [10,11]. Input data for the modified UBCSAND relied on correlations by Beaty and Byrne [36], drawn from a comprehensive numerical database introduced by Al-Jeznawi et al. [10], expanded in Al-Jeznawi et al. [11], encompassing 542 data points on pile settlement, featuring essential parameters like the soil Young's modulus (E), $(N_1)_{60}$, soil relative density (Dr), soil unit weight (γ), peak ground acceleration (PGA), and L/D.

The primary input data (void ratio, porosity, E, Ø, and γ) were initially obtained from the experimental work conducted by Mahmood et al. [37] and Hussein [38], and subsequently calibrated using the methodology proposed by Beaty and Byrne [36]. These correlations establish connections between various soil parameters:

$$K_G^e = 21.7 \times 20.0 \times (N_1)_{60}{}^{0.333} \tag{1}$$

$$K_G^P = (N_1)_{60}{}^2 \times 0.003 + 100 \tag{2}$$

$$Ø_p = Ø_{cv} + \frac{(N_1)_{60}}{10} \text{ for } (N_1)_{60} < 15 \tag{3}$$

$$Ø_p = Ø_{cv} + \frac{(N_1)_{60}}{10} + \max\left(0.0, \frac{(N_1)_{60} - 15}{5}\right) \text{ for } (N_1)_{60} \geq 15 \tag{4}$$

$$R_f = 1.1 \times (N_1)_{60}{}^{-0.15} \tag{5}$$

where K_G^e and K_G^P represent the elastic and plastic shear modulus values, respectively. $Ø_p$ and $Ø_{cv}$ indicate the peak and constant volume friction angles, respectively, and R_f represents the failure ratio. Hence, the data employed in this study were obtained from an extensive numerical database explicitly designed for assessing pile settlement in driven piles under seismic loadings.

The settlement behavior of piles under seismic shaking was initially investigated through experimental tests, specifically shaking table tests conducted by Hussein [38]. In these tests, a soil–pile model (scaled at 1:35, corresponding to model-to-prototype) was subjected to four recorded ground motions (Kobe, El Centro, Halabja, and Ali Algharbi). Subsequently, Al-Jeznawi et al. [10,11] conducted a comprehensive numerical study, incorporating various scales of soil–pile models, diverse earthquake histories, and different soil and pile properties. The numerical analysis utilized MIDAS GTS software, and the settlement of the pile was directly obtained as an output from the software. This combined approach, encompassing both experimental and numerical investigations, provides a wide range of values of dynamic pile settlement under seismic loading conditions. Table 2 provides a statistical overview of the dataset, where the dry or saturated soil condition is indicated by Dry or Sat, respectively. Figure 1 illustrates the data distribution for both the dry and saturated soils. The data tend to lean towards lower values, indicating a prevalence of softer or less rigid materials and conditions in the dataset. The friction angle (ϕ) and unit weight of soil (γ) are fairly normally distributed, although with a slight skew towards higher values, suggesting a moderate variation in shear strength and density across the samples. Overall, these data reveal a tendency towards more common occurrences of lower elasticity, shorter lengths, lower penetration resistance, ground acceleration, and less dense soil conditions, while maintaining a relatively consistent soil friction angle and unit weight.

Table 2. Statistical overview of the current data points.

Attribute		Mean	Std.	Min.	Max.
Corrected SPT test blow count $(N_1)_{60}$		14.5	3.2	10	18
PGA (g)		0.37	0.21	0.1	0.82
Soil unit weight (γ) (kN/m^3)		18.1	1.1	16	19.4
Closed-ended pile	$(\text{Pile settlement})_{Dry}$ (mm)	33.5	32.5	1	150
	$(\text{Pile settlement})_{Sat}$ (mm)	54.8	55.6	2.2	269
Open-ended pile	$(\text{Pile settlement})_{Dry}$ (mm)	44.2	41	1.6	211
	$(\text{Pile settlement})_{Sat}$ (mm)	64.6	65.5	3.3	423

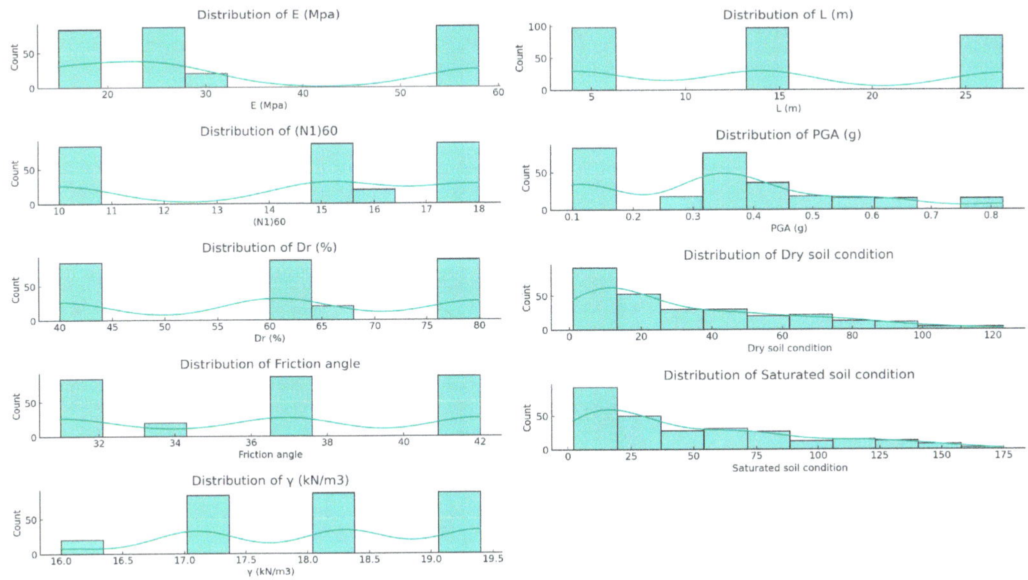

Figure 1. Data distribution.

2.2. Data Preparation

Appropriately preparing raw, assembled data is imperative before proceeding with predictive modeling [31]. Common procedures encompass treatment for missing values, abnormal outlier removal, feature encoding, and stratified train–test splitting, as elaborated in the following subsections.

2.2.1. Data Cleaning and Missing Values

Real-world observations frequently contain missing entries due to sensor limitations, equipment errors, data loss, or human oversight. Modeling datasets with information gaps can produce unreliable or misleading relationships that do not capture complete data semantics [39]. Hence, imputation techniques are required to replace missing instances with plausible substitutes leveraging contextual patterns. As only 1.1% of observations had partial voids, basic median and mode replacement was applied for numerical and categorical attributes based on their distribution statistics [40]. Sophisticated methods are warranted for larger missing proportions. Substitutions enabled the retention of the maximum raw data points.

2.2.2. Outlier Identification and Removal

With the cleaned complete data, statistical outlier detection was systematically conducted by computing z-scores (Equation (6)) and visually inspecting distributions. Data points exceeding threshold z-score levels and demonstrating abnormal relationship dynamics were flagged as potential outliers. Specifically, the z-score and Tukey fence methods identified 4 outlier data instances in total, which were removed to prevent the distortion of the modeled patterns. Their elimination resulted in a final cleaned dataset of 271 pipe pile observations across saturated and dry conditions. Figure 2 presents the box plot of the data:

$$Z = \frac{X - \mu}{\sigma} \tag{6}$$

where Z is the z-score, indicating how many standard deviations an element X is from the mean; X is the value of the element being standardized; μ is the mean of the population or sample; and σ is the standard deviation of the population or sample.

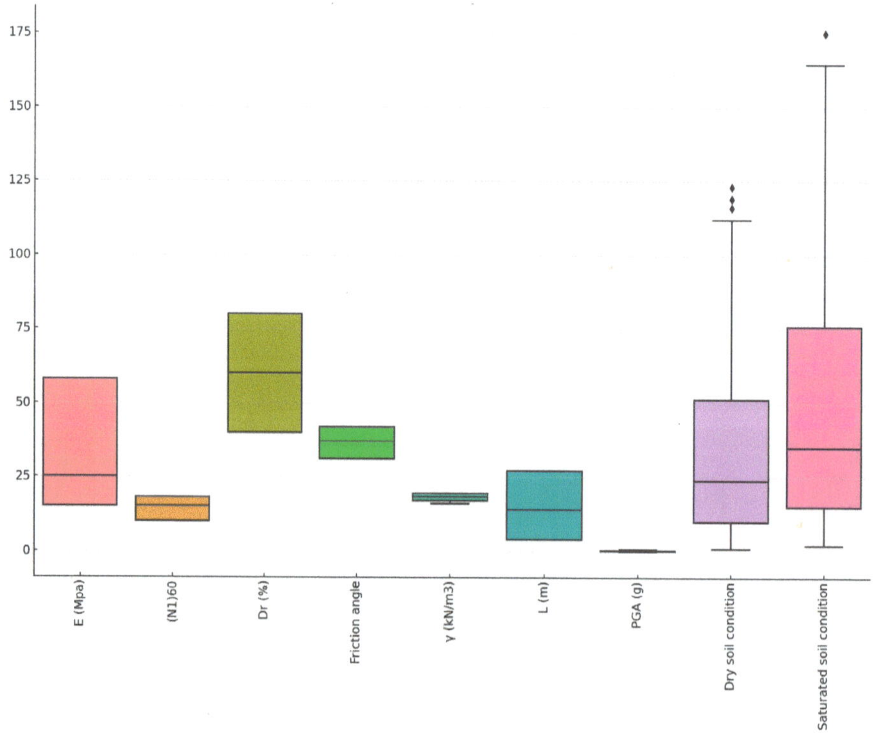

Figure 2. Box plot.

2.2.3. Feature Encoding

Categorical variables must be encoded into their numerical formats to enable mathematical coherence alongside continuous inputs during computation [41]. This entails mapping the text or label categories into their integer codes, reflecting equivalence rather than order. Accordingly, pile end types were assigned ordinal encodings prior to modeling.

2.2.4. Correlation Heatmap

Prior to conducting regression analysis, it is imperative to examine the presence and degree of collinearity among the feature variables, as strong collinearity can lead to instability in the modeling results. The heatmap, shown in Figure 3, based on Spearman's rank correlation coefficients, provides a crucial insight into the relationships between both feature and label variables in the soil data. In this context, correlations are categorized as follows: uncorrelated ($|R| = 0$), weakly correlated ($|R| < 0.4$), correlated ($0.4 < |R| < 0.75$), strongly correlated ($0.75 < |R| < 1$), and fully correlated ($|R| = 1$).

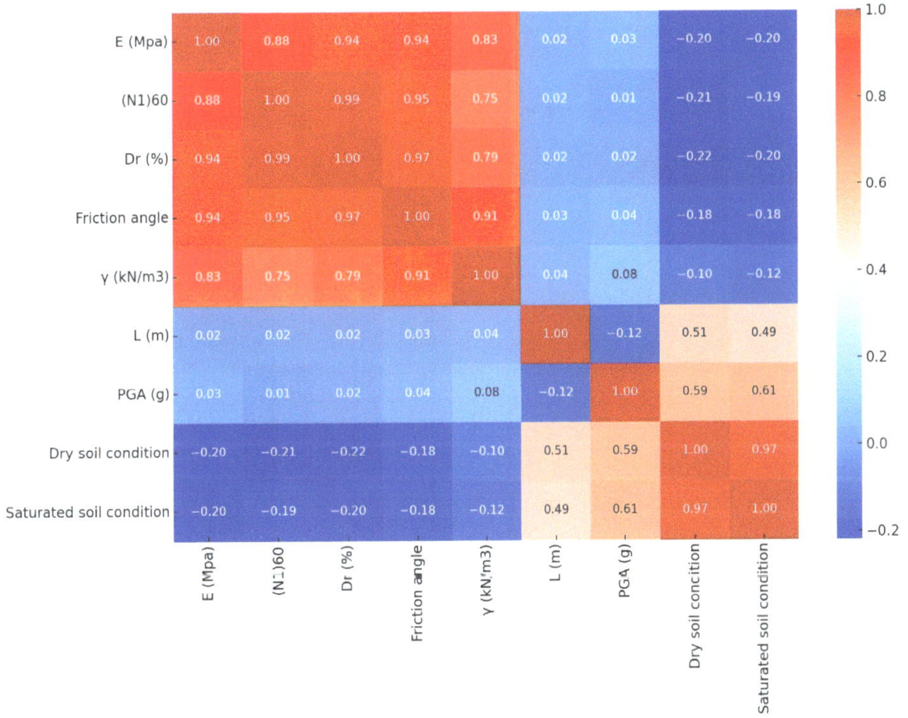

Figure 3. Correlation coefficient matrix heatmap.

Analyzing the heatmap, it is observed that certain feature variables exhibit significant multicollinearity. For example, the correlation between variables such as 'E (MPa)' and '$(N_1)_{60}$', as well as between 'PGA (g)' and both 'Dry soil condition' and 'Saturated soil condition', fall into the higher correlation brackets. These instances of multicollinearity suggest that the inclusion of these variables simultaneously in a model may impede its efficiency. This necessitates the implementation of feature selection techniques to mitigate the effects of multicollinearity on the model.

2.2.5. Stratified Train–Test Split

To objectively assess the model's generalization, the encoded dataset was randomly partitioned into mutually exclusive training data (70%) and holdout sets or validation data

(30%) for cross-validation based on stratification percentage optimization in preliminary experiments. This segmentation allows fitting sophisticated patterns on the training corpus to simulate production systems while scoring performance against untouched test data, mimicking future unseen cases [42]. Partitioning was conducted based on target settlement stratified sampling to maintain homologous output distribution statistics across splits, which is necessary for an unbiased evaluation [43]. Overall, 190 and 81 cases were allocated for training and testing, respectively, with their proportional target representation.

2.2.6. Model Optimization Scope

The problem scope targeted developing an accurate predictive model for pile settlement under seismic events based on key influencing variables identified from the literature and prior field evidence. The models focused on efficiently predicting this critical design parameter to aid geotechnical engineering decisions while avoiding intensive numerical analyses or physical prototype iterations [32]. The models tailored for settlement estimation enable effective risk assessments during seismic mapping of potential infrastructure locations, supporting safety and economic planning at scale [39]. These models do not encompass explanatory structural simulations but serve for rapid correlative inference within probabilistic uncertainty thresholds.

3. Model Development

The precise approach undertaken for the model's development encompassed sequential steps of appropriate machine learning algorithm selection based on empirical evaluations, hyperparameter tuning for optimization, followed by training, and testing iterative cycles to qualify the model's robustness and generalizability prior to finalization [31]. Each sub-process is elaborated in the following subsection.

3.1. Algorithm Selection

An ensemble RF regression model was selected as the principal supervised learning technique for predicting the seismic settlement of pile foundations based on a comparative assessment against other prevalent classifiers on a smaller prototype dataset. Ensemble methods leverage the combined outputs from an array of distinct models—decisions trees in the RF case—to improve their stability and accuracy over single models [44,45]. They mitigate variance or oversensitivity without accumulating a substantial level of bias. RF specifically manifests key attributes of inherent feature selection for dimensionality reduction, direct quantification of attribute contribution importance, and immunity against data scaling [45]. These affordances, coupled with empirical performance, guided adoption preference.

Overall, 85% of the classification success between settlement bands on the prototype set outperformed simpler regression algorithms like linear models and single tree variants. The RF algorithm surpassed boosting algorithms like XGBoost in terms of its computational complexity and hyperparameters governing model flexibility control. Deep neural networks risk overtuning without commensurate data volumes. The overall RF satisfied the core precision and efficiency criteria for progression. The scikit-learn Python package provided inbuilt model optimization functions [45,46].

3.2. Tuning Fundamentals

Tuning adjusts model configurations to discover the optimum combination of control parameters that return the highest accuracy or business value without materially compromising computational feasibility. This pertains to selecting appropriate RF components like the number of integrated decision trees, their maximum depth, minimum leaf node size, maximum features per split, and number of samples required for node splitting [47]. Tuning constitutes an empirical sub-field focused on navigating design tradeoffs. Grid search and Bayesian optimization are common approaches.

Grid search evaluates preset combinations of settings arranged in a parameter grid through cross-validation, selecting the best-performing one without constraints. Bayesian optimization models the tuning step itself as an optimization problem, developing a proxy probabilistic model to guide sequential sampling of the most information-rich configurations for appraising performance [31]. Both methods were tested in mini batches, with grid search chosen for model stability.

3.3. Hyperparameter Setting

In the conducted Random Forest analysis, the hyperparameters were selected to balance model complexity and computational efficiency, aiming for robust and interpretable results. The key hyperparameters include:

1. Number of trees (n_estimators): Set to 500, a value that offers a good balance between model performance and computational load. More trees generally improve accuracy but increase computation time.
2. Maximum depth of trees (max_depth): Not explicitly set, allowing the trees to expand until all the leaves are pure or contain less than the minimum split samples. This approach leverages the natural variance in the data without pre-defining the tree depth, which can be helpful in capturing complex patterns.
3. Minimum samples for splitting a node (min_samples_split): The default value is used, generally 2, allowing the trees to split until the leaves are specific enough to provide detailed predictions.
4. Random state (random_state): Set to a fixed value (e.g., 42) to ensure reproducibility of the results. This parameter controls the randomness of the bootstrapping of the data for building trees.

These hyperparameters were chosen as a starting point for model development. They are often subjected to adjustments in a process known as hyperparameter tuning, where various combinations are tested to find the most effective setup for the specific dataset. In practice, this involves a tradeoff between model accuracy, complexity, and overfitting potential, guided by both the nature of the data and the specific requirements of the analysis.

3.4. Hyperparameter Optimization

Comparing RF variants using grid search over key tuning factors produced a robust architecture with 500 integrated decision trees and an unlimited node depth and leaf size. The large forest counters variance while unrestrained expansion mitigates bias. To prevent the model from being overfitted, early stopping was used. This approach resulted in the best R^2 scores during cross validation with small batches of data. Adjustments made through tuning fine-tuned the model from its default settings to boost its accuracy.

3.5. Model Training and Validation

With optimized specifications, separate RF regression models were trained on dry and saturated observations from the encoded input dataset (training data) to determine variable relationships specific to each condition through multivariate correlation analysis. Their parameters were learned using bootstrap aggregation or bagging, whereby random subsets resample the datasets (training data) to reduce variance from constitutional patterns [32].

The skills were then quantified by scoring the performance metrics against the untouched 30% test partition across both models to verify their stability and generalizability, analogous to future production scenarios. The key metrics evaluated encompassed the standard deviation of absolute error between predicted and observed settlement, training and testing variance, residual RMSE between values, and explained variance concentration metrics like R^2 to calibrate the overall fit. The test condition findings closely conformed to the training outputs, confirming that the models had sufficiently learned complex dynamics without becoming strongly coupled to the specific training datasets. Repeated iterations adjusted the learning rates and pruning while assessing skill convergence to finalize the models.

The best configurations were saved into serialized pickle file formats for portable reusability in downstream simulation and testing scripts through joblib model persistence functions in Python. This avoided retraining computation [39].

4. Performance Evaluation

Performance metrics such as Mean Absolute Error (MAE), Mean Squared Error (MSE), and R^2 score (R^2) were calculated for both models (Equations (7)–(9)):

$$MAE = \frac{1}{n}\sum_{i=1}^{n}|y_i - \bar{y}_i| \tag{7}$$

$$MSE = \frac{1}{n}\sum_{i=1}^{n}(y_i - \bar{y}_i)^2 \tag{8}$$

$$R^2 = 1 - \frac{\sum_{i=1}^{n}(y_i - \bar{y}_i)^2}{\sum_{i=1}^{n}(y_i - \bar{y})^2} \tag{9}$$

In particular, for the dry soil condition, the obtained performance metrics were: MAE = 3.58 mm, MSE = 26.89 mm^2, and R^2 = 0.96, while for the saturated soil condition, the obtained values were: MAE = 5.96 mm, MSE = 83.5 mm^2, and R^2 = 0.95.

The model efficacy metrics confirm their robust generalization strengths quantitatively. These are further supported visually in the scatter plots (Figures 4 and 5) comparing the actual and predicted outputs for the dry and saturated conditions, respectively, with points closer to the diagonal line indicating higher prediction accuracy. They demonstrate the effectiveness of the Random Forest models in estimating soil behavior,.

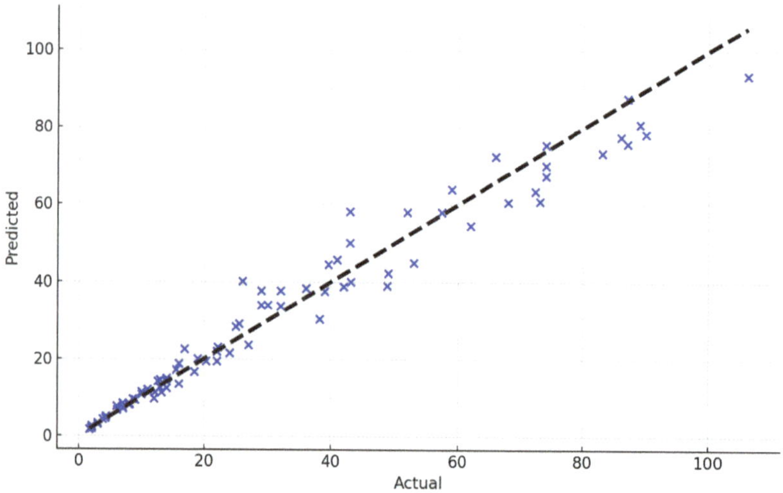

Figure 4. Actual vs. predicted pile settlement (mm) for the dry soil condition.

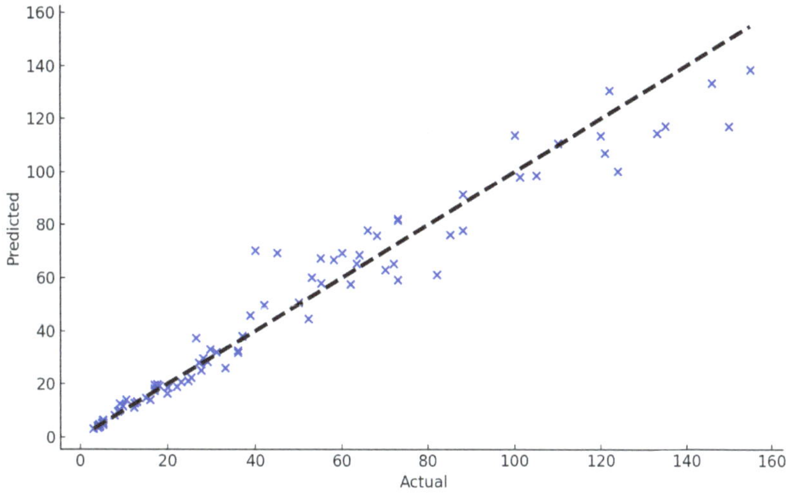

Figure 5. Actual vs. predicted pile settlement (mm) for the saturated soil condition.

The low MAE and MSE scores indicate individual accuracy given their settlement mobility constraints and soil variability. More critically, high enclosing R^2 values over 0.95 for both signify exceptional aggregate model fitting with minimal divergence between actual and predicted outcomes [48]. This verifies the precise seismic settlement capture capability. Almost all deviance is appropriately explained to apply predictions. Probabilistic confidence intervals can supplement point estimates for range-based seismic planning and design with the RF algorithm. Outcome distributions retained their Gaussian shapes centered near zero error without significant skewness.

Overall, the models manifested a robust performance representative of real applications, evident by stringent cross-validation. Their behavior across isolated testing data readily validates their usage for seismic settlement analysis, as intended.

Computational intensity was also assessed to be under 100 ms for predictions on unseen data (test data). This meets the expedited simulation criteria. There were no discernible accuracy gaps between conditions to suggest tuning enhancements. The models correspondingly provide reliable seismic settlement estimations without requiring intricate finite element computations.

5. Interpretability Assessment

While the RF algorithm delivers reliable predictions, its internal behavior as an ensemble of multiple decision trees hinders plain interpretability into the produced complexity, interactions, and feature contributions, frequently categorized as a 'black box' algorithm [31]. Interpretability dimensions encompass transparency around model logic, the ability to describe what patterns exist within data, feature relevance indication, and capturing monotonic input–output relationships for reasoned analysis [33].

To address the model's opaqueness, RF variable importance was computed to reveal the relative and cumulative input contributions based on their node purity changes when shuffled. Peak ground acceleration and pile diameter constituted the dominant inputs, collectively explaining over 81% of the variational influence on the observed settlement (Figures 6 and 7). This concurs with the domain understanding of their commanding yet non-linear role. Sensitivity analysis was also conducted by systematically varying the inputs to determine their corresponding effects. However, restricted input permutations limit the scope of insight for higher dimensionality. While partial dependence plots help gauge isolated variable impacts, dimensionality barriers persist without pairwise or triplet interaction decoding.

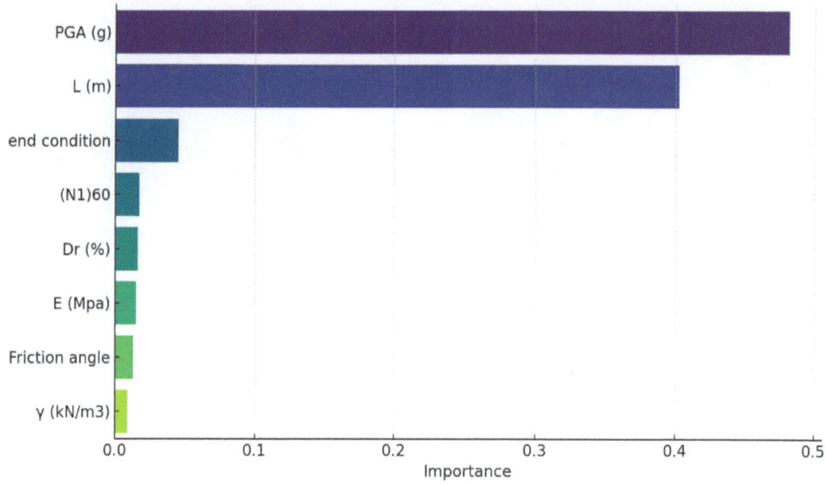

Figure 6. Feature importance for the dry soil condition.

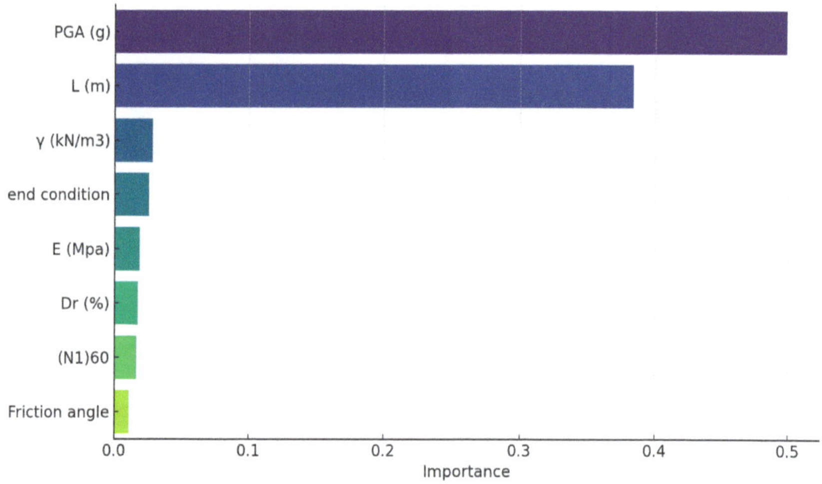

Figure 7. Feature importance for the saturated soil condition.

6. Partial Dependence Plots for the Top Features

The partial dependence plots created from the RF models (Figure 8) exhibit the distinct influence of selected soil parameters on the predicted settlement. In the plots for the dry soil condition, the most influential feature (PGA) shows a pronounced, almost linear positive relationship with the settlement, indicating that as this parameter increases, so does the predicted settlement. Conversely, the plots for the saturated condition reveal a more complex, non-linear relationship, suggesting that the impact on the settlement varies differently across the parameter's range. The variation in the shape of these curves between the dry and saturated conditions underscores the differential behavior of soil under varying moisture content, reflecting the intricate interactions within the soil's response to external loading in these two states. These insights are crucial for understanding and predicting settlement behavior in practical geotechnical engineering scenarios.

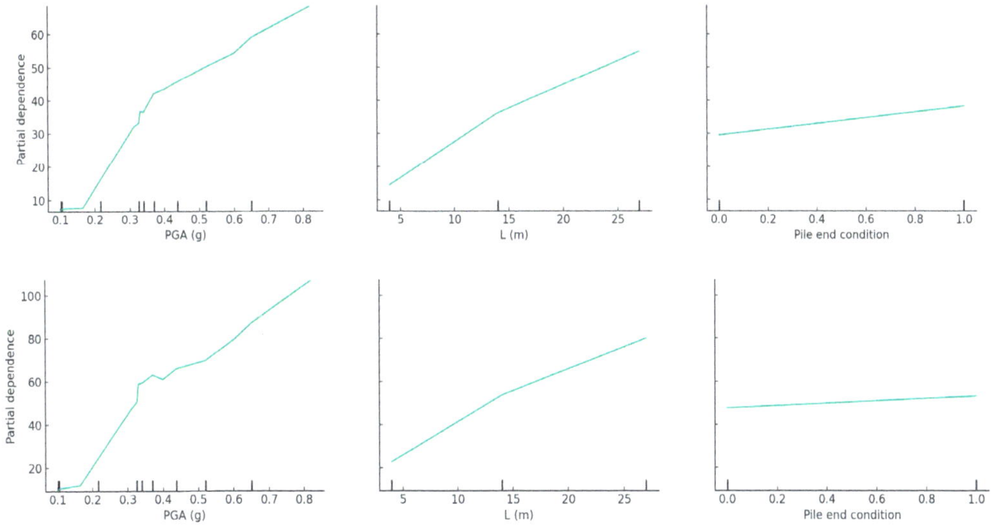

Figure 8. Partial dependence plots for the top features.

7. Conclusions and Recommendations

This study introduces and validates Random Forest (RF) regression models designed to predict seismic-induced settlements in pipe piles under both dry and saturated soil conditions. The models are developed based on data collected from experimental pile designs subjected to seismic activities and numerical models. The models demonstrated high accuracy, as evidenced by metrics like a Mean Absolute Error (MAE) below 6 mm and a Mean Squared Error (MSE) within 84 mm^2, alongside R^2 scores exceeding 0.95. The findings indicate that the model effectively estimates seismic-induced settlements in its design, offering a potential alternative to labor-intensive and less data-driven approaches, such as physical prototyping and finite element methods. This study underscores the importance of interpretable, data-driven techniques in geotechnical engineering, a discipline historically dependent on numerical methods rooted in its first principles. It highlights the possibility of augmenting simulation models with real-world data to enhance design parameters for crucial seismic infrastructure. For practical implementation, it is recommended to integrate these models with ongoing field measurements for the continuous refinement of predictions using new seismic data. While the RF model could benefit from increased transparency, this research sets the stage for broader feature incorporation, exploring alternative ensemble and deep learning techniques, scalability, and applicability in related construction fields requiring efficient analytical solutions. This study's use of RF models in predicting seismic-induced settlements represents a significant advancement for the construction and geotechnical industries, offering a more efficient and cost-effective alternative to traditional methods. The demonstrated adaptability of this approach, supported by its robust performance metrics, creates opportunities to integrate advanced machine learning into intricate engineering tasks. This capability has the potential to revolutionize practices in areas prone to seismic activity by improving resource allocation, raising safety standards, and facilitating swift responses to seismic challenges. Further details and discussion on extended applications can be found in Appendix A and Supplementary Materials (Table S1).

Supplementary Materials: The following supporting information can be downloaded at: https://www.mdpi.com/article/10.3390/jmse12020274/s1, Table S1: Pipe-piles datasets.

Author Contributions: Data collection, D.A.-J.; investigation, M.A.Q.A.-J.; writing—review and editing, D.A.-J. and S.E.R.; review and writing—original draft preparation, L.F.A.B. and S.E.R.; review, editing and visualization, M.A.Q.A.-J., D.A.-J. and L.F.A.B. All authors have read and agreed to the published version of the manuscript.

Funding: This research received no external funding.

Institutional Review Board Statement: Not applicable.

Informed Consent Statement: Not applicable.

Data Availability Statement: Data are contained within the article and supplementary materials.

Conflicts of Interest: The authors declare no conflicts of interest.

Appendix A. Random Forest Regression Code for Pile Settlement Prediction

```
# Import necessary libraries
from sklearn.ensemble import RandomForestRegressor
from sklearn.model_selection import train_test_split
from sklearn.metrics import mean_absolute_error, mean_squared_error, r2_score
import pandas as pd
import matplotlib.pyplot as plt
# Function to plot Actual vs. Predicted values
def plot_actual_vs_predicted(y_actual, y_predicted, title):
 plt.figure(figsize=(10, 6))
 plt.scatter(y_actual, y_predicted, c='blue')
 plt.plot([y_actual.min(), y_actual.max()], [y_actual.min(), y_actual.max()], 'k--', lw=3)
 plt.xlabel('Actual')
 plt.ylabel('Predicted')
 plt.title(title)
 plt.show()
# Load the dataset
file_path = 'path/to/your/excel/file.xlsx' # Replace with the actual path to your Excel file
df = pd.read_excel(file_path)
# Apply One-Hot Encoding to the 'Pile end condition ' column
df_encoded = pd.get_dummies(df, columns=['Pile end condition'])
# Features (common for both conditions)
X_encoded = df_encoded.drop(['Dry soil condition', 'Saturated soil condition'], axis=1)
# Targets
y_dry_encoded = df_encoded['Dry soil condition']
y_saturated_encoded = df_encoded['Saturated soil condition']
# Function to create, train, and evaluate a Random Forest model
def create_rf_model(X, y, title):
 # Split the data into training and test sets
 X_train, X_test, y_train, y_test = train_test_split(X, y, test_size=0.2, random_state=42)
 # Initialize and train the Random Forest model
 rf = RandomForestRegressor(n_estimators=100, random_state=42)
 rf.fit(X_train, y_train)
 # Make predictions on the test set
 y_pred = rf.predict(X_test)
 # Calculate and return performance metrics and plot
 mae = mean_absolute_error(y_test, y_pred)
 mse = mean_squared_error(y_test, y_pred)
 r2 = r2_score(y_test, y_pred)
 plot_actual_vs_predicted(y_test, y_pred, title)
 return mae, mse, r2
# Create, train, and evaluate the model for dry soil condition
mae_dry_encoded, mse_dry_encoded, r2_dry_encoded = create_rf_model(X_encoded, y_dry_encoded, 'Actual vs.
Predicted for Dry Soil Condition')
```

```
# Create, train, and evaluate the model for saturated soil condition
mae_saturated_encoded, mse_saturated_encoded, r2_saturated_encoded = create_rf_model(X_encoded, y_saturated_encoded, 'Actual vs. Pre-dicted for Saturated Soil Condition')
# Display the metrics
print("Metrics for Dry Soil Condition:", {'MAE': mae_dry_ encoded, 'MSE': mse_dry_encoded, 'R2': r2_dry_encoded})
print("Metrics for Saturated Soil Condition:", {'MAE': mae_saturated_encoded, 'MSE': mse_saturated_encoded, 'R2': r2_saturated_encoded})
```

References

1. Xu, S.H.; Li, Z.W.; Deng, Y.F.; Bian, X.; Zhu, H.H.; Zhou, F.; Feng, Q. Bearing performance of steel pipe pile in multilayered marine soil using fiber optic technique: A case study. *Mar. Georesources Geotechnol.* **2022**, *40*, 1453–1469. [CrossRef]
2. Abi, E.; Shen, L.; Liu, M.; Du, H.; Shu, D.; Han, Y. Calculation Model of Vertical Bearing Capacity of Rock-Embedded Piles Based on the Softening of Pile Side Friction Resistance. *J. Mar. Sci. Eng.* **2023**, *11*, 939. [CrossRef]
3. Wang, Y.; Qi, Z.; Wei, T.; Bao, J.; Zhang, X.; Zhou, Y. Numerical Study on the Responses of Suction Pile Foundations under Horizontal Cyclic Loading Considering the Soil Stiffness Degradation. *J. Mar. Sci. Eng.* **2023**, *11*, 2336. [CrossRef]
4. Wu, Q.; Ding, X.; Zhang, Y. Dynamic interaction of coral sand-pile-superstructure during earthquakes: 3D Numerical simulations. *Mar. Georesour. Geotechnol.* **2023**, *41*, 774–790. [CrossRef]
5. Bhattacharya, S. Pile Instability during Earthquake Liquefaction. Ph.D. Thesis, University of Cambridge, Cambridge, UK, 2003.
6. Barbosa, V.D.; Galgoul, N.S. Designing Piled Foundations with a Full 3D Model. *Open Constr. Build. Technol. J.* **2018**, *12*, 65–78. [CrossRef]
7. Tehrani, F.S.; Han, F.; Salgado, R.; Prezzi, M.; Tovar, R.D.; Castro, A.G. Effect of surface roughness on the shaft resistance of non-displacement piles embedded in sand. *Géotechnique* **2016**, *66*, 386–400. [CrossRef]
8. Al-Jeznawi, D.; Jais, I.B.M.; Albusoda, B.S.; Khalid, N. Numerical assessment of pipe pile axial response under seismic excitation. *J. Eng.* **2023**, *29*, 10, 1–11. [CrossRef]
9. Hussein, A.F.; Hesham El Naggar, M. Seismic axial behaviour of pile groups in non-liquefiable and liquefiable soils. *Soil Dyn. Earthq. Eng.* **2021**, *149*, 106853. [CrossRef]
10. Al-Jeznawi, D.; Khatti, J.; Al-Janabi, M.A.Q.; Grover, K.S.; Jais, I.B.M.; Albusoda, B.S.; Khalid, N. Seismic performance assessment of single pipe piles using three-dimensional finite element modeling considering different parameters. *Earthq. Struct.* **2023**, *24*, 455.
11. Al-Jeznawi, D.; Jais, I.B.M.; Albusoda, B.S.; Alzabeebee, S.; Keawsawasvong, S.; Khalid, N. Numerical study of the seismic response of closed-ended pipe pile in cohesionless soils. *Transp. Infrastruct. Geotechnol.* **2023**, 1–27. [CrossRef]
12. Sarkhani Benemaran, R.; Esmaeili-Falak, M.; Javadi, A. Predicting resilient modulus of flexible pavement foundation using extreme gradient boosting based optimized models. *Int. J. Pavement Eng.* **2022**, *24*, 2095385. [CrossRef]
13. Zhang, Q.; Afzal, M. Prediction of the elastic modulus of recycled aggregate concrete applying hybrid artificial intelligence and machine learning algorithms (retracted). *Struct. Concr.* **2021**, *23*, 2477–2495. [CrossRef]
14. Huang, L.; Jiang, W.; Wang, Y.; Zhu, Y.; Afzal, M. Prediction of long-term compressive strength of concrete with admixtures using hybrid swarm-based algorithms. *Smart Struct. Syst.* **2022**, *29*, 433–444.
15. Benemaran, R.S.; Esmaeili-Falak, M. Optimization of cost and mechanical properties of concrete with admixtures using MARS and PSO. *Comput. Concr.* **2020**, *26*, 309–316.
16. Qu, X.-Q.; Wang, R.; Zhang, J.-M.; He, B. Influence of Soil Plug on the Seismic Response of Bucket Foundations in Liquefiable Seabed. *J. Mar. Sci. Eng.* **2023**, *11*, 598. [CrossRef]
17. Poulos, H.G. *Tall Building Foundation Design*; CRC Press: Boca Raton, FL, USA, 2017.
18. Zhang, Y.; Hu, X.; Tannant, D.D.; Zhang, G.; Tan, F. Field monitoring and deformation characteristics of a landslide with piles in the Three Gorges Reservoir area. *Landslides* **2018**, *15*, 581–592. [CrossRef]
19. Lee, I.-M.; Lee, J.-H. Prediction of pile bearing capacity using artificial neural networks. *Comput. Geotech.* **1996**, *18*, 189–200. [CrossRef]
20. Che, W.F.; Lok, T.M.H.; Tam, S.C.; Novais-Ferreira, H. *Axial Capacity Prediction for Driven Piles at Macao using Artificial Neural Network*; AA Balkema Publishers: Leiden, The Netherlands, 2003.
21. Liu, H.; Li, T.J.; Zhang, Y.F. *The Application of Artificial Neural Networks in Estimating the Pile Bearing Capacity*; AA Balkema Publishers: Leiden, The Netherlands, 1997.
22. Hanna, A.M.; Morcous, G.; Helmy, M. Efficiency of pile groups installed in cohesionless soil using artificial neural networks. *Can. Geotech. J.* **2004**, *41*, 1241–1249. [CrossRef]
23. Shanbeh, M.; Najafzadeh, D.; Ravandi, S.A.H. Predicting pull-out force of loop pile of woven terry fabrics using artificial neural network algorithm. *Ind. Textila* **2012**, *63*, 37–41.
24. Xu, B.; Deng, J.; Liu, X.; Chang, A.; Chen, J.; Zhang, D. A Review on Optimal Design of Fluid Machinery Using Machine Learning Techniques. *J. Mar. Sci. Eng.* **2023**, *11*, 941. [CrossRef]
25. Wang, K.; Gaidai, O.; Wang, F.; Xu, X.; Zhang, T.; Deng, H. Artificial Neural Network-Based Prediction of the Extreme Response of Floating Offshore Wind Turbines under Operating Conditions. *J. Mar. Sci. Eng.* **2023**, *11*, 1807. [CrossRef]

26. Rumelhart, D.E.; Hinton, G.E.; Williams, R.J. Learning internal representation by error propagation. In *Parallel Distributed Processing*; Rumelhart, D.E., McClelland, J.L., Eds.; MIT Press: Cambridge, MA, USA, 1986; Chapter 8, Volume 1.
27. Albinmousa, J.; Peron, M.; Jose, J.; Abdelaal, A.F.; Berto, F. Fatigue of V-notched ZK60 magnesium samples: X-ray damage evolution characterization and failure prediction. *Int. J. Fatigue* **2020**, *139*, 105734. [CrossRef]
28. Marsavina, L.; Berto, F.; Radu, N.; Serban, D.A.; Linul, E. An engineering approach to predict mixed mode fracture of PUR foams based on ASED and micromechanical modelling. *Theor. Appl. Fract. Mech.* **2017**, *91*, 148–154. [CrossRef]
29. Al-Jeznawi, D.; Jais, M.; Albusoda, B.S. A Soil-Pile Response under Coupled Static-Dynamic Loadings in Terms of Kinematic Interaction. *Civ. Environ. Eng.* **2022**, *18*, 96–103. [CrossRef]
30. Song, W.; Liu, X.; Berto, F.; Razavi, S.M.J. Energy-based low cycle fatigue indicator prediction of non-load-carrying cruciform welded joints. *Theor. Appl. Fract. Mech.* **2018**, *96*, 247–261. [CrossRef]
31. Qian, G.; Lei, W.-S.; Yu, Z.; Berto, F. Statistical size scaling of breakage strength of irregularly-shaped particles. *Theor. Appl. Fract. Mech.* **2019**, *102*, 51–58. [CrossRef]
32. Lei, W.-S.; Qian, G.; Yu, Z.; Berto, F. Statistical size scaling of compressive strength of quasi-brittle materials incorporating specimen length to diameter ratio effect. *Theor. Appl. Fract. Mech.* **2019**, *104*, 102345. [CrossRef]
33. Breiman, L. Random forests. *Mach. Learn.* **2001**, *45*, 5–32. [CrossRef]
34. Raman, C.D.; Bhattacharya, S.; Blakeborough, A. Settlement Prediction of Pile-Supported Structures in Liquefiable Soils During Earthquake. In Proceedings of the 14th World Conference on Earthquake Engineering, Beijing, China, 12–17 October 2008.
35. Robinsky, E.I.; Morrison, C.F. Sand Displacement and Compaction around Model Friction Piles. *Can. Geotech. J.* **1964**, *1*, 81–93. [CrossRef]
36. Beaty, M.H.; Byrne, P.M. UBCSAND Constitutive Model Version 904aR. *Itasca UDM Web Site* **2011**, *69*, 71.
37. Mahmood, M.R.; Al-Helo, K.H.; AL-harbawee, A.M. Laboratory study of plug length development and bearing capacity of pipe pile models embedded within partially saturated cohesionless soils. In *Advances in Analysis and Design of Deep Foundations: Proceedings of the 1st GeoMEast International Congress and Exhibition, Egypt 2017 on Sustainable Civil Infrastructures*; Springer International Publishing: Cham, Switzerland, 2018; Volume 1, pp. 28–43. [CrossRef]
38. Hussein, R. Experimental and Numerical Modeling of Piles under Combined Loading in Liquefied Sandy Soil with Improvement by Nanomaterials. PhD Thesis, University of Baghdad, Baghdad, Iraq, 2021.
39. Namdar, A. Prediction of the settlement of a pile and assessment of seismic soil-pile interaction—An analytical investigation. *Procedia Struct. Integrity* **2020**, *28*, 311–322. [CrossRef]
40. Wang, L.; Zhang, X.; Qi, D. Indoor Thermal Stratification and Its Statistical Distribution. *Indoor Air* **2019**, *29*, 243–255. [CrossRef] [PubMed]
41. Cao, G.; Wang, S.; Yoo, E.-H. A Statistical Framework of Data Fusion for Spatial Prediction of Categorical Variables. *Stoch. Environ. Res. Risk Assess* **2014**, *28*, 1785–1799. [CrossRef]
42. Kohavi, R. Feature subset selection as search with probabilistic estimates. In Proceedings of the AAAI Fall Symposium on Relevance, Arlington, Virginia, 25–27 October 1994; pp. 122–126.
43. Geisser, F.; Eddy, W. A predictive approach to model selection. *J. Am. Stat. Assoc.* **1979**, *74*, 153–160. [CrossRef]
44. Ren, Z.; Sun, L.; Zhai, Q. Improved k-means and spectral matching for hyperspectral mineral mapping. *Int. J. Appl. Earth Obs. Geoinf.* **2020**, *91*, 102154. [CrossRef]
45. Zhou, J.; Li, X.; Mitri, H.S. Classification of rockburst in underground projects: Comparison of ten supervised learning methods. *J. Comput. Civ. Eng.* **2016**, *30*, 4016003. [CrossRef]
46. Tao, H.; Jingcheng, W.; Langwen, Z. Prediction of Hard Rock TBM Penetration Rate Using Random Forests. In Proceedings of the IEEE Control and Decision Conference, Osaka, Japan, 15–18 December 2015.
47. Rodriguez-Galiano, V.; Mendes, M.P.; Garcia-Soldado, M.J.; Chica-Olmo, M.; Ribeiro, L. Predictive modeling of groundwater nitrate pollution using Random Forest and multisource variables related to intrinsic and specific vulnerability: A case study in an agricultural setting (Southern Spain). *Sci. Total Environ.* **2014**, *476*, 189–206. [CrossRef]
48. Ching, J.; Phoon, K.K. Constructing site-specific multivariate probability distribution model using Bayesian machine learning. *J. Eng. Mech.* **2019**, *145*, 04018126. [CrossRef]

Journal of
Marine Science
and Engineering

MDPI

Article

Investigation of Soil Heaving and Penetration Resistance of Bucket Foundation with Inner Bucket and Cruciform Skirts

Zhong Xiao [1,*], Teng Ma [1], Hongwei Wang [2], Feng Bian [2], Haifeng Jin [2], Tengjiao Yu [2], Wei Zhang [1], Pan Hu [3,*] and Jinhui Li [4]

[1] State Key Laboratory of Hydraulic Engineering Simulation and Safety, Tianjin University, Tianjin 300072, China
[2] Shallow Water Development Company, PetroChina Dagang Oilfield Company, Tianjin 300280, China
[3] School of Engineering, Design and Built Environment, Western Sydney University, Sydney 2747, Australia
[4] Department of Civil and Environmental Engineering, Harbin Institute of Technology (Shenzhen), Shenzhen University Town, Shenzhen 518055, China
* Correspondence: tjuzhongxiao@tju.edu.cn (Z.X.); p.hu@westernsydney.edu.au (P.H.)

Abstract: Since concrete is cheaper and more resistant to corrosion than steel, the wide-shallow concrete bucket foundation is being used extensively in ocean engineering. By adding the inner bucket and cruciform skirts, both the bearing capacity and rigidity of the wide-shallow concrete bucket foundation increase significantly. When compared to the hollow steel bucket foundation, the inclusion of thicker skirts, as well as the addition of inner bucket and cruciform skirts, would cause changes to the soil flow mechanism, resulting in soil heave within each compartment and changes in soil strength evolution and penetration resistance during installation in clay. In order to study the influence of the addition of the inner bucket and cruciform skirts on the soil heaving inside each compartment, soil softening and penetration resistance, three-dimensional large deformation finite element (LDFE) models for the bucket foundation with and without inner bucket, and cruciform skirts considering soil remolding were established using the Coupled Eulerian–Lagrangian (CEL) approach. It was found that the inner bucket significantly changes the soil flow and softening of the soil during penetration of the bucket foundation. According to the theoretical analysis and numerical results, the diameter of the optimal inner bucket is equal to 5/8 of the outer diameter. The adhesion coefficient observed in this study falls within the range of 0.5 to 0.8, which is higher than the theoretical value of 0.25 that assumes the soil is fully remolded. The reason for this discrepancy is that the soil is only partially remolded during the actual installation of the bucket foundation. The neglect of the softening of the soil or considering the soil as completely softened will result in significant variation in the predicted penetration resistance; hence, partial softening of the soil should be taken into account.

Keywords: wide-shallow concrete bucket foundation; inner bucket and cruciform skirts; soil heaving; penetration resistance; soil remolding; optimal value

Citation: Xiao, Z.; Ma, T.; Wang, H.; Bian, F.; Jin, H.; Yu, T.; Zhang, W.; Hu, P.; Li, J. Investigation of Soil Heaving and Penetration Resistance of Bucket Foundation with Inner Bucket and Cruciform Skirts. *J. Mar. Sci. Eng.* **2023**, *11*, 996. https://doi.org/10.3390/jmse11050996

Academic Editor: Antoni Calafat

Received: 10 February 2023
Revised: 26 April 2023
Accepted: 5 May 2023
Published: 7 May 2023

1. Introduction

Due to the advantages of lower cost and better durability, the wide-shallow concrete bucket foundation has a great potential in coastal and ocean engineering, and it can be used as the foundation of an offshore deep-water wharf, breakwater, or offshore wind turbine. Normally, the rigidity of the wide-shallow concrete bucket foundation is relatively small due to its large diameter and shallow penetration depth. However, by adding the inner bucket and cruciform skirts, both the bearing capacity and rigidity of the bucket foundation will increase significantly [1]. A typical application of such a configuration is a deep-water wharf, as shown in Figure 1.

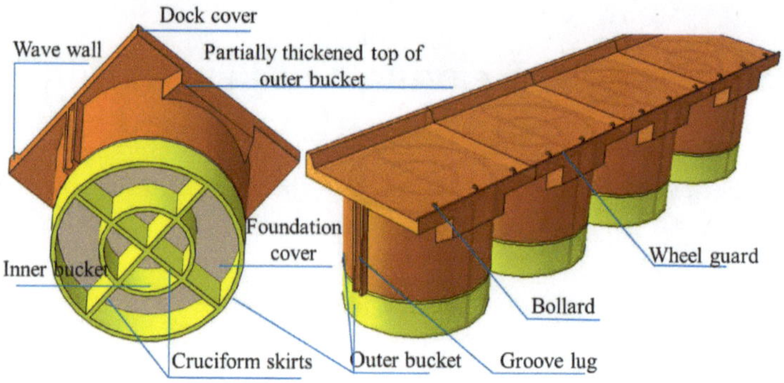

Figure 1. The deep-water wharf using bucket foundation with inner bucket and cruciform skirts.

The inner bucket and cruciform skirt significantly increase the total area of the bottom and skirt; therefore, the foundation will be subjected to larger penetration resistance during its installation process. In addition, soil heaving will occur in each compartment separated by the inner bucket and cruciform skirt, which may prevent the foundation from reaching the designed installation depth and affect its post-installation bearing capacity. However, current research on the installation of bucket foundations mainly focuses on the ones with no compartment. The bucket foundation was first applied in an actual engineering project in 1994, and Tjelta [2], through research on the measured data, found that installation methods, including self-weight penetration, pressing penetration, and suction-assisted penetration, have a significant impact on the soil flow mechanism. Iskander et al. [3] conducted 1g indoor experiments on the penetration and extraction behavior of suction bucket foundations in clay and sand and explained the variation of penetration resistance and soil heaving. Chen et al. [4] focused on the influence of different installation methods on the penetration process and installation effect of bucket foundations, studying the impact of different installation methods on the penetration process. Zhai et al. [5] also analyzed how to reduce penetration resistance through indoor experiments. Wang et al. [6] used the arbitrary Lagrangian–Eulerian (ALE) method to simulate the installation process of suction buckets and compared the difference in penetration resistance and soil heaving under different installation methods. Xiao et al. [7] studied the changes in penetration resistance and soil flow mechanism during the penetration process of a single bucket foundation, considering the strain softening and rate effect of soil using the coupled Eulerian–Lagrangian (CEL) method. The adverse effects of bucket penetration-induced soil heaving and softening on the ultimate bearing capacity were also studied [8]. Jin et al. [9] used the smoothed particle hydrodynamics (SPH) method to simulate the penetration process of bucket foundations in sand.

There have been few studies on bucket foundations with multiple compartments, mostly focusing on on-site or indoor experiments. Liu et al. [10] studied a new type of seven-compartment suction bucket shallow foundation (CBSF) used in a wind farm project in Jiangsu, China. Zhang et al. [11] conducted a large-scale model study on the installation speed, penetration resistance, and levelness of the foundation in typical saturated silty clay for the seven-compartment suction bucket.

However, overall, there is still a lack of relevant research using three-dimensional large deformation numerical simulation methods to simulate the penetration process of bucket foundations with internal buckets and cruciform skirts in soil and analyze the penetration resistance and soil flow characteristics with consideration of soil softening properties.

When the ratio of the diameter of the inner bucket to that of the outer bucket changes, the height of soil heave inside each compartment during the installation process may vary accordingly. As the foundation cover contacts the heaved soil inside the compartment, sinking of the foundation will be halted. In such case, the soils within some compartments are still not in contact with the foundation cover after installation, which will obviously reduce the post-installation vertical bearing capacity of the foundation. Therefore, it is necessary to study soil heaving and penetration resistance during the installation of a bucket foundation with inner bucket and cruciform skirts in clay considering soil remolding under different ratios of the diameter of the inner bucket to that of the outer bucket. The bucket foundation with inner bucket and cruciform skirts has the optimal ratio of the diameter of the inner bucket to that of the outer bucket when the height of the soil heaving in the inner and outer compartments is almost equal after installation.

Large deformation numerical analysis is required to model the installation of the bucket foundation with inner bucket and cruciform skirts in clay. In the literature, there are three large deformation numerical methods by which to simulate the installation of the bucket foundation in clay; these include the ALE method [7], the remeshing and interpolation technique by small strain (RITSS) [8], and the CEL method [7]. For the ALE method, a new mesh is created when the elements have obvious distortion. Then, the variables are mapped from the old mesh to the new mesh, which can accurately define material boundaries and complex contact interactions. However, it is mostly applied to solve plane strain or axisymmetric problems due to the limitation of computational efficiency. The RITSS method falls into the category of the ALE method in nature [12], but the topological relationship between its old mesh and new mesh can be changed. Nevertheless, it requires the user to write a program to implement the entire calculation process, and the realization of the interpolation of the variables from the old mesh to the new mesh is challenging. The CEL method is a new finite element analysis method for large deformations, which combines the Eulerian analysis method and Lagrangian analysis method. As the CEL method can better simulate the deformation of the material in the Eulerian mesh, it can effectively solve the large deformation problems such as mesh and element distortions in the traditional Lagrangian domain [13].

In this study, a three-dimensional large deformation finite element model incorporating the effect of soil remolding was established to model the installation of a bucket foundation with inner bucket and cruciform skirt in clay using the CEL method. The influences of the inner bucket and cruciform skirts on the soil heaving and penetration resistance during the installation of the bucket foundation in clay were analyzed and discussed. Meanwhile, the evolution of soil remolding at different penetration depths was demonstrated. The contribution of the resistance from each component of the foundation to the total penetration resistance was analyzed. The optimal ratio of the diameter of the inner bucket to that of the outer bucket was proposed based on theoretical analysis and validated by the numerical results. Key findings were obtained and discussed from the numerical results, and recommendations were made.

2. Method and Materials

2.1. Finite Element Mesh and Boundary Conditions

The CEL large deformation finite element analysis method in commercial software Abaqus is used to simulate the penetration of the concrete bucket foundation with inner bucket and cruciform skirts. It is commonly used to solve geomechanical boundary value problems involving large deformations. The CEL method adopts an explicit time integration scheme with the central difference rule for the solution of the non-linear system of differential equations. No iteration is needed as the unknown solution for the next time step can be found from the solution of the previous time step. In the CEL formulation, the Lagrangian domain deals with the deformations of the bucket foundation and the Eulerian domain deals with the displacement of the soil. The numerical model only contains the eight-node Eulerian elements (EC3D8R), which are the only available elements in CEL analyses. Displacement boundary conditions were applied to the bucket foundation through a reference point.

For concrete bucket foundation, the common sizes of D typically range from 5 to 25 m, with t falling within the range of 0.2 to 0.6 m. Therefore, the outer diameter of the outer bucket (D) is set as 10 m, while the diameter of the inner bucket (D_{in}) in the base is equal to

$D/2$. A typical value of t = 0.4 m is used for all the skirt thickness. In order to avoid the influence of soil boundary effect on the penetration process, the radius of the soil domain is set to $3D$ and the soil depth is set to $5D$. The height of the Eulerian mesh needs to exceed the original top boundary of the soil to avoid the loss of the material. A 5 m void layer was set above the soil field to allow sufficient flow space of the surface soil. In order to maintain the accuracy of the calculation of the penetration resistance and soil heaving in the bucket, a very fine mesh zone shaped in a square column with width of $1.25D$ was set surrounding the bucket foundation. The minimum mesh size was set as $t/6$, as shown in Figure 2, which is consistent with the setting in Xiao et al. (2019) [7]. Due to the symmetry, and in order to improve the computational efficiency, only a quarter of the domain was selected for the simulation, as shown in Figure 2. The soil base was fixed while only vertical displacements were allowed for side boundaries.

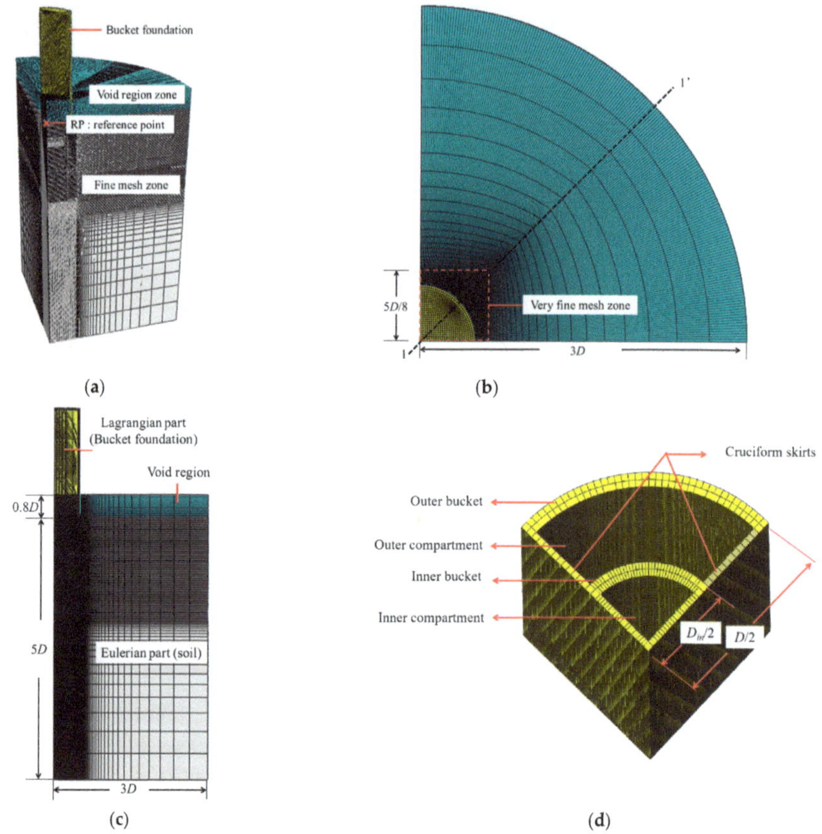

Figure 2. Geometry and mesh of the 3D CEL model. (**a**) 3D view. (**b**) Top view. (**c**) Side view. (**d**) Details of the bucket foundation.

2.2. Contact and Penetration Rate

In the CEL finite element model, the contact between the soil and the foundation adopted a universal contact surface based on the penalty contact method. The contact force is calculated using the normal and tangential components of the displacement and the friction coefficient. The normal component of the contact force acts to prevent penetration between the surfaces, while the tangential component of the contact force acts to prevent sliding between the surfaces. It aims to model an undrained behavior, as the soil always sticks to the bucket foundation while failure occurs within the soil. A sufficiently large friction coefficient μ was adopted so that the sliding failure occurs on the soil element adjacent to the bucket wall. The result of the case when μ = 0.2 is the same as that of the case when μ = 10. Therefore, μ is set as 0.2 to ensure that sliding failure occurs on the soil elements adjacent to the bucket wall.

The bucket foundation can be installed using suction or jacking after self-weight penetration. In this study, the caisson was penetrated into the soil at a constant rate using velocity control to model the jacking installation of the bucket foundation with inner bucket and cruciform skirts. A penetration rate of 1 m/s was adopted for the installation of the bucket foundation, which provides a good balance between the accuracy and efficiency of the simulation (Xiao et al., 2019) [7].

2.3. Material Properties

The bucket foundation with inner bucket and cruciform skirts was simulated as a rigid body as its rigidity is much greater than that of clay. The saturated clay was modeled by an ideal elastoplastic constitutive model based on the Tresca yield criterion. The gradient of the undrained shear strength of clay with depth z is usually 1–2 kPa/m [14–17]; hence, the undrained shear strength of the clay was prescribed with $s_u = 2 + 1.2z$ in this study. The effective unit weight of the clay was adopted as $\gamma' = 6.5$ kN/m^3, which is also a typical value for marine soft soil [15,17]. The elastic modulus E of the soil was set as $500s_u$ [18,19]. The Poisson's ratio υ was 0.499 [18,19], which is used to simulate the volume incompressibility of the saturated soft soil under undrained conditions.

In order to simulate the remolding of the soil, the softening model [20] and rate effect model [21] were combined with the Tresca yield criterion to describe the evolution of the undrained strength of clay. During the penetration, the soil is disturbed and therefore undergoes softening. Meanwhile, when the shear strain rate is higher than the reference strain rate, the shear strength of the soil will increase, which is termed the strain rate effect of the soil. Both of these factors act on the strength of the soil simultaneously. Therefore, the equation for calculating the undrained shear strength considering the strain softening and rate effects of marine soft soil is expressed as follows:

$$s_u = \beta_s \beta_r s_{u0} \tag{1}$$

where

$$\beta_s = \delta_{rem} + (1 - \delta_{rem}) e^{-3\xi/\xi_{95}} \tag{2}$$

and

$$\beta_r = \left[1 + \eta \left(\frac{\max(\dot{\gamma}, \dot{\gamma}_{ref})}{\dot{\gamma}_{ref}} \right)^{\beta} \right] / (1 + \eta) \tag{3}$$

where s_u is the shear strength of clay after considering the softening and rate effects; s_{u0} is the initial soil shear strength; β_s is the softening effect coefficient, while β_r is the rate effect coefficient; δ_{rem} is the ratio of the initial shear strength of the clay to the shear strength when the soil is completely disturbed, and its value is equal to the inverse of the soil sensitivity S_t; ξ is the accumulated absolute plastic shear strain; ξ_{95} is the accumulated shear strain corresponding to the 95% degradation in soil strength from intact to fully-remolded conditions; and η and β represent the viscosity characteristics and shear index of the soil, respectively, and the general values range from 0.1 to 2.0 and 0.05 to 0.2, respectively. The $\dot{\gamma}_{ref}$ is the reference shear strain rate, and $\dot{\gamma}$ is the maximum shear strain rate, which is calculated by:

$$\dot{\gamma} = \frac{\Delta\varepsilon_1 - \Delta\varepsilon_3}{\Delta t} \tag{4}$$

where $\Delta\varepsilon_1$ and $\Delta\varepsilon_3$ are the cumulative major and minor principal strains over the duration of Δt.

For marine clays, the commonly used values for soil sensitivity S_t and ductility coefficient ξ_{95} are 2–6 and 10–50, respectively [20,22]. The soil sensitivity S_t = 4 and the ductility coefficient ξ_{95} = 30 were taken in this study.

Some assumptions have been made on the numerical model: the distribution of soil strength is assumed to increase proportionally with depth; all lateral frictional resistance is believed to occur within the soil itself, as opposed to at the interface between the soil and the bucket wall; and finally, the bucket is assumed to be penetrated at a constant speed.

2.4. Model Validation

The CEL model was initially validated against the centrifuge test results reported for the jacking installation of the bucket foundation [16,23]. The sizes of the bucket foundations are shown in Figure 3. The undrained shear strength of the soil is s_{u0} = 1.25z kPa and s_{u0} = 10 + 2.8z kPa in validation case 1 and 2, respectively. The corresponding details of the foundation dimensions and soil properties are shown in Table 1. The comparison of the penetration resistance obtained from the CEL analysis in this study and the centrifuge testing results [16,23] can be seen in Figure 3, where d_p is the penetration depth of the bucket foundation relative to the mud surface. For both cases, reasonably good agreement can be found with the maximum difference less than 15%. The root mean square error (RMSE) for comparison of the penetration resistances in Chen et al. (2007) and Westgate et al. (2009) are approximately 0.2 and 0.05, respectively. The relatively close results confirm the validity of the CEL model in this study.

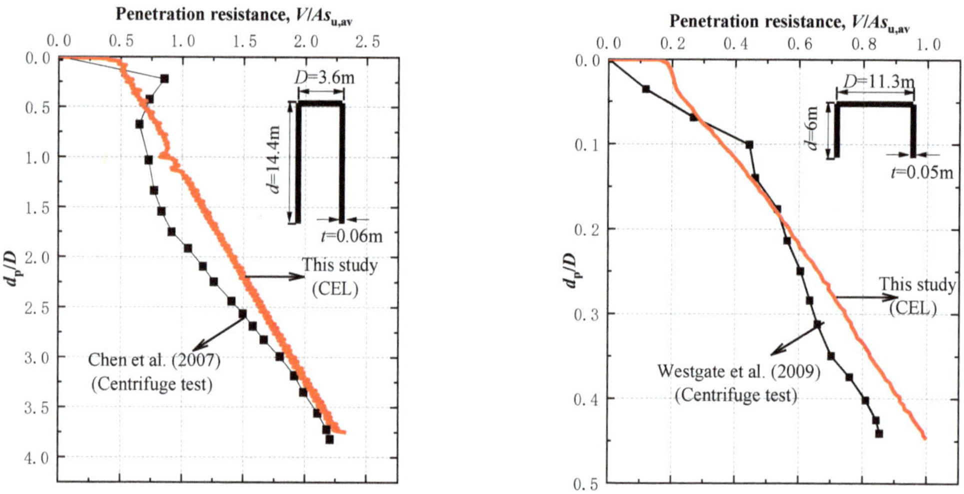

Figure 3. Comparison of the penetration resistance between the CEL analysis results and centrifuge test results [16,23].

Table 1. Details of the foundation dimensions and soil properties in validation cases.

Case	γ′(kN/m³)	S_t	ξ_{95}	η	β	E/s_u	s_{u0} (kPa)	D (m)	d (m)	t (m)	Reference
Case 1	6.7	2.6	10	1	0.1	500	1.25z	3.6	14.4	0.06	[Chen et al. (2007) [16]]
Case 2	5.9	3	10	1	0.1	500	10 + 2.8z	11.3	6	0.05	[Westgate et al. (2009) [23]]

3. Numerical Results

3.1. Soil Flow and Heaving in Each Compartment

During the penetration of the bucket foundation, the soil will be squeezed by the skirts; therefore, the height of the soil surface inside the bucket will vary. If soil heaving happens inside the bucket, the foundation cannot be installed to the predetermined design depth. As a consequence, the actual bearing capacity will be less than the design bearing capacity.

The average height of the soil surface inside the bucket foundation relative to the original mud surface is defined as H_{plug}. If it is positive, it indicates that the soil surface inside the bucket is higher than the original mud surface. Otherwise, the soil inside the bucket settles. Take the base case ($D_{in} = D/2$) as an example; the comparison of H_{plug} in the inner or outer compartment of the bucket foundation with and without inner bucket and cruciform skirts at different penetration depth ratio (d_p/D) is shown in Figure 4.

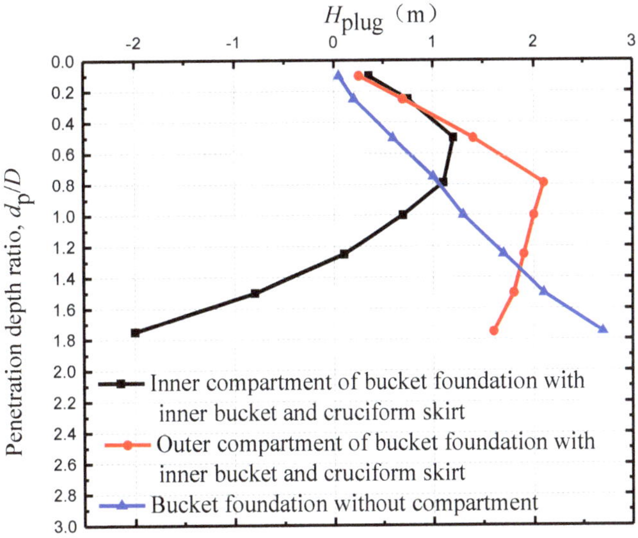

Figure 4. Average height of soil surface at different penetration depth ratio.

It can be seen from Figure 4 that the height of the soil surface in the bucket foundation without compartment continuously increases with the penetration depth ratio. With the increase of the penetration depth ratio, the soils displaced by the skirt are inclined to flow into the bucket. This is because the earth pressure within the bucket decreases with the increase of the penetration depth ratio due to the increase of the soil strength with depth. For the foundation with inner bucket and cruciform skirts during the shallow penetration, the H_{plug} is significantly higher than that of the bucket foundation without compartment. This is due to the added inner bucket and cruciform skirts squeezed more soil during the penetration, which gradually flowed into the inner or outer compartment. As the penetration depth ratio increases (i.e., $d_p/D > 0.5$ for the inner compartment and $d_p/D > 0.8$ for the outer compartment), the H_{plug} decreases with the increase of the penetration depth ratio. The main reason lies in the larger friction resistance caused by the inner bucket and cruciform skirt, which prevents the soil from flowing into the inner or outer compartment.

Meanwhile, it can be seen from Figure 4 that with further penetration, the difference of the H_{plug} between the inner and outer compartment becomes greater. Take $d_p/D = 1.5$ as an example; the H_{plug} in the outer compartment is about 1.7 m, which prevents the bucket foundation to penetrate into the predetermined depth and correspondingly its post-installation bearing capacity reduces. The H_{plug} in the outer compartment is about 1.0 m lower than the original mud surface, which greatly decreases the vertical bearing capacity of the foundation due to non-contact between the lid of the bucket and the soil.

The evolution of the soil surface height in each compartment has correlation with the soil flow. Figure 5 is the section view of the soil flow velocity vector at different positions and penetration depths (i.e., the section view is the soil flow along the section 1-1′ in Figure 2b). Figure 5a reflects the trend of soil flow during shallow penetration ($d/D = 0.1$). It can be seen that the soil around each skirt flows downwards and then evenly flows to both sides. As the penetration depth increases to $d/D = 0.5$ in Figure 5b, and since the outer boundary of the bucket is closer to the semi-infinite space and the inner boundary of the bucket is limited, the passive earth pressure inside the bucket is lower than that of

resistance of the bucket foundation in clay accounting for a partially-remolded condition in this study is more reasonable.

Although the bucket foundation studied in this paper incorporates the inner bucket and cruciform skirts, its penetration resistance can still be calculated according to the theory proposed by Houlsby and Byrne [24]. The ratio of the individual resistance to the total resistance of the bucket foundation with the inner bucket and cruciform skirts is shown in Figure 7.

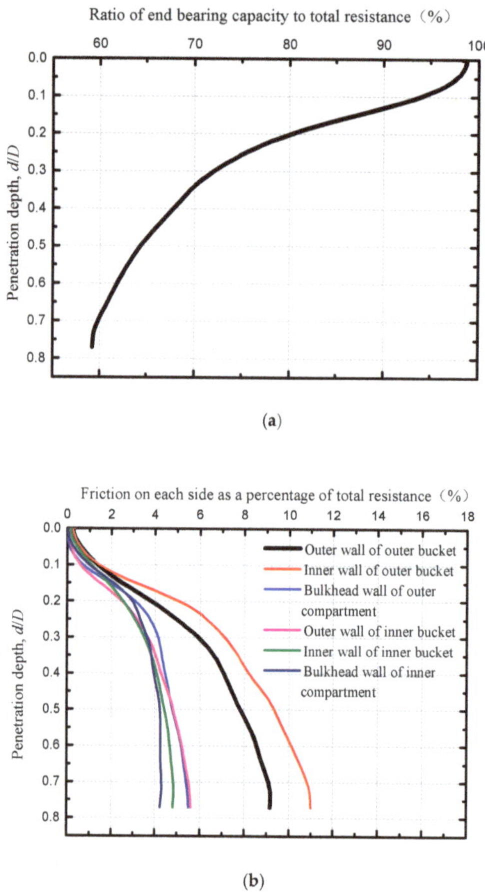

Figure 7. Proportion of resistance to total resistance. (**a**) Ratio of end bearing capacity to total resistance. (**b**) Percentage of friction on each part.

It can be seen that the total resistance of the concrete thick-wall bucket foundation with the inner bucket and cruciform skirts mainly comes from the end bearing capacity. The ratio of the friction resistance to total penetration resistance increases with the penetration depth.

3.2.1. End Bearing Capacity

The equation for calculating the end bearing capacity at the bottom of the bucket is:

$$F_N = (\gamma' d N_q + s_{u0,tip} N_c) A \quad (6)$$

where A is the bottom area. Therefore, the end bearing capacity coefficient N_c can be derived as:

$$N_c = (\frac{F_N}{A} - \gamma' d N_q)/s_{u0,tip} \tag{7}$$

For the hollow bucket foundation and the bucket foundation with inner bucket and cruciform skirts, the corresponding change of the end bearing capacity coefficient N_c with the penetration depth ratio is shown in Figure 8.

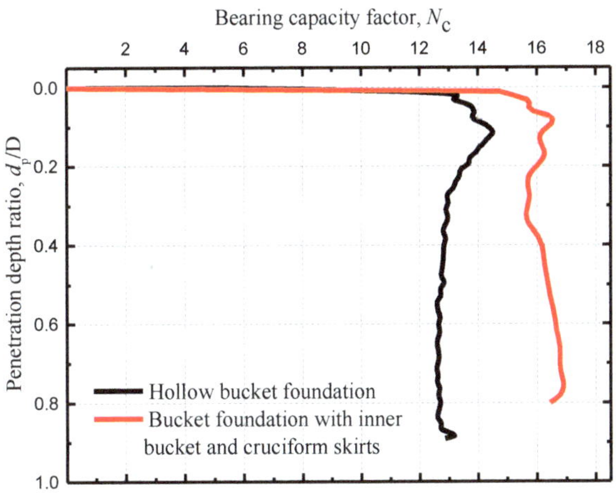

Figure 8. End bearing capacity coefficient N_c.

It can be seen from Figure 8 that the end bearing capacity coefficient of the hollow bucket foundation first increases with the penetration depth and then slightly decreases before stabilizing during further penetration. For bucket foundations with inner bucket and cruciform skirts, after the end bearing capacity coefficient increases and then slightly decreases with penetration depth, a slight increase occurs with penetration depth. For the investigated penetration depth, the end bearing capacity coefficient of the bucket foundation with inner bucket and cruciform skirts is obviously greater than that of the hollow bucket foundation.

3.2.2. Skirt Friction

According to Equation (5), the Equation for calculating skirt friction is shown as Equation (8):

$$F_f = d\alpha s_{u0,av} L \tag{8}$$

where L is the perimeter or width of each skirt. After conversion, the Equation for the adhesion coefficient α is:

$$\alpha = F_f / d s_{u0,av} L \tag{9}$$

The adhesion coefficient α between the skirting board and the soil on both sides are calculated and the results for each skirt are shown in Table 2.

Table 2. Adhesion coefficient for each skirt.

Outer Wall of Outer Bucket	Inner Wall of Outer Bucket	Bulkhead Wall ofOuter Compartment
0.51	0.68	0.60
Outer wall ofinner bucket	**Inner wall ofinner bucket**	**Bulkhead wall ofinner compartment**
0.70	0.79	0.63

The adhesion coefficient is generally considered as the inverse of the sensitivity of the soil, for which it is 0.25 in this case. However, it can be seen from the table that the value is much larger than 0.25 because the soil was partially remolded during the installation.

The adhesion coefficient α on the side of the skirting board is mainly determined by the degree of disturbance to the soil during the foundation installation, which can be expressed by the softening coefficient. However, for bucket foundations with inner bucket and cruciform skirts, the soil height in the compartment is too large due to the relatively small height of the foundation, which increases the adhesion coefficient.

Taking the penetration depth ratio $d/D = 0.5$ as an example, the section view of the softening factor of the soil is shown in Figure 9. The corresponding softening factor of the soil near different skirting boards is shown in Figure 10.

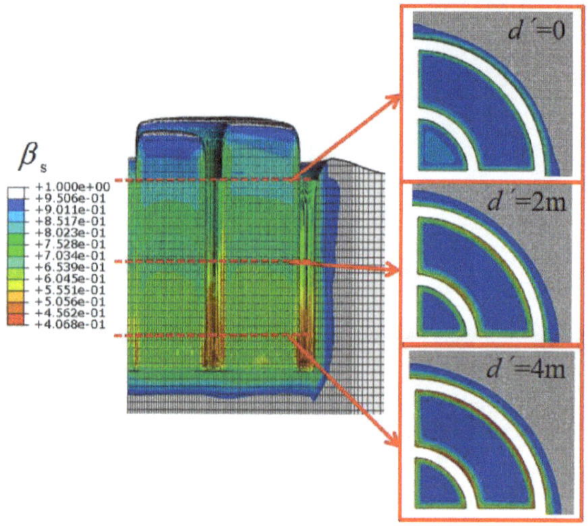

Figure 9. Section view of soil softening factor at different depths.

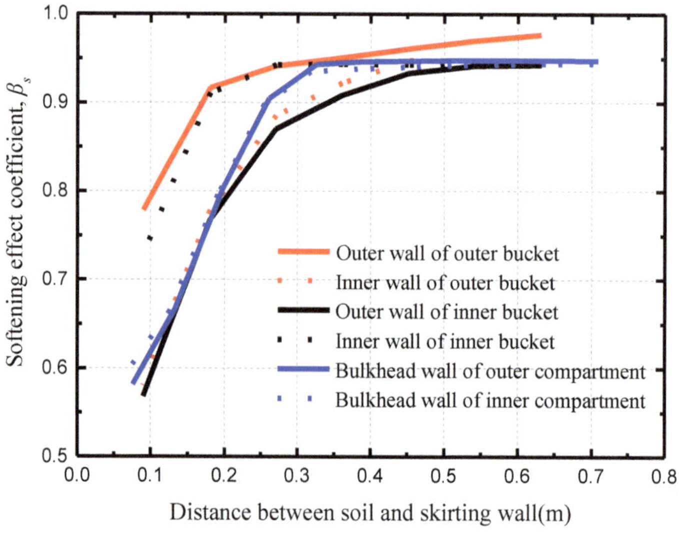

Figure 10. Softening factor of soil for each skirt.

It can be seen that the deeper the depth, the more severe the softening effect. Among the different skirting walls, most softening occurred in the outer wall of the inner bucket. However, due to the larger volume of soil in the outer compartment, the cohesion coefficient for the inner wall of the inner bucket is the largest while the cohesion coefficient of the outer wall of the outer bucket is the smallest.

3.3. The Optimal Inner Bucket Diameter

Through the analysis of soil flow and soil plug during the penetration, it is found that $D_{in} = D/2$ is not the optimal configuration based on the analysis of the soil plug inside the foundation after penetration, though the contribution of the inner bucket to improve the overall rigidity of the foundation can be maximized. When the penetration depth is relatively large, the excessive difference in soil height between the inner and outer compartments will adversely affect the bearing capacity of the foundations.

It is feasible to vary the diameter of the inner bucket to affect the soil flow, so that a part of the soil that originally flowed into the outer compartment can be diverted to the inner compartment and maintained similar height of soil. It can not only solve the negative height of the soil in the inner compartment, but also reduce the height of the soil in the outer compartment. It is more conducive to the installation of the foundation to the predetermined buried depth and to achieving the designed bearing capacity.

Based on previous analysis, for large penetration depth, the side friction resistance of the unit volume of the soil in the compartment will play a decisive role in the flow of the soil and the final height of the soil in the compartment. In order to ensure similar soil height for the inner and outer compartments, the total lateral friction resistance acting on the unit volume of the soil should be equal in the inner and outer compartments. This is expressed in Equation (10):

$$\frac{F_{fin}}{V_{in}} = \frac{F_{fout}}{V_{out}} \tag{10}$$

where F_{fin} and F_{fout} are the total friction of the soil in the inner and outer compartments, and it has correlation to the coefficient of side friction, the average soil shear strength over the penetration depth, the bottom perimeter of the compartment, and the penetration depth, as shown in Equation (8); V_{in} and V_{out} are the soil volumes in the compartments, and they can be expressed by the bottom area of the compartments and the penetration depth. During the penetration of the bucket foundation with inner bucket and cruciform skirts, the penetration depth and the average soil shear strength are equal in the inner and outer compartments. Based on previous analyses, the side friction coefficients of the skirts are similar. For the convenience of analysis, it can be assumed that the side friction coefficients of the skirts are equal, and Equation (10) is expressed as:

$$\frac{\alpha s_{u0,av} L_{in} d}{S_{in} d} = \frac{\alpha s_{u0,av} L_{out} d}{S_{out} d} \tag{11}$$

It can be simplified to:

$$\frac{L_{in}}{S_{in}} = \frac{L_{out}}{S_{out}} \tag{12}$$

where L_{in} and L_{out} are the bottom perimeter of the inner and outer compartment; and S_{in} and S_{out} are the bottom area of the inner and outer compartments. The diameter of the inner bucket should be changed so that the inner and outer compartments satisfy Equation (12). Further calculation concluded that when $a_{in} = a_{out}$, the diameter of the inner bucket is about 5/8 of the total diameter ($D_{in} = 5/8D$).

In addition, in order to better compare the influence of the diameter of internal cylinder on the soil flow and the soil surface in the inner and outer compartments during the foundation penetration process, it is necessary to calculate the penetration model when the diameter of the inner bucket is 3/4 of the total diameter.

When D_{in} = $1/2D$, $5/8D$, and $3/4D$, the average height of the soil surface and soil flow in the inner and outer compartments at different penetration depths (the 1-1′ section in Figure 2b) are shown in Figure 11. It can be seen that by changing the diameter of the inner bucket, the trend of soil flow changes significantly, and at the theoretically optimal inner bucket diameter (i.e., D_{in} = $5/8D$), the inner and outer compartments can maintain similar soil surface. However, if the diameter of the inner cylinder is too large (i.e., D_{in} = $3/4D$), the soil in the outer compartment will appear to have a negative height due to the excessive friction when the penetration depth is relatively large.

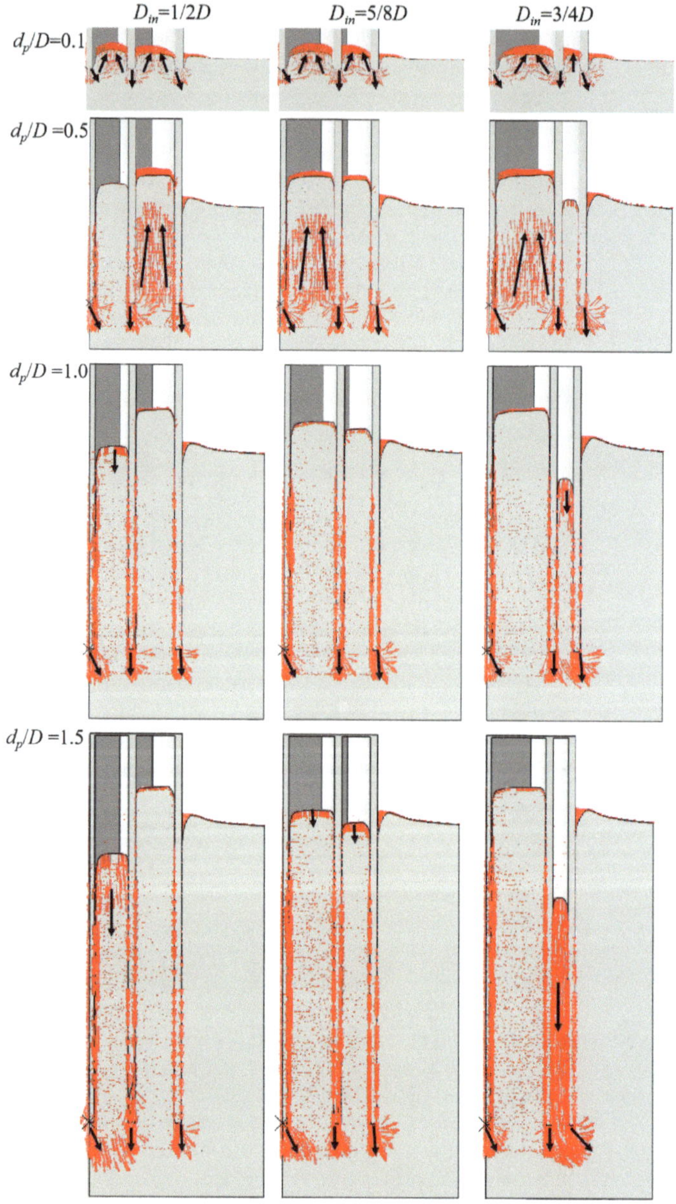

Figure 11. The height of the soil surface and soil flow in the inner and outer compartments under different inner bucket diameters and penetration depths.

4. Conclusions

In this paper, a CEL large-deformation finite element model was established to simulate the penetration of the bucket foundation with inner bucket and cruciform skirts. The soil strain softening effect was considered. The soil flow, soil strength evolution, and penetration resistance during the installation were investigated. Then, the optimal inner bucket diameter was suggested based on the theoretical analysis and numerical results. The main findings are as follows:

(1) Due to the addition of inner bucket and cruciform skirts, the average height of the soil surface inside the inner or outer compartment is significantly higher than that of the hollow bucket foundation during shallow penetration. However, if the adopted diameter of the inner bucket is half the diameter of the outer bucket, the larger frictional resistance in the outer compartment hinders the upward flow of the soil with the increase of the penetration depth, while that in the inner compartment a large amount of soil is pushed downward. The significant difference of soil height between the outer and inner compartment will affect the post-installation bearing capacity of the foundation.

(2) The total resistance of a concrete thick-walled bucket foundation with the inner bucket and cruciform skirts is equal to the sum of the friction resistance of all skirts and the end bearing capacity. The end bearing capacity coefficient of the bucket foundation with the inner bucket and cruciform skirts was significantly larger than that of the hollow bucket foundation. Meanwhile, the end bearing capacity accounts for a large proportion of the total resistance. Due to the partial remolding of the soil during the penetration, the friction coefficient for the bucket skirt was larger than the inverse of the sensitivity of the soil.

(3) The optimal inner bucket diameter is equal to 5/8 of the outer diameter obtained from the theoretical analysis and numerical results. Under the circumstances, the height of the soil surface in the inner and outer compartments is almost equal because the friction resistance acting on the unit volume of the soil in the inner compartment is equal to that in the outer compartment. It can be seen that the trend of soil flow is effectively varied by expanding the ratio of the diameter of the inner bucket to the outer diameter of the bucket foundation.

The findings add further value to the application of multi-compartment foundations such as bucket foundations with internal cylindrical and cruciform skirts. Further investigation is necessary to study the penetration of the bucket foundation, equipped with an inner bucket and cruciform skirts, into sandy soil, while taking into account the staged installation and variation in pore pressure.

Author Contributions: Conceptualization, Z.X.; methodology, Z.X. and T.M.; software, Z.X. and T.M.; validation, Z.X. and T.M.; formal analysis, Z.X. and T.M.; investigation, Z.X. and T.M.; resources, Z.X.; data curation, Z.X.; writing—original draft preparation, Z.X. and T.M.; writing—review and editing, Z.X. and P.H.; visualization, Z.X. and T.M.; supervision, Z.X., H.W., F.B., H.J., T.Y., W.Z., P.H. and J.L.; project administration, Z.X. and T.Y.; funding acquisition, Z.X., H.W., F.B., H.J., P.H. and J.L. All authors have read and agreed to the published version of the manuscript.

Funding: This research was funded by the National Key Research and Development Program of China (No. 2021YFB2600200) and the National Natural Science Foundation of China (No. 51879187; No. 51979067; No. 51890915).

Institutional Review Board Statement: Not applicable.

Informed Consent Statement: Not applicable.

Data Availability Statement: Data is contained within the article.

Conflicts of Interest: The authors declare no conflict of interest.

Notation

D	Outer diameter of bucket foundation
d	Skirt depth
d_p	Penetration depth
s_u	Undrained soil shear strength
γ'	Submerged unit weight
E	Young's modulus
v	Poisson's ratio
S_t	Soil sensitivity
β_s, β_r	Soil strain softening and rate factors
δ_{rem}	Reciprocal of soil sensitivity, = $1/S_t$
ξ	Current accumulated absolute plastic shear strain
ε_1, ε_3	Major and minor principal total strains
H_{plug}	Average height of soil surface relative to original mud surface in bucket foundation
V'	Penetration resistance
N_c	End bearing capacity factor
α	Frictional coefficient
α_0, α_i	Outer skin and inner skin frictional coefficient
ξ_{95}	Soil relative ductility (value of ξ for the undrained shear strength to achieve 95% remoulding)
$s_{u0,av}$	Soil average intact undrained shear strength over the penetration depth
$s_{u0,tip}$	Soil intact undrained shear strength at the skirt tip level
d'	Depth for a position under the same penetration depth
D_{in}	Diameter of the inner bucket
F_{fin}, F_{fout}	Total friction of the soil in the inner and outer compartments
S_{in}, S_{out}	Bottom areas of the inner and outer compartments
L_{in}, L_{out}	Bottom perimeter of the inner and outer compartments

References

1. Xiao, Z.; Liu, Y.; Ge, B.; Fu, D.F.; Zhou, Z.F.; Yan, Y. Bearing performance of offshore bucket foundation with internal cruciform skirt under combined loading. *Mar. Georesour. Geotechnol.* **2020**, *38*, 1209–1222. [CrossRef]
2. Tjelta, T.I. Geotechnical Experience from the Installation of the Europipe Jacket with Bucket Foundations. In Proceedings of the Offshore Technology Conference, Houston, TX, USA, 1–4 May 1995; pp. 897–908.
3. Iskander, M.; El-Gharbawy, S.; Olson, R. Performance of Suction Caissons in Sand and Clay. *Can. Geotech. J.* **2002**, *39*, 576–584. [CrossRef]
4. Chen, W.; Randolph, M. Radial Stress Changes around Caissons Installed in Clay by Jacking and by Suction. In Proceedings of the Fourteenth International Offshore and Polar Engineering Conference, Toulon, France, 23–28 May 2004; pp. 493–499.
5. Zhai, H.; Zhang, P.; Ding, H.; Le, C.; Wei, W. Laboratory tests on measures of reducing penetration resistance in bucket foundation installation. *Mar. Georesour. Geotechnol.* **2021**, *39*, 929–936. [CrossRef]
6. Wang, Y.; Zhu, X.; Lv, Y.; Yang, Q. Large Deformation Finite Element Analysis of the Installation of Suction Caisson in Clay. *Mar. Georesour. Geotechnol.* **2018**, *36*, 883–894. [CrossRef]
7. Xiao, Z.; Fu, D.; Zhou, Z.; Lu, Y.; Yan, Y. Effects of strain softening on the penetration resistance of offshore bucket. *Ocean. Eng.* **2019**, *193*, 106594. [CrossRef]
8. Xiao, Z.; Lu, Y.M.; Wang, Y.Z.; Tian, Y.H.; Zhao, Y.B.; Fu, D.F.; Zhang, D.H. Investigation into the influence of caisson installation process on its capacities in clay. *Appl. Ocean. Res.* **2020**, *104*, 102370. [CrossRef]
9. Jin, Z.; Yin, Z.; Kotronis, P.; Jin, Y.F. Numerical Investigation On Evolving Failure of Caisson Foundation in Sand Using the Combined Lagrangian-SPH Method. *Mar. Georesour. Geotechnol.* **2019**, *37*, 23–35. [CrossRef]
10. Liu, R.; Chen, G.; Lian, J.; Ding, H. Vertical Bearing Behaviour of the Composite Bucket Shallow Foundation of Offshore Wind Turbines. *J. Renew. Sustain. Energy* **2015**, *7*, 13123. [CrossRef]
11. Zhang, P.; Guo, Y.; Liu, Y.; Ding, H. Experimental Study On Installation of Hybrid Bucket Foundations for Offshore Wind Turbines in Silty Clay. *Ocean. Eng.* **2016**, *114*, 87–100. [CrossRef]
12. Ghosh, S.; Kikuchi, N. An Arbitrary Lagrangian-Eulerian Finite Element Method for Large Deformation Analysis of Elastic-Viscoplastic Solids. *Comput. Methods Appl. Mech. Eng.* **1991**, *86*, 127–188. [CrossRef]
13. Simulia, D. *ABAQUS Version 6.13. Analysis User's Manual*; Abaqus 6.13 Documentation; Dassault Systèmes Inc.: Providence, RI, USA, 2013.
14. Chen, W.; Zhou, H.; Randolph, M.F. Effect of installation method on external shaft friction of caissons in soft clay. *J. Geotech. Geoenviron. Eng.* **2009**, *135*, 605–615. [CrossRef]

15. Zhou, H.; Randolph, M.F. Large deformation analysis of suction caisson installation in clay. *Can. Geotech. J.* **2006**, *43*, 1344–1357. [CrossRef]
16. Chen, W.; Randolph, M.F. External radial stress changes and axial capacity for suction caissons in soft clay. *Géotechnique* **2007**, *57*, 499–511. [CrossRef]
17. Zhang, Y.; Wang, D.; Cassidy, M.J.; Bienen, B. Effect of installation on the bearing capacity of a spudcan under combined loading in soft clay. *J. Geotech. Geoenviron. Eng.* **2016**, *140*, 04014029. [CrossRef]
18. Vulpe, C.; Gourvenec, S.; Power, M. A generalised failure envelope for undrained capacity of circular shallow foundations under general loading. *Géotech. Lett.* **2014**, *4*, 187–196. [CrossRef]
19. Vulpe, C. Design method for the undrained capacity of skirted circular foundations under combined loading: Effect of deformable soil plug. *Géotechnique* **2015**, *65*, 669–683. [CrossRef]
20. Einav, I.; Randolph, M.F. Combining upper bound and strain path methods for evaluating penetration resistance. *Int. J. Numer. Methods Eng.* **2005**, *63*, 1991–2016. [CrossRef]
21. Liu, H.; Xu, K.; Zhao, Y. Numerical investigation on the penetration of gravity installed anchors by a coupled Eulerian—Lagrangian approach. *Appl. Ocean. Res.* **2016**, *60*, 94–108. [CrossRef]
22. Randolph, M.F. Characterization of soft sediments for offshore applications. In Proceedings of the 2nd International Conference on Site Characterization, Millpress Science, Rotterdam, The Netherlands, 19–22 September 2004; pp. 209–231.
23. Westgate, Z.J.; Tapper, L.; Lehane, B.M.; Gaudin, C. Modelling the installation of stiffened caissons in overconsolidated clay. In Proceedings of the ASME 28th International Conference on Ocean, Offshore and Arctic Engineering, Honolulu, HI, USA, 31 May–5 June 2009; pp. 119–126.
24. Houlsby, G.T.; Byrne, B.W. Design procedures for installation of suction caissons in clay and other materials. *Proc. Inst. Civ. Eng. Geotech. Eng.* **2005**, *158*, 75–82. [CrossRef]
25. Det Norske Veritas. Geotechnical design and installation of suction anchors in Clay. In Proceedings of the Offshore Technology Conference-Asia, Houston, TX, USA, 2–5 May 2005. Recommended practice E303.

Journal of
Marine Science and Engineering

MDPI

Article

Borehole Instability in Decomposed Granite Seabed for Rock-Socketed Monopiles during "Drive-Drill-Drive" Construction Process: A Case Study

Bo Sun [1], Qi Zhang [1,*], Wenxuan Zhu [1], Jian Leng [2] and Guanlin Ye [1]

1 State Key Laboratory of Ocean Engineering, Department of Civil Engineering, Shanghai Jiao Tong University, Shanghai 200240, China; bo.sun@sjtu.edu.cn (B.S.); zhuwx1992@sjtu.edu.cn (W.Z.); ygl@situ.edu.cn (G.Y.)
2 CCCC Third Harbor Engineering Co., Ltd., Shanghai 200032, China; keepwalkinglj@163.com
* Correspondence: zhangqi33@sjtu.edu.cn

Abstract: Monopiles are commonly used in the construction of offshore wind turbines. However, implementing drive-drill-drive construction techniques in decomposed granite seabed may lead to borehole instability during the window period between drilling and pile driving, resulting in significant project losses. This study provides a comprehensive understanding and approach to address the causes of borehole instability in rock-socketed monopiles in decomposed granite seabed. Using the Pinghai Bay offshore wind farm project in Fujian, China as an example, the details of drive-drill-drive and reverse-circulation drilling techniques employed in monopile construction were introduced. An improved sampling method was utilized to obtain decomposed granite samples, and a series of in situ and laboratory tests were conducted to analyze the physical and mechanical properties of marine-decomposed granite. By examining three cases of monopile construction, the factors contributing to borehole instability during rock-socketed monopile construction in decomposed granite seabed were identified, and corresponding recommendations were proposed. The results indicated that construction technology and unfavorable geological characteristics of decomposed granite are the primary causes of borehole instability. Collapses occurred mainly in highly and moderately decomposed granite layers. Employing smaller boreholes can reduce the likelihood and impact of borehole instability.

Keywords: borehole instability; decomposed granite; rock-socketed monopile; field case

Citation: Sun, B.; Zhang, Q.; Zhu, W.; Leng, J.; Ye, G. Borehole Instability in Decomposed Granite Seabed for Rock-Socketed Monopiles during "Drive-Drill-Drive" Construction Process: A Case Study. *J. Mar. Sci. Eng.* **2023**, *11*, 990. https://doi.org/10.3390/jmse11050990

Academic Editor: M. Dolores Esteban

Received: 19 March 2023
Revised: 20 April 2023
Accepted: 6 May 2023
Published: 6 May 2023

1. Introduction

Offshore wind power is an important contributor to mitigating the effects of climate change and reducing the global carbon footprint. The construction of offshore wind turbines has seen a significant increase in recent years to meet the growing demand for renewable energy [1]. However, offshore wind turbine construction presents a unique set of challenges, one of them being the elevated cost compared to onshore wind turbine construction [2]. Among the contributing factors to the elevated cost of offshore wind turbine construction are the design and installation of the foundation account for a substantial portion, potentially reaching 20–30% [3].

The foundations for offshore wind turbines can take various forms, such as gravity, monopiles, tripods, jackets, and suction bucket foundations [4]. Among these, monopiles have become a popular choice for offshore wind power due to their simplicity and cost-effectiveness [5–7]. Despite their popularity, monopile construction in certain geological formations can pose problems that result in increased foundation installation costs. In southeastern China, granites with varying degrees of weathering are widely present on the seabed [8,9]. The construction of monopiles in this area requires passing through decomposed granite, which led to frequent incidents of borehole instability and pile running during construction. These incidents have a significant impact on the safety and economics of offshore wind turbine construction.

Borehole instability is a major concern in drilling operations and has been studied in different formations. In anisotropic rocks, when the borehole is perpendicular to the bedding plane, the influence of the bedding plane on the wellbore instability is negligible, while the borehole instability is prone to occur when the borehole is oriented along the bedding plane [10]. Borehole instabilities in shale involve multiple mechanisms, including pore pressure diffusion, plasticity, anisotropy, capillary effects, permeability, and physicochemical changes [11]. In sandstones with moderate to high porosity and rich in quartz, borehole instabilities result in the formation of elongated, slender fractures that extend to substantial distances and are oriented perpendicular to the far-field principal compressive stress [12]. The failure mode of sandy soil in the borehole was found to be primarily governed by shear failure [13]. Extensive model borehole tests showed that in clay formations, over-consolidation can greatly improve the stability of the borehole, and for normally consolidated clay, the hole wall becomes unstable when the volume change exceeds 5% [14]. However, borehole stability performance in decomposed granite seabed is still not fully understood and is an area of scarce research.

To fully understand the causes of borehole instability of rock-socketed monopile in decomposed granite seabed, a series of in situ and laboratory tests were carried out to reveal the engineering geological characteristics of the decomposed granite in the southeastern coastal region of China. Combining three cases of monopile construction, the reasons for borehole instability during the construction process of rock-socketed monopiles in the decomposed granite seabed were clarified, and corresponding countermeasures were proposed. The results of this study can provide a reference for the construction of offshore wind power projects in similar geological formations.

2. Background

2.1. Project Information

As part of the Pinghai Bay Offshore Wind Power Project, the offshore wind farm is located near Luci Island in the southeastern region of China, as shown in Figure 1. The site has been designated for the installation of 50 wind turbines, each with a capacity of 5 MW, yielding a total installed capacity of 250 MW. The water depth within the site varied between 10~20 m, with the closest distance to the coastline estimated at approximately 6.0 km. The primary foundation forms utilized for the wind turbines are steel tube monopiles, group pile foundations, and suction bucket foundations, with an initial plan to utilize monopiles for 28 wind turbines. However, due to construction challenges such as borehole collapse and pile running, the number of wind turbines employing monopiles was reduced to 16.

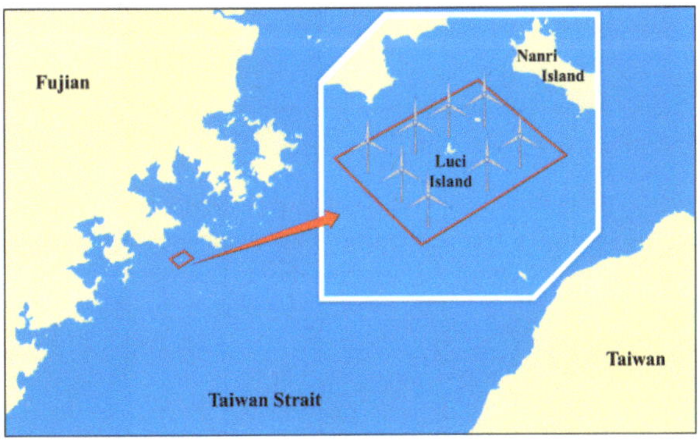

Figure 1. The location of the offshore wind farm.

The geological characteristics of this area, identified by drilling surveys conducted in the study area, are characterized by a soft top layer and a hard bottom layer. The overlying soil is the result of extensive weathering of offshore rocks and sediments, while the bottom is primarily composed of moderately decomposed granite (MDG) to slightly decomposed granite (SDG) with a relatively intact rock mass and high hardness. Between the two layers are completely decomposed granite (CDG) and highly decomposed granite (HDG), which are sensitive to disturbance and tend to soften when exposed to water.

The construction of offshore wind power foundations is significantly influenced by the marine environment. Typically, construction activities are carried out for 3~4 consecutive days under ideal weather conditions. However, the area is frequently affected by unfavorable weather conditions from October to March of the subsequent year, and the continuity of construction is significantly affected. The optimal construction period occurs from April to September annually, although this timeframe is subject to typhoon disruptions. Data obtained from historical records of the construction site and local meteorological stations show that the construction site is typically impacted by 5~6 typhoons each year, with each typhoon resulting in 3~7 days of suspension. The restricted effective construction time necessitates stringent demands on the construction of offshore wind power foundations and the installation of offshore wind turbines.

2.2. Construction Technology

The construction of monopiles typically employs pile-driving construction technology, which involves driving the pile to the required depth using a hydraulic hammer. During the construction process in the area of Pinghai Bay, the monopile could easily penetrate the surface sedimentary soil layers. However, the pile-driving method is no longer effective due to the high soil shear strength when reaching the decomposed granite stratum. To address this issue, the construction method was altered to "drive-drill-drive" construction technology [15], as shown in Figure 2. The process is as follows:

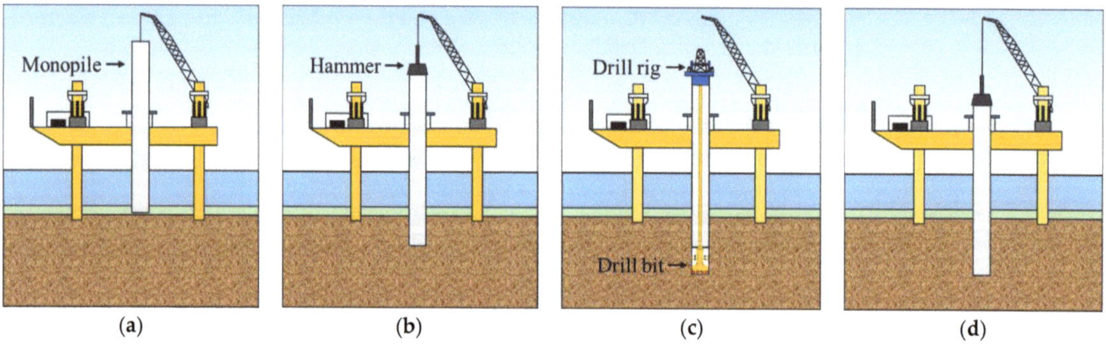

Figure 2. Construction process of monopile: (**a**) locating monopile on the seabed, (**b**) pile driving, (**c**) drilling hole, and (**d**) pile driving.

The monopile is first positioned into the cage mouth using a crane and then pressed down with a hydraulic hammer. When the monopile reaches a certain depth and encounters harder formation, it is difficult to move downward. At this stage, a drilling rig is utilized to hollow out the soil in the monopile, and drilling is continued for a set distance to reduce the resistance for driving the monopile. After the drilling rig is drilled to the desired depth, the piling equipment is replaced to further drive the monopile into a deeper formation. This process of drilling and pile driving is repeated until the pile foundation reaches the design depth. The on-site construction of pile driving and drilling rig hoisting is shown in Figure 3.

(a) (b)

Figure 3. The on-site construction of: (**a**) monopile driving and (**b**) drilling rig hoisting.

To complete the installation of the offshore wind turbine foundations within the limited construction window, the highly efficient and rapid gas-lift reverse-circulation construction technology was employed during the drilling phase, as illustrated in Figure 4. The reverse-circulation construction process entails directing compressed air into the drill bit through the annular space of the double wall drill pipe, causing the compressed air to mix with seawater and the drill cuttings generated during drilling to form a low-density gas–liquid mixture [16]. The resulting difference in liquid density between the inside and outside of the pipe drives the cuttings in the drill hole to be transported from the inner pipe to the surface via water flow, creating a cyclical reaction. This process enables a significant amount of drill cuttings to be efficiently discharged in a short period, thereby significantly increasing the construction speed.

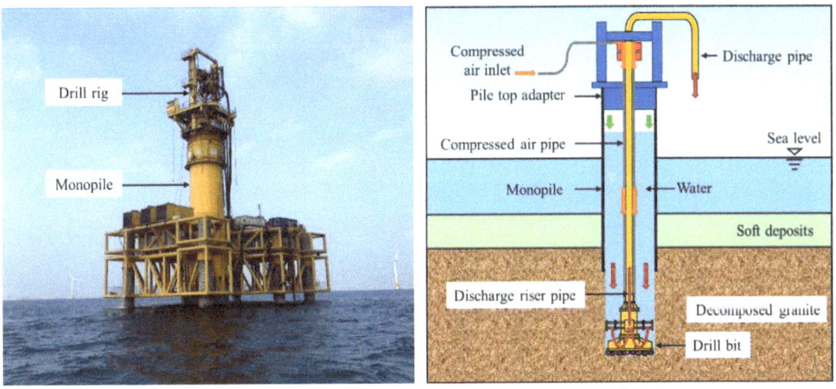

(a) (b)

Figure 4. Reverse-circulation drilling: (**a**) photo of drilling rig and (**b**) schematic diagram of drilling.

2.3. Borehole Instability

During the construction of the monopile at the Pinghai Bay offshore wind farm, borehole instability problems occurred several times, which occurred during the transitional period between drilling and pile driving, leading to issues such as pile running and deviation from the desired pile verticality. These problems resulted in prolonged construction periods and increased costs.

Borehole instability often occurs during the window between the drilling and pile driving processes, which required the transportation and replacement of large-scale construction machinery by offshore engineering vessels. This transition period, which often lasted several days, caused the soil near the exposed hole wall to collapse and fall, triggering a cascading collapse of the hole wall, as shown in Figure 5. During the construction process of the monopiles for the 16 wind turbines at the wind farm, varying degrees of collapse occurred. Among them, pile running occurred during the construction of three monopiles due to the large range of hole wall collapse. In the construction process of the remaining monopiles, the bottom elevation of the holes increased by 0.5 to 5 m compared to the pre-collapse level due to the hole wall collapse.

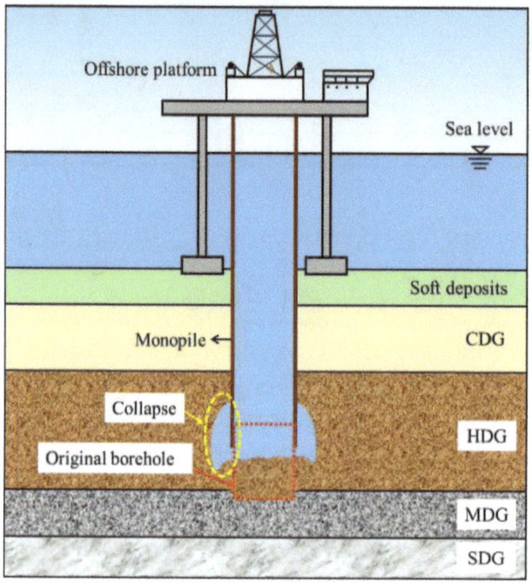

Figure 5. Schematic diagram of hole collapse after drilling.

2.4. Typical Borehole Collapse Cases

2.4.1. Case 1

The geological conditions and collapse situation of Case 1 are shown in Figure 6. The water depth is 15.45 m, and the overlying layer is 16.4 m thick, primarily consisting of silt-fine sand and medium-fine sand. The lower part of the surface seabed sediments is composed of CDG, HDG, and MDG. The CDG layer has a thickness of 3.6 m, the HDG has a thickness of 19.5 m, and the MDG is located below it. At first, the monopile was driven to a depth of 33.25 m, with the bottom of the monopile located in the HDG layer, followed by drilling operations. After five days of drilling, the depth reached 6.25 m below the monopile tip, with the exposed rock thickness being 6.15 m in the HDG layer and 0.1 m in the MDG layer. Subsequently, the plan was to replace the driving equipment for further pile driving, but the equipment arrived 12 days after the completion of drilling, and a collapse problem occurred in the meantime. The elevation of the pile bottom had increased by 5 m due to the borehole collapse, which mainly occurred in the HDG layer.

2.4.2. Case 2

The geological conditions and collapse situation of Case 2 are presented in Figure 7. The overlying layer is 6.8 m thick and consists mainly of silt-fine sand. The lower part of the surface seabed sediments consists of HDG, and there is no CDG stratum at this turbine location. The HDG stratum has a thickness of 12.75 m, and the MDG stratum below it has a thickness of 13.7 m. After the completion of monopile driving, the monopile entered the strata at a depth of 26.85 m, followed by drilling operations. The drilling depth reached 8.4 m below the monopile tip, with 2.1 m of the exposed rock being in the SDG stratum. Subsequently, the plan was to replace the piling equipment for further piling, with the equipment arriving 10 days after the completion of the drilling. However, it was discovered that a collapse had occurred. Upon testing, it was found that the elevation of the monopile bottom had increased by 4.2 m due to the collapse, which mainly occurred in the MDG stratum.

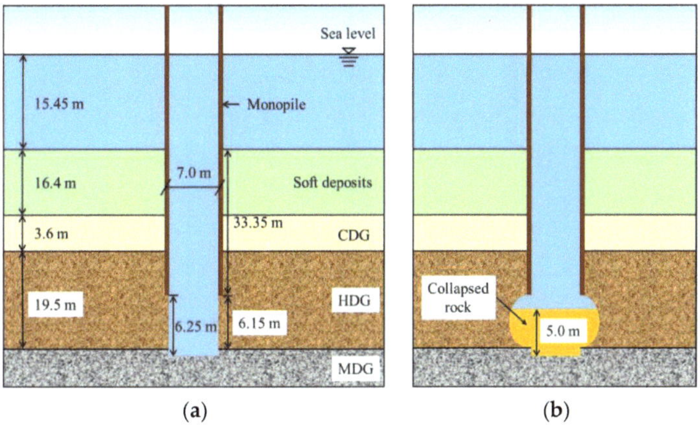

Figure 6. Borehole instability of Case 1: (**a**) original borehole and (**b**) borehole after the collapse.

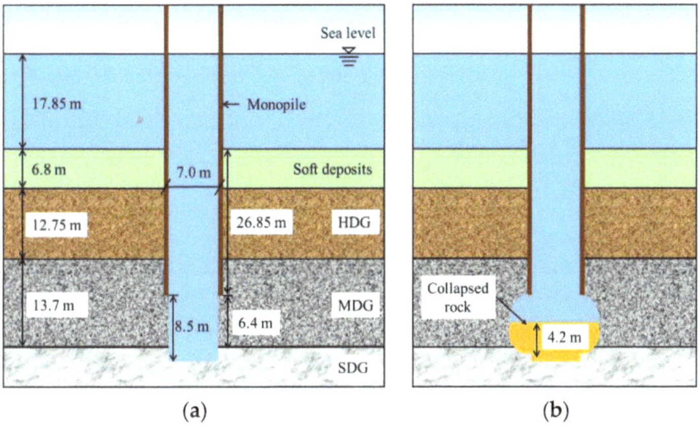

Figure 7. Borehole instability of Case 2: (**a**) original borehole and (**b**) borehole after the collapse.

2.4.3. Case 3

The geological conditions and collapse situation in Case 3 are depicted in Figure 8. The overlying layer is 24.6 m thick, consisting mainly of silt-fine sand and powdery clay. The lower part of the surface seabed sediments comprises HDG, with the CDG stratum missing at this location, similar to Case 2. The HDG stratum has a thickness of 7.9 m, with the MDG stratum below it. The monopile had a depth of 33.2 m in the strata, with the bottom of the monopile located in the MDG stratum, followed by drilling operations. After 10 days of drilling, the drilling depth reached 4.9 m below the monopile tip, with all exposed rock being in the MDG stratum. Subsequently, the plan was to replace the piling equipment for further piling, with the equipment arriving 12 days after the completion of the drilling. However, a collapse problem occurred, and it was found that the elevation of the monopile bottom had increased by 4.8 m due to the collapse, which mainly occurred in the MDG stratum.

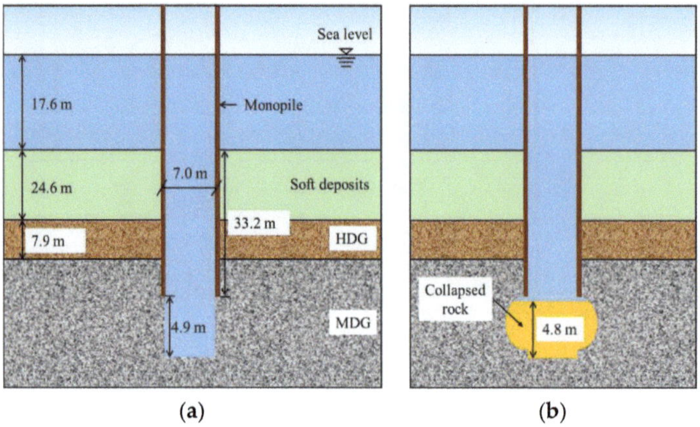

Figure 8. Borehole instability of Case 3: (**a**) original borehole and (**b**) borehole after the collapse.

3. Methodology

Since the borehole instability of rock-socketed monopile in decomposed granite seabed is closely related to the physical and mechanical properties of the granite, a series of in situ and laboratory tests were first carried out.

3.1. Sampling Method

Due to the poor mechanical properties of the soil, especially for the completely decomposed granite (CDG) and highly decomposed granite (HDG), a less disturbed field sampling method was adopted. The sampling process is depicted in Figure 9. Firstly, the casing was lowered to the seabed surface and the drilling rig was secured on the platform. The soil sampler was then installed on the drill pipe and lowered from within the casing to the seabed surface. The drilling rig rotated the drill pipe and extracts soil samples while a slurry pit and a pump were employed on the platform to circulate drilling fluid in the borehole and remove drilling debris.

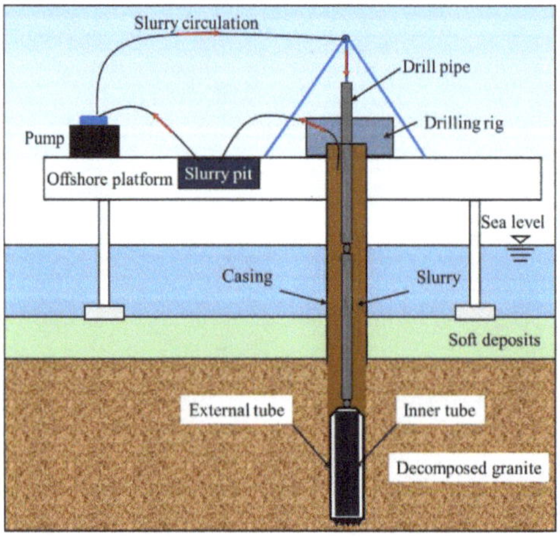

Figure 9. Schematic diagram of the soil sampling method.

As shown in Figure 10, the soil sampler utilized was a double-tube rotary sampler, with the outer tube rotating during the soil retrieval process, while the inner core tube remained stationary [17]. This method reduces the impact of vibration on the soil samples during the sampling process, minimizes scouring of the soil samples by the drilling liquid, and effectively reduces the disturbance to soil samples [18]. To enhance the performance of the slurry and ensure the quality of the retrieved soil, a certain amount of biopolymer was added to the drilling fluid. Biopolymer is a natural, organic high molecular polymer that comes in the form of a light red powder and is soluble in water [19]. The drilling fluid prepared with biopolymer exhibits improved viscoelasticity, lubricity, and rheology compared to the traditional slurry, effectively reducing the mechanical vibration of the drill pipe, protecting the core and hole wall, and being non-toxic and biodegradable. To gain a comprehensive understanding of the soil properties in the sea region of the offshore wind farm, a total of 22 sampling locations within the sea region were drilled, and samples were collected.

Figure 10. Double-tube rotary sampler sampling of marine-decomposed granite: (**a**) drilling rig, (**b**) double-tube rotary sampler, (**c**) biopolymer, and (**d**) sample.

3.2. In Situ Tests

In this research, various techniques were utilized for geological surveying, including the standard penetration test (SPT), heavy cone dynamic probing test (DPT), shear wave velocity tests, and resistivity tests. Standard penetration tests were carried out on the overlying soil layers such as silt-fine sand, powdery clay, and the CDG and HDG soil layers. An automatic hammer device with a hammer weight of 63.5 kg and a fall distance of 7 cm was used for the test. During the test, after pre-drilling 15 cm, the probe was kept vertical, and the hammer was struck at a uniform speed. The number of standard penetrations for each 10 cm and the cumulative penetration of 30 cm were recorded as one test point. The penetration depth at the 50th blow was recorded when the number of standard penetrations reached 50 within 30 cm. Standard penetration tests were conducted on each layer inside the borehole with a spacing of 2~3 m within a depth of 0~20 m.

In the MDG stratum, heavy dynamic probing tests were conducted using an automatic disengagement free-fall hammer for penetration, with continuous penetration required, and the number of blows were recorded for every 10 cm. Heavy dynamic probing tests were carried out every 2~3 m, with a continuous penetration of no less than 1 m per layer in the same geological unit.

In this study, shear wave velocity tests were conducted on each soil layer using a suspended wave velocity logger. The equipment included a host machine, an in-well suspended probe, and connecting cables, such as signal cables and trigger cables. The in-well suspended probe was composed of a fully sealed electromagnetic excitation source, two independent fully sealed detectors, and high-strength connection hoses, with a distance of 1 m between the two receiving detectors. The shear wave testing was conducted after the completion of drilling and before the stratum shrinkage, with a test point spacing of 1 m in the uncased borehole. Electrical resistivity testing was also conducted after sampling, with an electrode spacing of 0.1 m and 0.95 m.

3.3. Laboratory Tests

After the completion of the sampling work, samples were promptly sealed with wax, labeled, placed in shockproof boxes, and sent to the laboratory. To ensure the quality of undisturbed samples, efforts were made to minimize human disturbances during the packaging, transportation, storage, protection, and delivery of soil samples. Based on ASTM D2487 (2011) standard [20], the laboratory conducted tests to determine the basic physical indicators of rock and soil, including natural density (ρ_n), dry density (ρ_d), void ratio (e), water content (ω), specific gravity of solid particles (G_s), and permeability coefficient (k). Mechanical indicators of rock and soil, including compression coefficient $a_{v0.1\sim0.2}$ (the slope of the tangent of the e-p compression curve when the pressure p is between 0.1 and 0.2 MPa), compression modulus $E_{s1\sim2}$ (calculated when the additional pressure p equals 1 and 2 MPa), and elastic modulus (E_e), were also obtained. The cohesive strength (c) and internal friction angle (φ) of the CDG and HDG were determined through direct simple shear tests, while the point load strength index (I_s) of the MDG was obtained through point load tests. The uniaxial compressive strength (R) of the MDG and SDG was determined through a uniaxial compression test.

CDG and HDG exhibit disintegration properties when immersed in water, which can accelerate the collapse of borehole walls. To investigate this characteristic, disintegration tests were performed by immersing samples in a saline solution.

4. Results and Discussion

4.1. Physical and Mechanical Properties of Decomposed Granite

Typical samples of four types of weathered granite are displayed in Figure 11. Table 1 summarizes the typical characteristics of four types of weathered granite classified according to the degree of weathering and physical properties, and the weathering classification system for granite and volcanic rocks in Hong Kong [21,22] is referenced. Based on data from 22 drilling sites in Pinghai Bay and laboratory tests, Table 2 summarizes the physical properties of the four types of weathered granite, including shear wave velocity (V_s), resistivity (ρ), and the standard penetration test number (N). The sample morphology in Figure 11, combined with the qualitative indicators of the samples in Table 2 and the general description of the sample properties in Table 1, can be used to conveniently classify weathered granite according to the degree of weathering in engineering.

Table 1. Classification of differently decomposed granite.

Type	Grade	General Characteristics	Typical Characteristics
CDG	V	Gray-yellow with brown yellow Destroyed completely, soil like	High content of clay minerals Can be crumbled by hand
HDG	IV	Gray-white with black	Mainly quartz and feldspar
		Original rock texture preserved, sandy soil like	Can be broken by hand
MDG	III	Brown yellow	Part of feldspar is weathered
		Clear structure of parent rocks, clastic blocks	Not easily broken by hand, easily broken by hammer
SDG	II	Gray-white with black	Mainly quartz, feldspar, and mica
		Distinct structure of parent rocks, column-like	Cannot usually be broken by hand, not easily broken by hammer

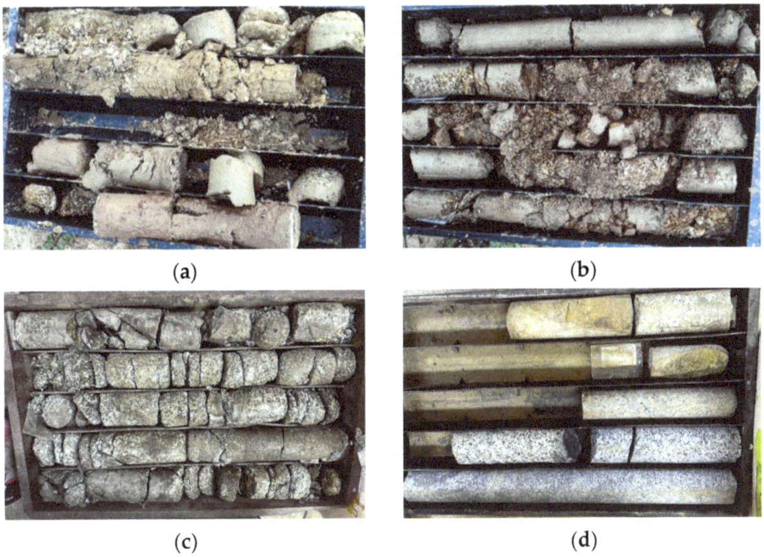

Figure 11. Typical images of different decomposed granite: (**a**) CDG, (**b**) HDG, (**c**) MDG, and (**d**) SDG.

Table 2. Main physical and mechanical properties of decomposed granite.

Parameter	CDG	HDG	MDG	SDG
$\rho_n/(\mathrm{g/cm^3})$	1.83~2.04	1.87~2.21	/	/
$\rho_d/(\mathrm{g/cm^3})$	1.50~1.73	1.46~1.92	2.68~2.72	2.62~2.81
e	0.552~0.814	0.397~0.850	/	/
$\omega/\%$	17.7~29.9	14.3~30.6	/	/
G_s	2.69~2.72	2.68~2.72	/	/
$a_{v0.1\sim0.2}/\mathrm{MPa^{-1}}$	0.30~0.54	0.23~0.28	/	/
$E_{s1\sim2}/\mathrm{MPa}$	3.35~5.49	5.53~11.61	/	/
E_e/GPa	/	/	/	16.8~51.9
c/kPa	21.3~25.3	18.3~23.4	205.9~6900	/
$\varphi/°$	27.4~28.8	28.4~35.2	34.9~52.4	/
I_s/MPa	/	/	0.33~0.68	/
R/MPa	/	/	5.38~18.20	30.63~107.20
$k/(10^{-6}\ \mathrm{cm/s})$	58.2~87.5	61.2~426.0	/	/
$V_s/(\mathrm{m/s})$	311~346	316~495	447~681	610~936
$\rho/(\Omega\cdot\mathrm{m})$	1.46~1.91	2.18~11.36	10.73~44.26	23.66~152.55
N	33~47	55~67	/	/

Overall, both CDG and HDG have undergone significant weathering processes, resulting in poor engineering geological properties similar to residual soil. Some physical indicators of CDG and HDG, such as natural density (ρ_n), dry density (ρ_d), void ratio (e), water content (ω), specific gravity of solid particles (G_s), and permeability coefficient (k), are comparable in magnitude and overlap. The compression coefficient ($a_{v0.1\sim0.2}$) and the compression modulus ($E_{s1\sim2}$) of CDG are significantly smaller than those of HDG, indicating that CDG is more compressible. The strengths of CDG and HDG are similar and relatively low, with cohesion (c) around 20 kPa, an internal friction angle (φ) of approximately 28° for CDG, and approximately 32° for HDG. There are noticeable differences in the field test indicators of CDG and HDG, manifested as higher shear wave velocity (V_s), resistivity (ρ), and standard penetration test number (N) for CDG than HDG.

In contrast to CDM and HDG, MDG exhibits overall characteristics of fragmented rock mass. MDG rock blocks retain part of the original rock strength, such as block cohesion (c) and point load strength (I_s) reaching several thousand kPa, and uniaxial compressive strength (R) around 10 MPa. The shear wave velocity (V_s) and resistivity (ρ) of MDG are significantly higher than those of CDG and HDG. Although the strength of individual MDG rock blocks is substantial, the structure of the MDG rock mass is extremely fragmented, with an average spacing of structural surfaces less than 0.4 m, making the rock mass prone to failure along these structural surfaces.

Among the four rock types, SDG exhibits the lowest degree of weathering, with core strength closer to fresh rock and the highest uniaxial compressive strength (R) reaching approximately 100 MPa. Additionally, the rock mass structure of SDG is relatively intact, with structural surface spacing greater than 1 m. Moreover, the shear wave velocity (V_s) and resistivity (ρ) of SDG are higher than those of MDG.

4.2. Disintergration Properties of Decomposed Granite

The typical images of the samples during disintegration are shown in Figure 12, where t represents the immersion time and R_d represents the percentage of disintegration of the sample. The disintegration process of CDG due to water immersion shows a gradual disintegration from the outside to the inside. Surface particles detach continuously from the sample, falling to the bottom of the container. The sample disintegrates rapidly, and a significant quantity of soil particles enter the solution, forming a suspension in a short time, leading to complete disintegration. During the disintegration process of CDG, multiple cracks occur, causing small samples to disintegrate quickly and fall to the bottom of the screen, while larger samples remain relatively stable for a short time before additional cracking occurs.

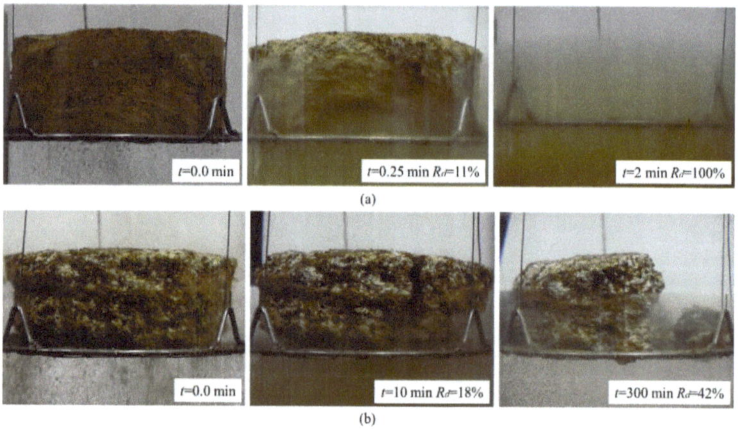

Figure 12. Typical images of samples during disintegration: (**a**) CDG with a buried depth of 3.5 m and (**b**) HDG with a buried depth of 9.6 m.

The results of the disintegration tests are presented in Figure 13. The disintegration rate of CDG was relatively fast, and it completely disintegrated in approximately 2 min. The disintegration process occurred gradually from the outside to the inside, resulting in a smooth disintegration curve. In contrast, HDG had a slower disintegration rate and demonstrated resistance to disintegration. During the process, multiple cracks were observed, and the disintegration curve showed a step-like increase. The disintegration rate and ratio varied based on the depth of burial, with the sample buried at a depth of 6.5 m taking approximately 10 min to reach a stable disintegration state with a final disintegration ratio (R_d) of approximately 80%. The sample buried at 9.6 m took around 2 h to reach a stable disintegration state with R_d of 42%, while the sample buried at a depth of 12.8 m showed only a small amount of disintegration after 2 to 3 h and took 10 h to reach a stable state, with R_d of 16%.

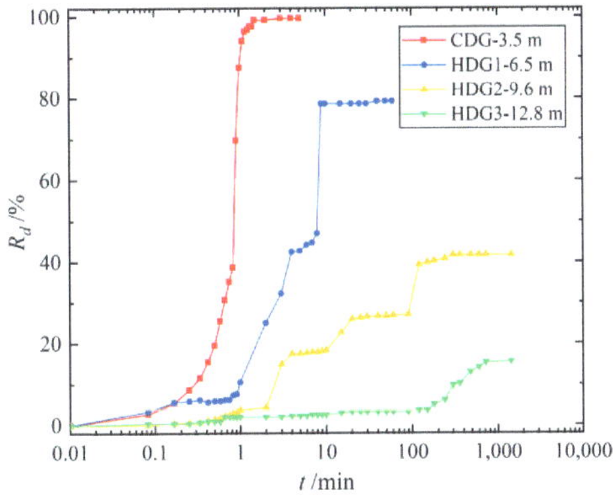

Figure 13. Disintegration curves of granite with different weathering degrees.

4.3. Causes for Borehole Instability

During the monopile construction, it was observed that hole wall collapse was primarily an issue in the HDG or MDG stratum. This was due to the high resistance encountered during monopile driving in this layer, which requires switching from piling to drilling methods. In contrast, the sedimentary layer on the seabed surface and the CDG formation has low soil strength, allowing the monopile to penetrate without the need for drilling equipment. However, after drilling is completed, the HDG and CDG tend to collapse before the next piling. When entering the SDG, the switch between piling and drilling is also required, but the low degree of weathering of SDG results in the stability of the hole wall, with rare incidents of instability.

Upon considering the construction technology characteristics, instances of borehole collapse, and the rock and soil mass properties in the formation, it is clear that inappropriate construction technology (including a lack of hole wall protection measures, extended exposure time of the hole wall, excessive drilling depth, and large wall diameter) and poor rock mass properties (including low strength, high in situ stress, susceptibility to disintegration upon water immersion, and high permeability) are the primary and direct causes of borehole collapse. Additionally, engineering experience gained from monopile construction in wind farms indicates that external factors such as fractures, faults, and the operating environment also impact borehole stability to some extent. The causes of borehole instability are summarized and ranked according to their degree of influence, as shown in Figure 14.

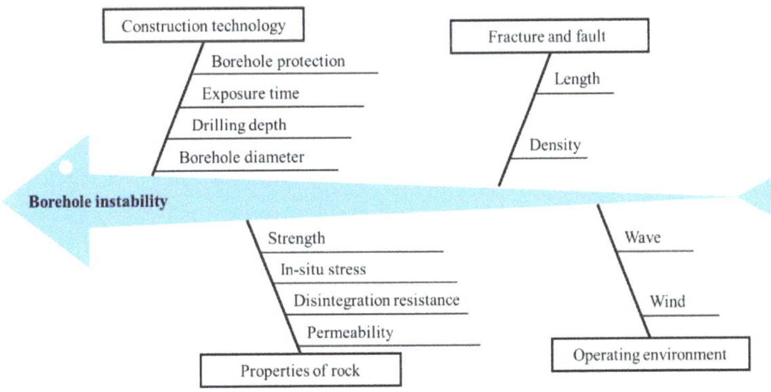

Figure 14. Causes of borehole instability.

(1) The construction of monopiles in this wind farm used the drill-drive-drill method, but the processes between drilling and monopile driving were not rapid enough, resulting in a long-time interval between the two operations. In addition, to shorten the construction time and reduce costs, reverse-circulation drilling with seawater was used during construction instead of using slurry to protect the hole wall, resulting in direct exposure of decomposed granite at the hole wall to water. The long window period between drilling and monopile driving operations and the lack of hole wall protection measures directly led to borehole instability.

Additionally, if the depth of drilling is too large at one time, the increased extent of exposed soil exacerbates the problem of borehole instability. The use of large-diameter monopiles is also an important factor that contributes to borehole instability. The diameter of the monopile is 7 m in the Pinghai Bay offshore wind farm, and the size of the borehole is close to the monopile diameter. The large diameter of the borehole results in a broad range of soil being affected by unloading. Based on the soil arching effect [23], it is evident that the larger the borehole size, the higher the likelihood of hole wall collapse issues.

(2) The insufficient soil shear strength and broken structure of HDG and MDG are the main causes of borehole instability. Drilling and excavation activities change the original in situ stress state of the soil, leading to the loss of horizontal restraint at the hole wall and subsequent changes in the stress state of the soil. The greater the depth of the soil, the higher the original in situ stress and the more pronounced the unloading effect of drilling excavation, exacerbating the change in a stress state. Evidence from the construction of the Pinghai Bay wind farm suggests that collapsed holes tend to occur at depths of 30–40 m, where excavation causes a significant unloading effect on the hole wall. Most of these borehole instabilities occur in MDG and HDG layers, where the shear strength of the soil is not sufficient to resist the shear deformation caused by excavation, resulting in deformation and borehole collapse.

As observed from the three construction cases, the stability of the borehole is influenced by the adjacent strata, but it is fundamentally controlled by the strength of the rock mass. In Case 2, MDG is adjacent to the less weathered and higher strength SDG. According to the previously mentioned stratum variation, the strength of the rock mass increases with depth. As a result, the exposed part of the MDG in this case has a relatively higher strength, leading to a collapse height of only 49% of the original exposed height. In contrast, in Case 3, the exposed rock mass is entirely MDG, and the exposure location has just entered the MDG layer from the HDG layer at 7.3 m. This indicates that the exposed MDG strength is not as high as that in Case 2, which is also reflected in the collapse ratio reaching 98%.

Additionally, CDG and HDG exhibit disintegration properties when immersed in water. Therefore, if the exposed rock after drilling is CDG or HDG, the gradual disintegration due to water immersion can accelerate the collapse of the borehole wall and sustain damage continuously. The duration of soil soaking in water corresponds to the extent of hole collapse. Additionally, the higher the permeability coefficient of HDG, the easier it is for water to seep into the rock, leading to soil mass disintegration and borehole wall collapse.

(3) Geological fractures and faults discovered can aggravate hole wall collapse problems. The MDG formation exhibits many structural planes, which are evident in the block-like nature of the MDG samples collected. These structural planes significantly reduce the overall strength of the rock mass, and the length and density of these planes are crucial factors that affect rock mass strength. Besides, HDG formation also contains geological cracks that, if present in the hole wall, can cause the surrounding rock mass to slide rapidly along these cracks, resulting in a hole wall collapse.

(4) Adverse offshore weather conditions, such as strong winds and large waves, can affect the stability of drilling to some extent. Based on engineering experience, strong winds and large waves make it challenging to maintain vertical drilling during the drilling process, potentially causing slight tilting and increasing the likelihood of borehole instability. Furthermore, dynamic loads from strong winds and large waves can induce small vibrations

in the offshore platform and drilling equipment. These vibrations are transmitted to the hole wall through the drill rod and monopile foundation, exerting a disturbance on the soil and affecting borehole stability. However, although these effects can be detrimental, their impact is relatively small compared to construction technology and rock mass properties.

4.4. Judgement of the Possibility of Borehole Instability

Accurate identification of the type of exposed rock is essential to determine the likelihood of borehole instability during the construction of a rock-socketed monopile in similar formations. This can be achieved through sampling or field tests, such as shear wave velocity testing, and the different types of decomposed granite can be identified by referring to Figure 11, Tables 1 and 2. Based on construction experience at the Pinghai Bay offshore wind farm, for the monopile with a diameter of 7 m, if the exposed rock is of the HDG or MDG, the height of the exposed rock exceeds 4 m, and the monopile installation is not completed within five days after the borehole excavation, serious instability issues are likely to occur in the borehole.

4.5. Improvement Measures to Prevent Borehole Instability

Considering the unfavorable engineering characteristics of decomposed granite, implementing the drive-drill-drive technology for constructing monopiles in multi-layered decomposed granite strata inevitably faces the challenge of borehole instability. However, various measures can be taken to minimize the extent of hole wall collapse and delay its occurrence, mitigating the negative impact on monopile construction. Efforts can be made to drive the monopile foundation deeper into strata with higher soil strength to avoid drilling or to minimize the time interval and drilling depth between drilling and pile driving. Additionally, construction processes can be improved to avoid borehole instability issues, as shown in Figure 15. Specific suggestions for improvement include:

(1) Using a smaller diameter drilling rig during drilling. In this case, the range of exposed soil is reduced, and considering the soil arching effect, the possibility of borehole instability is significantly reduced. Even if borehole instability occurs, the buffer effect provided by the distance between the borehole wall and the monopile can prevent pile running. Additionally, constructing a hollow cylindrical structure of soil within the monopile can also significantly reduce pile driving resistance.

(2) Using a casing that can be lifted and lowered freely. After drilling is completed, the casing can be lowered to protect the borehole wall and then lifted back up after the next step of pile driving is completed. This method can provide support to the borehole wall to prevent borehole instability.

(3) Using biopolymer drilling fluid during drilling to form a filter cake on the borehole wall. This can prevent decomposed granite from directly contacting seawater, avoiding disintegration. It can also exert static water pressure on the borehole wall to counteract the stress release caused by excavation and maintain borehole stability. Compared with traditional slurry, biopolymer drilling fluid causes minimal pollution to the marine environment, but its disadvantage is its high cost.

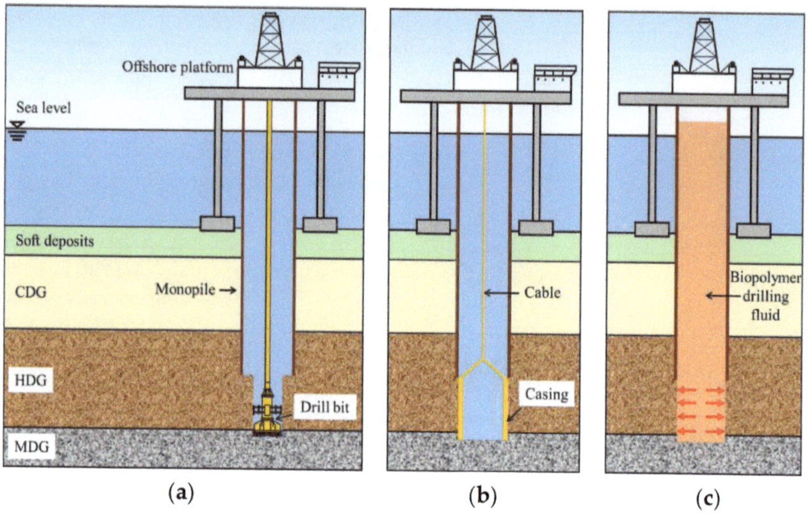

Figure 15. Construction recommendations: (**a**) drilling smaller borehole, (**b**) using casing, and (**c**) using biopolymer drilling fluid.

5. Conclusions

In this research, the causes of borehole instability in the construction of monopiles for offshore wind turbines in Pinghai Bay in decomposed granite seabed were investigated. A series of field sampling, in situ tests, and laboratory tests were conducted to reveal the geotechnical characteristics of marine-decomposed granite in the area. Combining three cases of monopile construction, the reasons for borehole instability during the construction process of rock-socketed monopiles in the decomposed granite seabed were clarified, and corresponding recommendations were proposed. The main conclusions are as follows:

(1) The results of the sampling operation demonstrate that employing a double-tube rotary sampler with biopolymer drilling fluid significantly minimizes disturbance to decomposed granite, enabling the acquisition of complete samples. A series of in situ and laboratory tests revealed an increase in the weathering degree of granite from SDG to MDG, HDG, and CDG, while mechanical strength and integrity declined considerably. These four types of decomposed granite can be effectively classified based on apparent characteristics, in situ tests, and laboratory tests.

(2) The long window period between drilling and monopile driving operations, the insufficient soil shear strength, and the lack of hole wall protection are the main causes of borehole instability. Due to HDG's insufficient strength and the highly fragmented structure of MDG, hole collapse predominantly occurs within CDG and MDG strata. The water-immersion disintegration properties of HDG exacerbate the collapse rate in HDG strata. Additional factors, such as large pile foundation dimensions, geological fractures or faults, and adverse construction environments, may worsen this issue.

(3) The occurrence of hole collapse can be comprehensively determined by considering factors such as stratum type, soil exposure range, and exposure duration. The key to preventing hole wall instability lies in reducing exposed soil range and duration or implementing suitable retaining wall measures. Drilling smaller holes has proven effective in engineering practice. Employing casing or plant-based drilling fluid for hole wall protection may also be considered a potential improvement measure.

Author Contributions: Conceptualization, B.S. and Q.Z.; formal analysis, B.S.; investigation, G.Y. and W.Z.; resources, J.L.; data curation, W.Z. and J.L.; writing—original draft preparation, B.S.; writing—review and editing, Q.Z.; visualization, B.S.; supervision, Q.Z. and G.Y. All authors have read and agreed to the published version of the manuscript.

Funding: This research was funded by National Natural Science Foundation of China (Grant Nos. 42207186, 42072317), the Natural Science Foundation of Shanghai (Grant No. 22ZR1435000), and the Shanghai Pujiang Talent Program (22PJ1408800).

Institutional Review Board Statement: Not applicable.

Informed Consent Statement: Not applicable.

Data Availability Statement: Data is available on request from the authors.

Conflicts of Interest: The authors declare no conflict of interest.

References

1. Global Wind Energy Council. *Global Wind Report 2022*; Global Wind Energy Council: Brussels, Belgium, 2022.
2. Zhang, J.; Wang, H. Development of offshore wind power and foundation technology for offshore wind turbines in China. *Ocean Eng.* **2022**, *266*, 113256. [CrossRef]
3. Gasch, R.; Twele, J. *Wind Power Plants: Fundamentals, Design, Construction and Operation*; Springer Science & Business Media: Berlin/Heidelberg, Germany, 2011.
4. Wu, X.; Hu, Y.; Li, Y.; Yang, J.; Duan, L.; Wang, T.; Adcock, T.; Jiang, Z.; Gao, Z.; Lin, Z. Foundations of offshore wind turbines: A review. *Renew. Sustain. Energy Rev.* **2019**, *104*, 379–393. [CrossRef]
5. Byrne, B.W.; Houlsby, G.T.; Burd, H.J.; Gavin, K.G.; Igoe, D.J.; Jardine, R.J.; Martin, C.M.; McAdam, R.A.; Potts, D.M.; Taborda, D.M. PISA design model for monopiles for offshore wind turbines: Application to a stiff glacial clay till. *Géotechnique* **2020**, *70*, 1030–1047. [CrossRef]
6. Byrne, B.W.; McAdam, R.A.; Burd, H.J.; Beuckelaers, W.J.; Gavin, K.G.; Houlsby, G.T.; Igoe, D.J.; Jardine, R.J.; Martin, C.M.; Muir Wood, A. Monotonic laterally loaded pile testing in a stiff glacial clay till at Cowden. *Géotechnique* **2020**, *70*, 970–985. [CrossRef]
7. Fan, S.; Bienen, B.; Randolph, M.F. Effects of monopile installation on subsequent lateral response in sand. I: Pile installation. *J. Geotech. Geoenviron. Eng.* **2021**, *147*, 04021021. [CrossRef]
8. Chiu, C.; Ng, C.W. Relationships between chemical weathering indices and physical and mechanical properties of decomposed granite. *Eng. Geol.* **2014**, *179*, 76–89. [CrossRef]
9. Liu, X.; Zhang, X.; Kong, L.; Li, X.; Wang, G. Effect of cementation on the small-strain stiffness of granite residual soil. *Soils Found.* **2021**, *61*, 520–532. [CrossRef]
10. Duan, K.; Wu, W.; Kwok, C.Y. Discrete element modelling of stress-induced instability of directional drilling boreholes in anisotropic rock. *Tunn. Undergr. Space Technol.* **2018**, *81*, 55–67. [CrossRef]
11. Fink, J. *Petroleum Engineer's Guide to Oil Field Chemicals and Fluids*; Gulf Professional Publishing: Houston, TX, USA, 2021.
12. Haimson, B. Micromechanisms of borehole instability leading to breakouts in rocks. *Int. J. Rock Mech. Min. Sci.* **2007**, *44*, 157–173. [CrossRef]
13. Lan, H.; Moore, I.D. Experimental investigation examining influence of pump rates on horizontal borehole stability in sand. *Tunn. Undergr. Space Technol.* **2022**, *127*, 104608. [CrossRef]
14. Abdulhadi, N.O.; Germaine, J.T.; Whittle, A.J. Experimental study of wellbore instability in clays. *J. Geotech. Geoenviron. Eng.* **2011**, *137*, 766–776. [CrossRef]
15. Castillo, J.G.; Nagula, S.; Schallück, C.; Grabe, J. Numerical Simulation of Drive-Drill-Drive Techniques for Open-Ended Pile Installations. In *Recent Advancements in Civil Engineering: Select Proceedings of ACE 2020*; Springer: Berlin/Heidelberg, Germany, 2022; pp. 539–548.
16. Li, L.; Ru, D.; Li, K.; Yishan, W. Research and Application of Reverse Circulation Drilling Technology. In Proceedings of the International Oil & Gas Conference and Exhibition in China, Beijing, China, 5–7 December 2006.
17. Viana da Fonseca, A.; Pineda, J. Getting high-quality samples in 'sensitive'soils for advanced laboratory tests. *Innov. Infrastruct. Solut.* **2017**, *2*, 34. [CrossRef]
18. Mayne, P.W.; Christopher, B.R.; DeJong, J. *Manual on Subsurface Investigations*; National Highway Institute Publication No. FHWA NHI-01-031; Federal Highway Administration: Washington, DC, USA, 2001.
19. Xia, S.; Zhang, L.; Davletshin, A.; Li, Z.; You, J.; Tan, S. Application of polysaccharide biopolymer in petroleum recovery. *Polymers* **2020**, *12*, 1860. [CrossRef] [PubMed]
20. *ASTM D2487*; Standard Practice for Classification of Soils for Engineering Purposes. ASTM International (ASTM): West Conshohocken, PA, USA, 2011.
21. Hencher, S.R.; Martin, R.P. The Description and Classification of Weathered Rocks in Hong Kong for Engineering Purposes. In Proceedings of the 7th South-East Asian Geotechnical Conference, Hong Kong, China, 22–26 November 1982; pp. 125–142.

22. Ng, C.W.; Yau, T.L.; Li, J.H.; Tang, W.H. Side resistance of large diameter bored piles socketed into decomposed rocks. *J. Geotech. Geoenviron. Eng.* **2001**, *127*, 642–657. [CrossRef]
23. Li, Z.; Lin, W.; Ye, J.; Chen, Y.; Bian, X.; Gao, F. Soil movement mechanism associated with arching effect in a multi-strutted excavation in soft clay. *Tunn. Undergr. Space Technol.* **2021**, *110*, 103816. [CrossRef]

Journal of
Marine Science and Engineering

MDPI

Article

Development and Influence of Pore Pressure around a Bucket Foundation in Silty Soil

Xue-Liang Zhao [1,*], Xin Wang [1], Peng-Cheng Ding [2], Shu-Huan Sui [1] and Wen-Ni Deng [1]

1 Key Laboratory of Concrete and Prestressed Concrete Structure of Ministry of Education, Southeast University, Nanjing 210096, China
2 Shanghai Investigation, Design & Research Institute Co., Ltd., Shanghai 200335, China
* Correspondence: zhaoxl@seu.edu.cn; Tel.: +86-15661451058

Abstract: Silty soil is common in the seabed of eastern coastal areas of China. The behaviors of the silty soil and a bucket foundation installed within it need more study. In this work, model tests of a bucket foundation in silty soil were performed. The development of the excess pore water pressures in the different positions around the bucket was measured. Different loading conditions, with a change in the horizontal cyclic load amplitude ratio, horizontal cyclic frequency, and vertical load ratio, were considered. The effects of the pore water pressure on the shear strength of the soil around the bucket and the horizontal bearing capacity of the bucket foundation were investigated. The results show that the normalized pore water pressures close to the bucket wall at depths between 0.1 L and 0.3 L exhibit distinct change under the cyclic load. Consistent with the distribution of the pore water pressure, the degradation of the undrained shear strength is more obvious with a greater load amplitude ratio, a greater load frequency, and a smaller vertical load. The degradation rate of the static horizontal ultimate bearing capacity is in a range of 1.57% to 14.90%, under different loading conditions.

Keywords: silt; bucket foundation; pore pressure; cyclic load; bearing capacity

Citation: Zhao, X.-L.; Wang, X.; Ding, P.-C.; Sui, S.-H.; Deng, W.-N. Development and Influence of Pore Pressure around a Bucket Foundation in Silty Soil. *J. Mar. Sci. Eng.* **2022**, *10*, 2020. https://doi.org/10.3390/jmse10122020

Academic Editor: Michele Arienzo

Received: 16 November 2022
Accepted: 14 December 2022
Published: 17 December 2022

1. Introduction

Offshore wind energy is considered to be one of the most important renewable energy resources due to its rich reserve, wide distribution, and freedom from pollution. In recent years, a number of wind farms have been planned and constructed in the eastern coastal areas of China. In this region, silty soil is widely distributed in the seabed. The physical properties and mechanical behaviors of silty soils are quite different from those of clay and sand. Using wave tank tests, Clukey et al. [1] found that silts were more susceptible to wave-induced liquefaction due to the build-up of excess pore-water pressures. The accumulation of the pore pressure in silty soil may affect the bearing capacity of the foundation of the wind turbine, which will bring challenges to the design and operation of wind farms [2,3].

A suction bucket foundation can be installed into the seabed with self-weight and suction pressure, which is thought to be a cost-effective fountain type due to the speed of installation and reduction in material costs comparing with other commonly used foundations. The cyclic lateral loadings induced by wind, waves, and currents in a marine environment may cause the accumulation of excess pore pressure in the soil around the bucket foundation, which may result in a decrease in the effective stress of the soil. Consequently, the bearing capacity of the suction bucket foundation will be decreased.

The development of pore pressure in clay under cyclic loads has been extensively studied over the past few decades. Several numerical models have been proposed to study the oscillatory and residual soil response under vertical wave loading by solving the Biot equation [4–6]. The development of pore water pressure and a liquefaction zone around the monopile in the seabed can be predicted and analyzed. In recent years, the soil–foundation interaction under vertical cyclic loads was also investigated by some researchers. Shen [7] proposed a finite element model of a suction anchor in the porous seabed subjected to cyclic uplifting loads. The results indicated that soil properties and the cyclic loading condition affected the accumulation of pore pressure around the suction anchor and the degradation of the anchor–soil frictional resistance was estimated. Some researchers [8–10] investigated the influence of the axial cyclic loading on the behavior of the foundation, considering the installation process. For the suction bucket foundation, the suction during installation may lead to seepage and cause a reduction in the soil strength. Some centrifuge tests [11,12], small scale tests [13,14], and numerical simulations [10,15,16] were also conducted to study the bearing capacity of the bucket foundation. In the previous work, more attention was paid to the influence of the L/D ratio (L is height and D is diameter of the bucket), loading regime, soil properties, and installation process on the horizontal bearing capacity of the foundation. Lazcano et al. [17] proposed a method to calculate the pore water pressure considering the in situ effective vertical stress, static shear stress, cyclic shear stress, and void ratio, and it was implemented to a numerical model, which could be used to calculate the bearing capacity of a shallow foundation under cyclic loading. Previous researchers also paid attention to the effect of cyclic loading on the soil strength degradation and bearing capacity degradation separately. Hodder et al. [18] proposed an analytical framework to estimate the degradation of undrained shear strength due to gross disturbance by a T-bar penetrometer. Ajmera et al. [19] investigated the effect of plasticity characteristics and mineralogical compositions on post-cyclic shear strength degradation and found that the post-cyclic undrained strength was related to clay minerals, the plasticity index (PI), and the cyclic stress ratio. Moses et al. [20] studied the strength and deformation behaviors of cemented marine clay under cyclic loading and hypothesized that the effect of cyclic loading can be summarized in terms of a degradation parameter. Chu et al. [21] proposed a linear equation which indicated that the soil stiffness degradation ratio was a function of the excess pore water pressure ratio in a cyclic triaxial system. Hanna et al. [22] investigate the effect of cyclic loading on sensitive clays under undrained and drained conditions and introduced the concept of a safe zone for practical design. Mao et al. [23] investigated the foundation-bearing capacity degradation of an offshore jacket platform under cyclic loading, and the results showed that the motion amplitude and the frequency of the piles exerted significant effects.

However, up to now, few studies have comprehensively and causally focused on the development of pore water pressure, the degradation of the undrained shear strength of the soil, and the bearing capacity of the bucket in silty soil under cyclic loadings. In this study, model tests were performed to investigate the development of pore pressure in the silty soil around the bucket foundation. The strength of the soil with the accumulation of the pore pressure induced by the cyclic loading was studied, and the consequent bearing capacity of the bucket foundation with the changed soil strength was analyzed.

2. Model Test

2.1. Bucket Model and Instrumentation

In the model test, the bucket was made of steel. The diameter of the bucket was 100 mm, and the height was 150 mm. Two holes were set on the top of the bucket. One hole was connected to the vacuum device to provide suction pressure during installation, and the other was used to lead out the test element circuits, such as the pore water pressure gauges. The tests were conducted in a model box with a size of 1 m × 1 m × 1.2 m (length × width × height). In order to study the accumulation of pore water pressure at different positions around the bucket, five pore water pressure gauges were installed at different depths, 0.1 L, 0.2 L, 0.3 L, 0.5 L, and 1.0 L, respectively, close to the wall of the bucket. Two gauges were installed at depths of 0.2 L and 0.3 L and a distance of 0.8 D from the bucket wall, and one gauge was also placed at a depth of 0.2 L and 1.2 D from the bucket wall. The arrangement of the pore water pressure gauges is shown in Figure 1. Other equipment to provide the load, to control the test, and to measure the data included the servo motor, the LVDT displacement gauges, the tension gauge, and the dynamic and static data acquisition instruments.

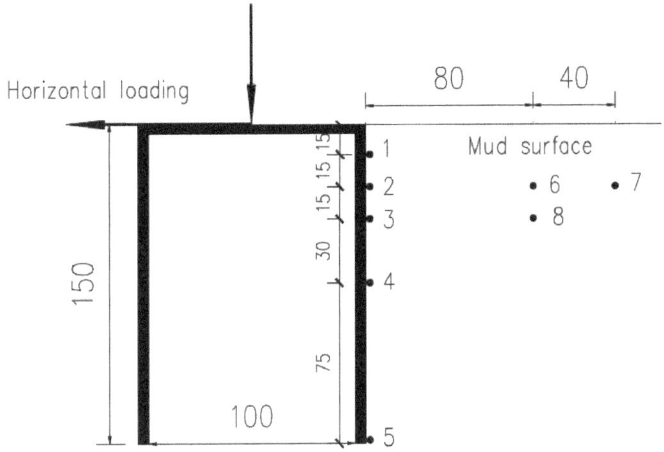

Figure 1. Arrangement of pore pressure gauges (mm).

2.2. Test Soil

The test soil in this study was taken from a wind farm in Lianyungang, China. The contents of silt, clay, and sand are 60%, 10%, and 30%, respectively, which is considered as the makeup of a typical silty soil in this region. The soil was screened to ensure the uniformity of the sample, and the soil sample was made into a slurry with a water content of about 60%. A sealable rubber cloth was placed on the box bottom, a gravel drainage layer with thickness of 0.2 m was laid, and a layer of geotextile was installed for drainage. The slurry was then poured into the test box. The vacuum preloading method was used to speed up the consolidation of the test soil, and the vacuum preload was kept at 50 kPa. After 30 days, more slurry was added to the model box to continue preloading, if settlement occurred. Finally, the thickness of the soil layer was maintained at 0.9 m. The soil parameters in the model test are listed in Table 1.

Table 1. Soil parameters.

Particle Content (%) (Silt–Clay–Sand)	Density (kN·m^{-3})	Consolidation Coefficient (m^2/s)	Permeability (m/s)	Porosity Ratio
60–10–30	19.0	1.26×10^{-6}	8.6×10^{-6}	0.814

2.3. Test Program

A total of 10 tests were conducted to study the pore pressure response around the suction bucket foundation and its effects on the soil strength and bearing capacity of the bucket foundation. The loading conditions with change in three parameters, horizontal cyclic load amplitude ratio H_d/H_j, horizontal cyclic frequency f, and vertical load ratio V_d/V_j, are considered. V_d and H_d are the applied vertical and horizontal loads, and V_j and H_j are the vertical and horizontal ultimate static bearing capacities of the foundation. A summary of the tests is shown in Table 2.

Table 2. Summary of all the tests.

Test NO.	Vertical Load Ratio, V_d/V_j	Cyclic Frequency, f (Hz)	Cyclic Load Amplitude Ratio, H_d/H_j
Test 1	0.1	0.2	0.3
Test 2	0.4	0.2	0.05
Test 3			0.1
Test 4			0.2
Test 5			0.3
Test 6			0.4
Test 7			0.5
Test 8		0.3	0.1
Test 9		0.5	0.1
Test 10	0.2	0.2	0.3

3. Test Results and Analyses

3.1. Development of Pore Water Pressure

Using the measured pore pressure at different depths and different distances, the development and distribution of the pore water pressures under different loading conditions are investigated, and the effects of the horizontal cyclic load amplitude ratio, the horizontal cyclic load frequency, and the vertical load ratio on the accumulation of pore water pressure are analyzed. The distribution of pore water pressure is consistent with that on the external side of the suction anchors in Shen's [7] numerical model.

3.1.1. Effect of the horizontal cyclic load amplitude ratio

A series of tests with different $\frac{V_d}{V_j}$ values (0.05, 0.1, 0.2, 0.3, 0.4, 0.5) were conducted to study the effect of horizontal cyclic load amplitude on the accumulation of the pore water pressure. In these tests, the load frequency was set to be 0.2 Hz and the vertical load was set to be 40% of the ultimate static bearing capacity (vertical load ratio $\frac{V_d}{V_j} = 0.4$). The measured pore pressures were normalized with the effective self-weight of the soil, p/σ_0'. Figure 2 shows the normalized pore water pressures at different depths close to the bucket wall, 0.8 D and 1.2 D from the wall, with different horizontal dynamic load amplitudes.

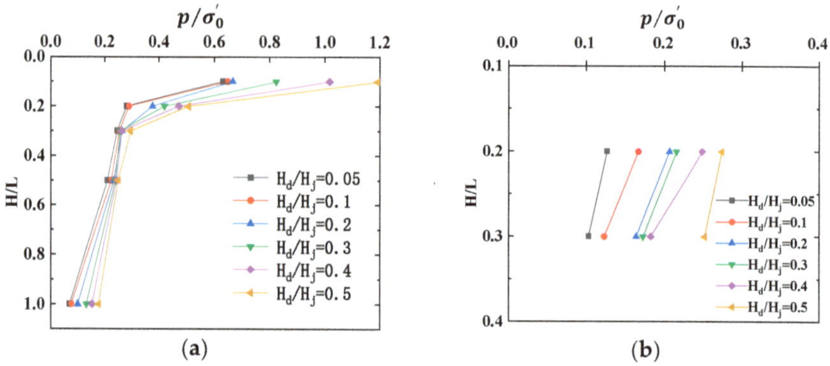

Figure 2. *Cont.*

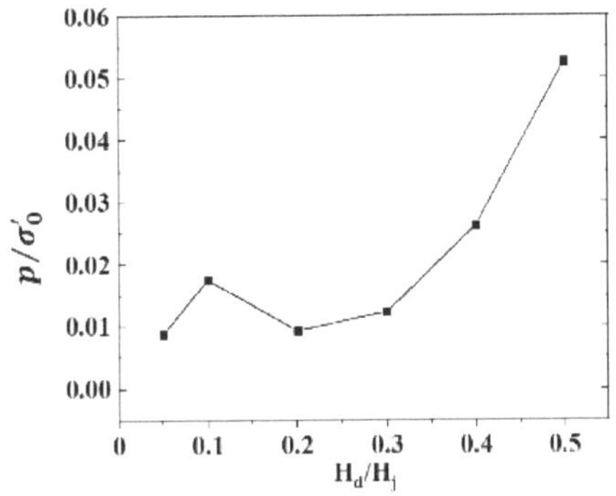

(c)

Figure 2. Normalized pore pressures with different horizontal cyclic load amplitudes: (**a**) close to the wall; (**b**) 0.8 D from the bucket; (**c**) 1.2 D from the bucket(0.2 L).

The pore water pressures at a distance of 0.8 D from the bucket wall and at a depth of 0.2 L and 0.3 L are shown in Figure 2b. The normalized pore pressures at 0.2 L are greater than those at 0.3 L, which is consistent with the distribution of pore water pressure close to the bucket wall, as shown in Figure 2a. Comparing the results at 0.2 L in Figure 2b with those in Figure 2a, the pore pressures measured close to the bucket foundation are about 2 times those at 0.8 D from the bucket wall. However, as the load amplitude increases, the difference at 0.3 L becomes smaller.

Figure 2c shows that the normalized pore water pressures at a distance of 1.2 D from the bucket wall are very small, varying from 0.0088 to 0.0526, which is even smaller than 10% of those measured close to the bucket wall. With an increase in the horizontal cyclic load amplitudes, the trend of normalized pore water pressures increases.

3.1.2. Effect of the Horizontal Cyclic Frequency

With the vertical load ratio of 0.4 and the horizontal cyclic load amplitude ratio of 0.1, different horizontal cyclic frequencies of 0.2 Hz, 0.3 Hz, and 0.5 Hz were applied in Test 3, Test 8, and Test 9 to study the effect of the horizontal cyclic frequency on the dynamic response of the pore water pressure around the bucket foundation. Normalized pore water pressures, with different cyclic load frequencies at different positions, are shown in Figure 3.

As shown in Figure 3a, the distribution of pore pressure around the bucket under different horizontal cyclic frequencies is consistent with that under the vertical cyclic loads [7]. After cyclic loading, the accumulation of pore pressure above 0.5 L close to the bucket wall is obvious, and the normalized pore pressures continue to increase with the increasing load frequency. Figure 3b shows the measured pore pressures at 0.8 D from the bucket wall. Compared with Figure 3a, the measured pore pressures close to the bucket foundation are also about 2 times those at 0.8 D from the bucket foundation.

From Figure 3c, it can be seen that the pore water pressures at a distance of 1.2 D from the bucket wall can still be observed, with the vertical load ratio of 0.4 and the load amplitude ratio of 0.1. The measured pore pressures at 0.8 D from the bucket foundation are about 2.5–3 times those at 1.2 D from the foundation. Moreover, the results of pore water pressure under a higher load frequency (0.5 Hz, $p/\sigma'_0 \approx 0.116$) are greater than those under a lower load frequency (0.2 Hz, $p/\sigma'_0 \approx 0.018$). The reason may be that the lower the frequency applied, the better the drainage in the test soil, so that the pore water pressure is easy to dissipate.

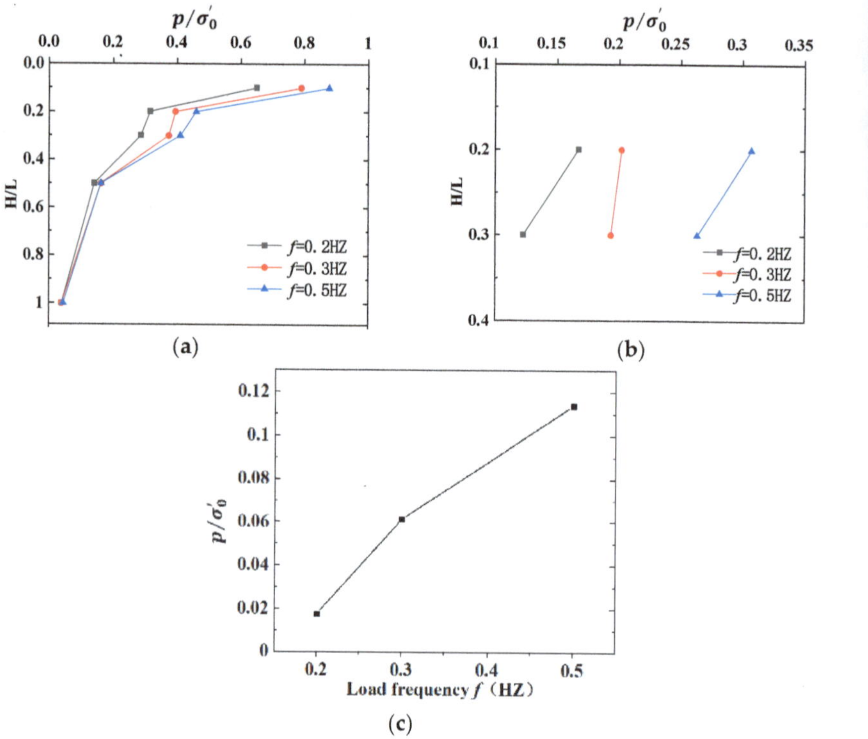

Figure 3. Normalized pore water pressures with different cyclic load frequencies: (**a**) close to the wall; (**b**) 0.8 D from the wall; (**c**) 1.2 D from the wall (at 0.2 L).

3.1.3. Effect of the Vertical Load Ratio

With the horizontal cyclic load amplitude ratio of 0.3 and a load frequency of 0.2 Hz, different vertical load ratios of 0.1, 0.2, and 0.4 were applied in Test 1, Test 5, and Test 10 to study the effect of the vertical load on the dynamic response of the pore water pressure at different positions around the bucket foundation.

Figure 4a shows that the normalized pore pressures change significantly above a depth of 0.3 L. In addition, the greater the vertical load ratio, the smaller the pore pressure ratio will be, which is similar to the results shown in Figure 4b,c. The reason may be that increasing the vertical load ratio will limit the vibration of the foundation, and the soil will be less disturbed by the horizontal cyclic loading. Therefore, the measured normalized pore pressures with a higher vertical load ratio (0.4) are smaller than those with a lower vertical load ratio in the surface soil, while below 0.5 L, the normalized pore pressures are approximate at the vertical load ratio of 0.1 and 0.2, which are smaller than those at the vertical load ratio of 0.4.

Figure 4c shows the result of measured pore water pressure at 1.2 D from the bucket wall. The smaller vertical load ratio (0.1) causes larger pore-pressure accumulation, and there is no pore water pressure measured at all when the vertical load ratio is 0.4 at a depth of 0.2 L.

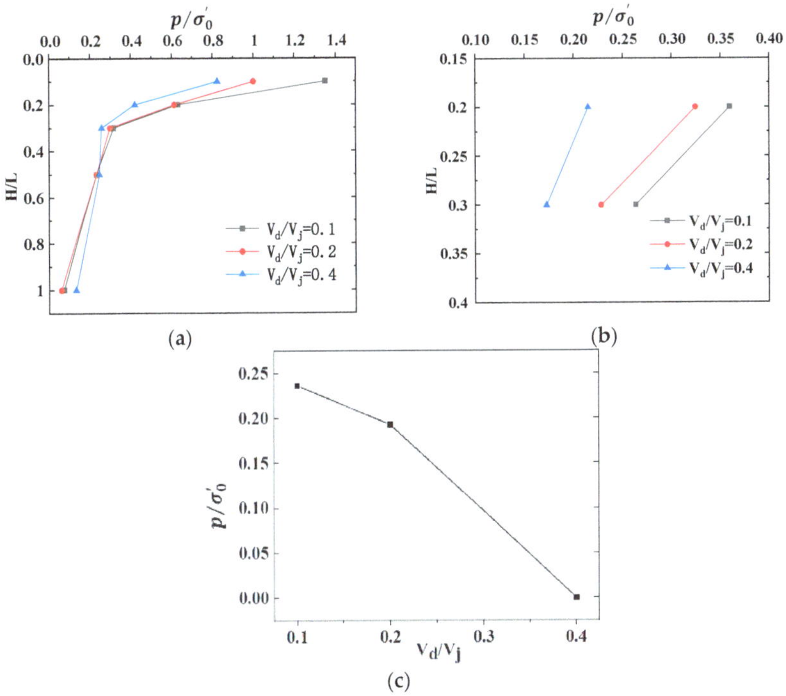

Figure 4. Normalized pore pressures with different vertical load ratios: (**a**) close to the bucket; (**b**) 0.8 D from the bucket; (**c**) 1.2 D from the wall (at 0.2 L).

3.2. Effects on the Undrained Shear Strength

A micro-penetrometer is a portable geotechnical test instrument that can be used to measure the undrained shear strength of the soil in situ or in laboratory. In this study, the micro-penetrometer was used to measure the undrained shear strength around the bucket foundation before and after the cyclic loading. At a given depth and distance from the bucket wall, three values were measured, and the distance between the three penetration points was less than 3 times the probe diameter. The penetration force and undrained shear strength can be calculated from the readings of the micro-penetrometer.

For sandy soil, it may become denser and the shear modulus may increase under cyclic loads, and no obvious excess pore water pressure will be observed because the pore water pressure can dissipate immediately in sand [11]. Thus, the strength may not degrade. However, for silty seabed soil, due to the content of fine particles, the excess pore water pressure induced by the cyclic load cannot dissipate immediately. The accumulation of excess pore pressure will cause a decrease in effective stress, and the shear strength of the soil will consequently decrease. If the fine particle content is within a certain range, and the excess pore water pressure is equal to the effective stress, the soil will be liquified, and the shear strength will be lost completely. Moreover, compared with the sand under cyclic loading, the fine particles in the silt may be lost during the seepage due to the amplification of excess pore pressure induced by sustained horizontal cyclic loading [24,25]. Liu [26] found that the strength of the silty seabed sediments decreased first, then recovered, and finally enhanced under progressive waves in the wave flume experiment, and the change rate became larger when wave height increased.

The shear strength of the test soil before the cyclic loading was measured, with a value of S_{uq} = 19.739 kPa. The undrained shear strength S_{uh} after each cyclic loading test was then measured. The values of undrained shear strength at a depth of 0.2 L after the tests and degradation rates are listed in Table 3. The degradation of undrained shear strength is consistent with the results of previous research [18,19].

Table 3. Degradation rate of undrained shear strength.

	Distance from Bucket Wall	S_{uh}	Degradation Rate (%)
Test 1	0 0.8 D 1.2 D	18.947 19.014 19.173	4.01 3.67 2.87
Test 2	0 0.8 D 1.2 D	19.162 19.254 19.417	2.93 2.45 1.63
Test 3	0 0.8 D 1.2 D	19.146 19.252 19.470	3.01 2.47 1.36
Test 4	0 0.8 D 1.2 D	19.110 19.221 19426	3.18 2.62 1.58
Test 5	0 0.8 D 1.2 D	18.937 19.002 19.196	4.06 3.73 2.75
Test 6	0 0.8 D 1.2 D	18.900 19.011 19.140	4.25 3.69 3.03
Test 7	0 0.8 D 1.2 D	18.889 19.036 19.097	4.31 3.56 3.25
Test 8	0 0.8 D 1.2 D	18.977 19.002 19.178	3.86 3.73 2.84
Test 9	0 0.8 D 1.2 D	18.861 19.026 19.162	4.45 3.61 2.92
Test 10	0 0.8 D 1.2 D	18.960 19.026 19.133	3.95 3.61 3.07

The results show that the undrained shear strength degradation is the most obvious close to the bucket wall, and the farther away from the bucket wall, the lower the undrained shear strength degradation will be. At the same time, the rate of degradation is larger when the load frequency and amplitude are higher. The undrained shear strength degradation rate decreases with the increase in the vertical load ratio, which is consistent with the distribution of the measured pore pressure around the bucket foundation, as shown in Figure 4.

The relationship between normalized pore water pressures p/σ_0' and the undrained shear strength degradation rate at the depth of 0.2 L close to the bucket foundation is shown in Figure 5. It is displayed in Figure 5a,b that the variation in the undrained shear strength degradation rate is consistent with that of the pore water pressure. Under given load conditions, when normalized pore water pressures close to the bucket foundation approach 0.5 ($p/\sigma_0' \approx 0.5$), the degradation rate of the undrained shear strength is about 4%. With a change in the vertical load ratio (V_d/V_j), Figure 5c shows that the degradation rate of the undrained shear strength exhibits little change, although the normalized pore water pressure decreases from about 0.6 to about 0.4. This means that the vertical load ratio may have little effect on the undrained shear strength of the soil.

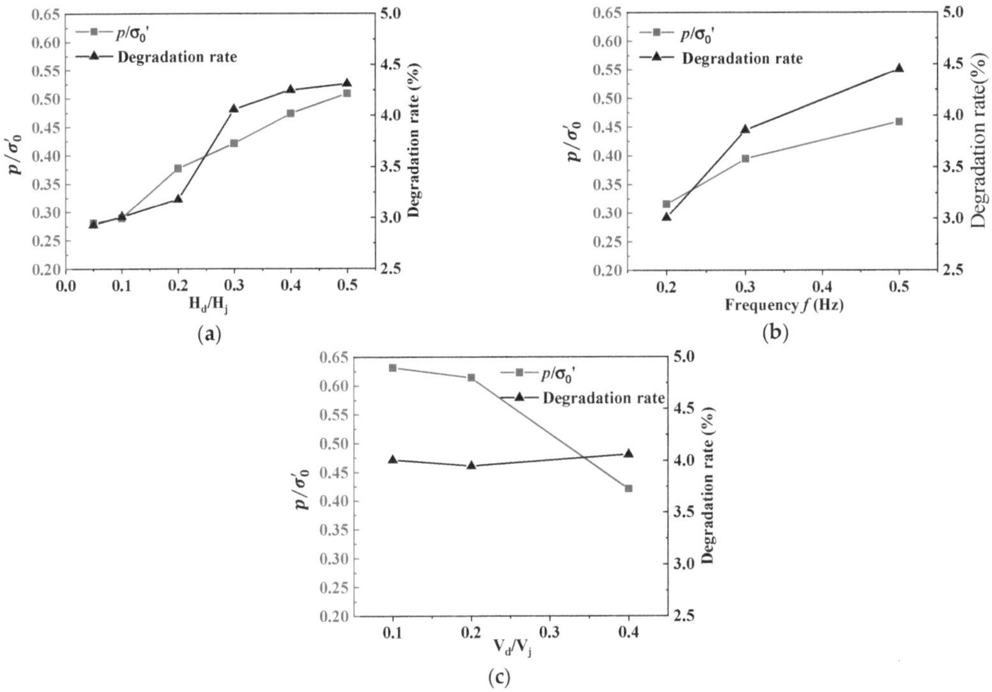

Figure 5. Relationship between p/σ_0' and undrained shear strength degradation rate: (**a**) horizontal cyclic load amplitude ratio; (**b**) horizonal cyclic load frequency; (**c**) vertical load ratio.

3.3. Effects on the Static Horizontal Bearing Capacity of Bucket Foundation

Lateral loads induced by wind and waves in a marine environment are considered to be the main load for offshore wind turbines. The horizontal bearing capacity of the bucket foundation is mainly provided by soil resistance inside and outside of the bucket skirts and the friction between the soil and the bucket wall [27]. Under cyclic loading, the pore water pressures in silty soil cannot dissipate immediately, and the strength of the soil is reduced, as analyzed in Section 3.2. Consequently, the horizontal static bearing capacity of the bucket foundation will be influenced by the cyclic loadings. In this study, the static horizontal bearing capacity of the bucket foundation after 1500 cycles was measured. The loading point of the horizontal static load was located at the top of the bucket to avoid a bending moment. According to the data obtained in this study and the method detailed in the literature, when the lateral displacement reaches 3% of the bucket diameter, the corresponding lateral load is regarded as the ultimate lateral bearing capacity of the bucket foundation [11].

The ultimate horizontal bearing capacity of the suction bucket foundation, before and after the cyclic loadings, and the attenuation rate under different load conditions, are listed in Table 4. It can be seen that after cyclic loading, the static horizontal bearing capacities degrade in a range of 1.57% to 14.90%, under different loading conditions. The larger the cyclic load amplitude and frequency, the higher the degree of horizontal bearing capacity degradation will be, and the smaller the vertical load ratio, the larger the degradation rate will be. These conclusions are consistent with the effects of the load parameter on the normalized pore pressures and the degradation of the undrained shear strength, described in Sections 3.1 and 3.2. This means that the change of capacity of the foundation can be explained by the change in pore water pressure and soil strength, comprehensively.

Table 4. Degradation rate of undrained shear strength.

Test NO.	Vertical Load Ratio, V_d/V_j	Cyclic Frequency, f (Hz)	Load Amplitude Ratio, H_d/H_j	Bearing Capacity (N)	Attenuation Rate (%)
Before loading	-	-	-	210	0
Test 1	0.1	0.2	0.3	193.6	7.81
Test 2	0.4	0.2	0.05	206.7	1.57
Test 3			0.1	205.7	2.05
Test 4			0.2	203.9	2.91
Test 5			0.3	197.9	5.76
Test 6			0.4	181.0	13.80
Test 7			0.5	178.7	14.90
Test 8		0.3	0.1	196.4	6.48
Test 9		0.5	0.1	192.0	8.57
Test 10	0.2	0.2	0.3	198.3	5.57

The load displacement curves of the bucket foundation, before and after the cyclic loadings, and the bearing capacity attenuation rates under different load conditions are shown in Figures 6–11.

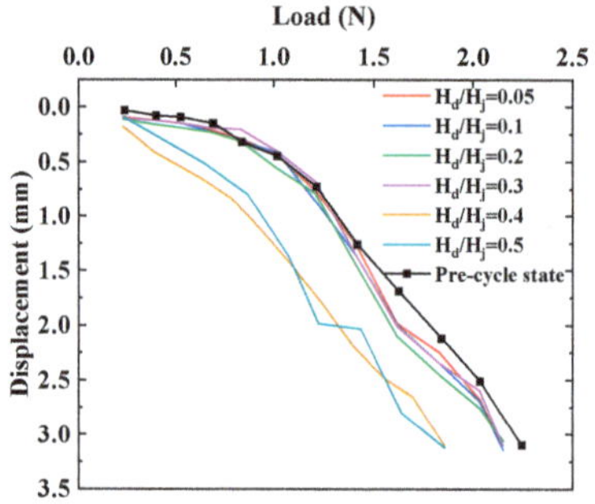

Figure 6. Load displacement curves with different cyclic load amplitudes ratios.

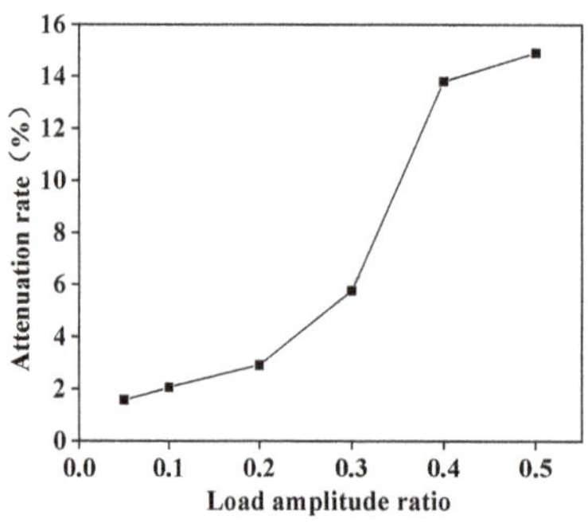

Figure 7. Bearing capacity attenuation rates with different cyclic load amplitude ratios.

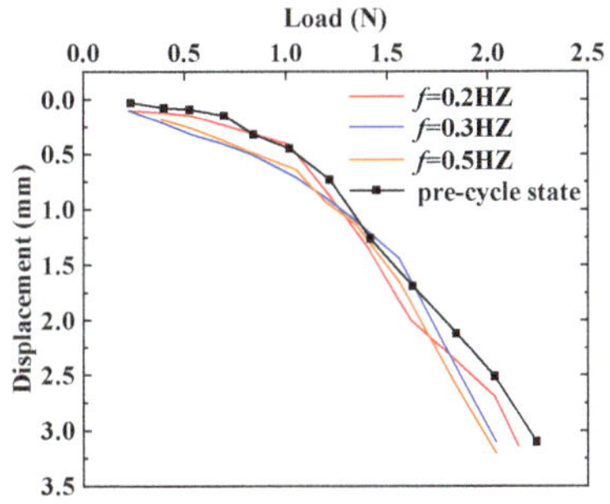

Figure 8. Load displacement curves with different cyclic frequencies.

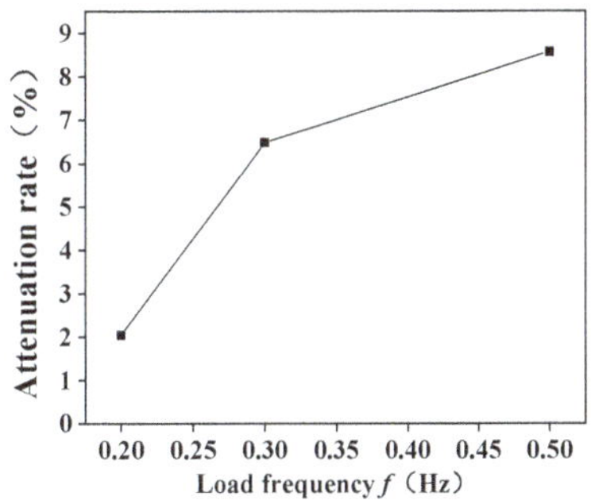

Figure 9. Degradation ratios of bearing capacity with different cyclic frequencies.

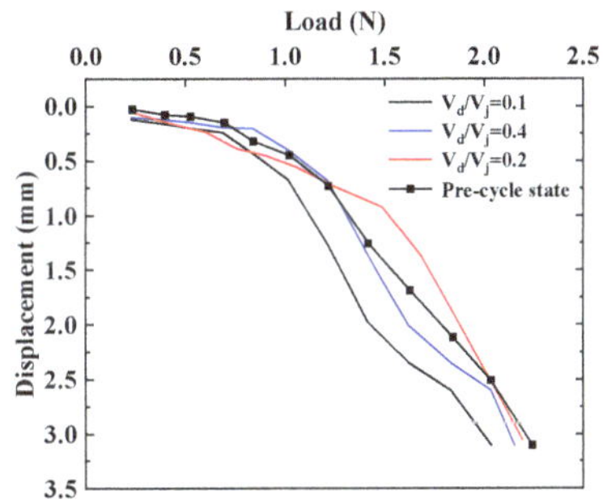

Figure 10. Load displacement curves with different vertical load ratios.

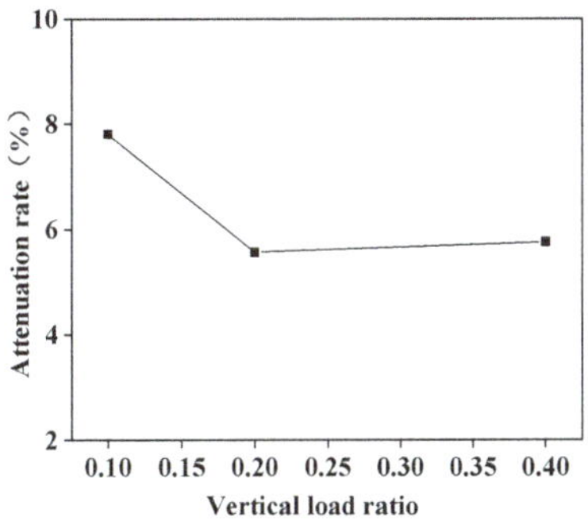

Figure 11. Degradation ratios of bearing capacity with different vertical load ratios.

Figure 6 shows the load displacement curves of the tests, with different cyclic load amplitudes ratios from 0.05 to 0.5 in tests 2–7, which are compared with the load displacement curve of test before cyclic loading, and the rates of bearing capacity degradation are given in Figure 7. From both the load displacement curves and the bearing capacity degradation rates, it can be seen that when the cyclic load amplitude ratios are smaller than 0.3 in tests 2–5, the curves are approximate, and the degradation ratios show no big differences, but when the cyclic load amplitude ratios are 0.4 and 0.5 in tests 6 and 7, both the load displacement curves and the bearing capacity degradation rates show significant changes. As analyzed previously in Section 3.1, at a depth of 0.2 L, the pore pressure ratios p/σ'_0 are between 0.2 to 0.5 when the cyclic load amplitude ratios are smaller than 0.3, as in tests 2–5, but in tests 6 and 7, when the load amplitude ratios are 0.4 and 0.5, the pore pressure ratios p/σ'_0 at a depth of 0.2 L are larger than 0.5, and they are larger than 1 at a depth of 0.1 L (see Figure 2). A pore pressure ratio larger than 1 means that the soil may be liquified, and the bearing capacity may be lost at this given location. In summary, a larger load amplitude ratio causes larger pore water pressure, and the displacement and consequently, the bearing capacity attenuation rate under certain load conditions will be higher.

The load displacement curves and the bearing capacity degradation ratios of the tests with different cyclic load frequencies of 0.2, 0.3, and 0.5 Hz in tests of 3, 8, and 10 are shown in Figures 8 and 9, respectively. This shows that the load frequency has little influence on the load displacement curve. The attenuation rates of the bearing capacity are 2.05%, 6.48%, and 8.57%, respectively, with cyclic frequencies of 0.2, 0.3, and 0.5 Hz. This means that the degradation increases with the increase in load frequency. Analyses of the pore pressure at a depth of 0.2 L show that there are some accumulations of pore pressure, but that the pore pressure ratios are all smaller than 0.5 (see Figure 3).

Figures 10 and 11 show the load displacement curves and the bearing capacity attenuation rates of the tests, with different vertical load ratios of 0.1, 0.4 and 0.2 in Test 1, Test 5, and Test 10, respectively. It can be seen in Figure 10 that the displacement is larger under the same load at $V_d/V_j = 0.1$ compared with the pre-cycle state, and the bearing capacity attenuation rate is also relatively larger, as shown in Figure 11. Considering the development of pore pressure, the increase in the vertical load may restrict the accumulation of pore pressure (see Figure 4), but Figure 5c shows that the degradation rate of the undrained shear strength of the soil shows no big difference (about 4% close to the bucket at the depth of 0.2 L) under different vertical load ratios (see Figure 5c), so the bearing capacity attenuation rate is approximate (about 6%) when the vertical load ratios are 0.2 and 0.4.

4. Discussion

The main goal of this paper is to investigate the bearing capacity of the bucket foundation in silty soil under cyclic loading conditions. It is well known that the reduction of the bearing capacity of the bucket foundation may be caused by the strength degradation of the surrounding soils, which may be a consequence of the development of the pore water pressure during cyclic loading. The development of the pore water pressure due to cyclic loading, the strength degradation of the soil due to the pore water pressure accumulation, and the reduction of the bearing capacity of the bucket foundation have all been analyzed extensively, but separately, in the literature, but these three topics have seldom been studied comprehensively and causally. This is the main concept and content of this paper.

In this paper, from the experimental model test, the accumulation of pore pressure at depths between 0.1 L and 0.3 L was found to noticeably change, and the values close to the bucket foundation were shown to be about 2 times those at 0.8 D from the bucket wall. The reduction in the undrained shear strength of the soil is consistent with the distribution of the pore water pressure under different cyclic conditions. Consequently, the ultimate static horizontal bearing capacity was attenuated, with rates in a range of 1.57% to 14.90%, under different loading conditions. This can be explained by the corresponding distribution of the pore pressure and the degradation of the soil strength.

It should be noted that the results of this paper are based on specific soils from a wind farm in China. The contents of the silt, clay, and sand are given, but the particle contents are known to have a significant effect on the soil behaviors, including the development of the pore water pressure and the soil strength. Thus, the conclusions of this paper are supposed to be qualitatively reasonable, but the results may change quantitatively with different soils. Additional study of the effects of particle contents on the development of the pore water pressure and consequently, on the soil strength degradation, may be meaningful. Moreover, the analysis of the bearing capacity is provided for a bucket foundation with the given geometry. The foundation may have different sizes and length-diameter ratios, so the results regarding the bearing capacity are given mainly for a qualitative analysis.

Author Contributions: Conceptualization, X.-L.Z. and W.-N.D.; methodology, X.W. and P.-C.D.; data curation, X.W.; writing—review and editing, S.-H.S.; supervision, X.-L.Z. All authors have read and agreed to the published version of the manuscript.

Funding: This research received no external funding.

Institutional Review Board Statement: Not applicable.

Informed Consent Statement: Not applicable.

Data Availability Statement: Not applicable.

Conflicts of Interest: The authors declare no conflict of interest.

References

1. Clukey, E.C.; Kulhawy, F.H.; Liu, P.L.F.; Tate, G.B. The impact of wave loads and pore-water pressure generation on initiation of sediment transport. *Geo-Mar. Lett.* **1985**, *5*, 177–183. [CrossRef]
2. Wang, H.; Liu, H.-J. Evaluation of storm wave-induced silty seabed instability and geo-hazards: A case study in the Yellow River delta. *Appl. Ocean. Res* **2016**, *58*, 135–145. [CrossRef]
3. Xu, G.; Sun, Y.; Wang, X.; Hu, G.; Song, Y. Wave-induced shallow slides and their features on the subaqueous Yellow River delta. *Can. Geotech. J.* **2009**, *46*, 1406–1417. [CrossRef]
4. Zhao, H.; Jeng, D.-S.; Zhang, H.; Zhang, J.; Zhang, H. 2-D integrated numerical modeling for the potential of solitary wave-induced residual liquefaction over a sloping porous seabed. *J. Ocean Eng. Mar. Energy* **2015**, *2*, 1–18. [CrossRef]
5. Jeng, D.S.; Zhao, H.Y. Two-Dimensional Model for Accumulation of Pore Pressure in Marine Sediments. *J. Waterw. Port Coast. Ocean Eng.* **2015**, *141*, 04014042. [CrossRef]
6. Liao, C.; Chen, J.; Zhang, Y. Accumulation of pore water pressure in a homogeneous sandy seabed around a rocking mono-pile subjected to wave loads. *Ocean Eng.* **2019**, *173*, 810–822. [CrossRef]
7. Shen, K.; Wang, L.; Guo, Z.; Jeng, D.S. Numerical investigations on pore-pressure response of suction anchors under cyclic tensile loadings. *Eng. Geol.* **2017**, *227*, 108–120. [CrossRef]
8. Zhang, J.-F.; Zhang, X.-N.; Yu, C. Wave-induced seabed liquefaction around composite bucket foundations of offshore wind turbines during the sinking process. *J. Renew. Sustain. Energy* **2016**, *8*, 023307. [CrossRef]
9. Tasan, H.E.; Yilmaz, S.A. Effects of installation on the cyclic axial behaviour of suction buckets in sandy soils. *Appl. Ocean Res.* **2019**, *91*, 101905. [CrossRef]
10. Le, V.H.; Remspecher, F.; Rackwitz, F. Development of numerical models for the long-term behaviour of monopile foundations under cyclic loading considering the installation effects. *Soil. Dyn. Earthq. Eng.* **2021**, *150*, 106927. [CrossRef]
11. Wang, X.; Yang, X.; Zeng, X. Centrifuge modeling of lateral bearing behavior of offshore wind turbine with suction bucket foundation in sand. *Ocean Eng.* **2017**, *139*, 140–151. [CrossRef]
12. Ma, P.; Liu, R.; Lian, J.; Zhu, B. An investigation into the lateral loading response of shallow bucket foundations for offshore wind turbines through centrifuge modeling in sand. *App. Ocean Res.* **2019**, *87*, 192–203. [CrossRef]
13. Wang, Y.; Lu, X.; Wang, S.; Shi, Z. The response of bucket foundation under horizontal dynamic loading. *Ocean Eng.* **2006**, *33*, 964–973. [CrossRef]
14. Li, D.; Zhang, Y.; Feng, L.; Gao, Y. Capacity of modified suction caissons in marine sand under static horizontal loading. *Ocean Eng.* **2015**, *102*, 1–16. [CrossRef]
15. Jia, N.; Zhang, P.; Liu, Y.; Ding, H. Bearing capacity of composite bucket foundations for offshore wind turbines in silty sand. *Ocean Eng.* **2018**, *151*, 1–11. [CrossRef]
16. Liu, M.; Yang, M.; Wang, H. Bearing behavior of wide-shallow bucket foundation for offshore wind turbines in drained silty sand. *Ocean Eng.* **2014**, *82*, 169–179. [CrossRef]
17. Lazcano, D.R.P.; Aires, R.G.; Nieto, H.P. Bearing capacity of shallow foundation under cyclic load on cohesive soil. *Comput. Geotech.* **2020**, *123*, 103556. [CrossRef]

18. Hodder, M.S.; White, D.J.; Cassidy, M.J. Analysis of soil strength degradation during episodes of cyclic loading, illustrated by the T-Bar penetration test. *Int. J. Geomech.* **2010**, *10*, 117–123. [CrossRef]
19. Ajmera, B.; Brandon, T.; Tiwari, B. Characterization of the reduction in undrained shear strength in fine-grained soils due to cyclic loading. *J. Geotech. Geoenviron.* **2019**, *145*, 04019017. [CrossRef]
20. Moses, G.G.; Rao, S.N. Degradation in cemented marine clay subjected to cyclic compressive loading. *Mar. Georesour. Geotechnol.* **2003**, *21*, 37–62. [CrossRef]
21. Chu, M.C.; Ge, L. Stiffness degradation of coarse and fine sand mixtures due to cyclic loading. *Eng. Geol.* **2021**, *288*, 106155. [CrossRef]
22. Hanna, A.M.; Javed, K. Experimental investigation of foundations on sensitive clay subjected to cyclic loading. *J. Geotech. Geoenviron.* **2014**, *140*, 04014065. [CrossRef]
23. Mao, D.; Zhong, C.; Zhang, L.; Chu, G. Dynamic response of offshore jacket platform including foundation degradation under cyclic loadings. *Ocean Eng.* **2015**, *100*, 35–45. [CrossRef]
24. Tzang, S.-Y.; Ou, S.-H. Laboratory flume studies on monochromatic wave-fine sandy bed interactions. *Coast. Eng.* **2006**, *53*, 965–982. [CrossRef]
25. Tzang, S.-Y.; Ou, S.-H.; Hsu, T.-W. Laboratory flume studies on monochromatic wave-fine sandy bed interactions Part 2. Sediment suspensions. *Coast. Eng.* **2009**, *56*, 230–243. [CrossRef]
26. Liu, X.-L.; Jia, Y.-G.; Zheng, J.-W.; Hou, W.; Zhang, L.; Zhang, L.P.; Shan, H.X. Experimental evidence of wave-induced inhomogeneity in the strength of silty seabed sediments: Yellow River Delta, China. *Ocean Eng.* **2013**, *59*, 120–128. [CrossRef]
27. Wang, X.; Zeng, X.; Yu, H.; Wang, H. Centrifuge Modeling of Offshore Wind Turbine with Bucket Foundation under Earthquake Loading. In Proceedings of the IFCEE 2015, San Antonio, TX, USA, 17–21 March 2015; pp. 1741–1750.

Journal of
*Marine Science
and Engineering*

MDPI

Article

Lateral Bearing Capacity of a Hybrid Monopile: Combined Effects of Wing Configuration and Local Scour

Biao Li [1,2], Yifa Wang [3], Wengang Qi [1,2,*], Shunyi Wang [1,2] and Fuping Gao [1,2]

1 Institute of Mechanics, Chinese Academy of Sciences, Beijing 100190, China
2 School of Engineering Science, University of Chinese Academy of Sciences, Beijing 100049, China
3 Department of Infrastructure Engineering, Faculty of Engineering and Information Technology, The University of Melbourne, Melbourne 3010, Australia
* Correspondence: qiwengang@imech.ac.cn

Abstract: Pile foundations for offshore wind turbines are subjected to large lateral loads. By mounting wings on the perimeter of regular monopiles, winged monopiles have shown better performance in resisting deformation under horizontal loading. However, the hazardous effect of local scour on the lateral bearing capacity of winged monopiles installed in the sandy seabed has not been systematically evaluated. In this study, a modified Mohr–Coulomb model considering the pre-peak hardening and post-peak softening behavior of dense sand is adopted to simulate laterally loaded winged monopiles in the locally scoured sandy seabed, using three-dimensional finite element analyses. The effect of local scour depth on the lateral capacity of winged monopiles is examined and explained by soil failure mechanisms. The enhancement of lateral capacity with wings attached to the monopile is demonstrated to be more effective than extending pile embedment length. The effects of the relative density of sand and the wing load orientation are also discussed. Finally, the wing efficiency is evaluated to determine the optimal configuration of winged monopiles.

Keywords: winged monopile; lateral capacity; local scour; sandy seabed; modified Mohr–Coulomb model

Citation: Li, B.; Wang, Y.; Qi, W.; Wang, S.; Gao, F. Lateral Bearing Capacity of a Hybrid Monopile: Combined Effects of Wing Configuration and Local Scour. *J. Mar. Sci. Eng.* **2022**, *10*, 1799. https://doi.org/10.3390/jmse10121799

Academic Editor: Domenico Curto

Received: 22 October 2022
Accepted: 19 November 2022
Published: 22 November 2022

1. Introduction

The development and utilization of renewable energy play an indispensable role in the process of "carbon neutrality". Offshore wind power has been attracting attention because of its abundant reserves, high energy density and minor disturbance to the surrounding environment. By the end of 2021, the global installed capacity of offshore wind power had reached 57.2 GW [1]. The offshore wind industry continues to grow rapidly, with the global installed capacity expected to reach 203 GW by 2030 [2].

Foundations account for approximately 35% of the total cost of an offshore wind project [3]. Monopiles are the most commonly used foundation solution for offshore wind turbines (OWTs) in shallow to medium water depths due to their ease of design, manufacture and installation [4]. To improve energy harvesting efficiency, the offshore energy industry is currently making intense efforts to upscale offshore wind turbines and advance into deeper waters [5]. The bigger turbine size and deeper water depth bring out larger loads, which require more reliable and powerful foundations to resist them. Several new hybrid monopile foundations are proposed to support large-capacity wind turbines operating in deeper waters at a reasonable cost [6].

As innovative hybrid monopiles with wings welded orthogonally to the perimeter of regular monopiles, winged monopiles have shown excellent lateral bearing capacity in both field and centrifuge tests [7–9]. Existing studies on the lateral bearing capacity of winged monopiles mainly focus on the optimized arrangement of the wings. Considering the multidirectional loading in the marine environment, the wings are usually evenly attached to the pile perimeter. With the same height and width of wings, the numerical analysis of Babu and Viswanadham [10] showed that the star-wing monopiles (i.e., monopile with eight wings) exhibit a higher lateral bearing capacity than four-wing monopiles. Abongo [11] further found that the lateral bearing capacity of a four-wing monopile is slightly higher than that of a three-wing monopile by conducting numerical simulations and 1g model tests. Winged monopiles with three or four wings are commonly preferred because of the greater installation costs associated with an increased number of wings. In addition, the shape of wings has a significant influence on the lateral response of a hybrid monopile. Rectangular wings, compared with triangular or trapezoidal wings, demonstrate superior performance in reducing the lateral deformation of pile foundations [12,13]. Furthermore, the aspect ratio of the wings (i.e., the height-to-width ratio of the wing) and wing load orientation may affect the lateral bearing capacity of winged monopiles. Nasr [13] suggested that the ultimate lateral bearing capacity of winged monopiles depends on the height of the wings to a large extent. As the height of the wing approaches 0.4 times the embedded length of the monopile, the displacement of the pile head can be reduced by up to 70% compared with regular monopiles. Pei and Qiu [14] compared the effect of the wing width of winged monopiles on the wing efficiency with different embedded depths and identified that the optimal wing width was equal to the pile diameter for both long and short piles. Variations in wing load orientations can also alter the lateral bearing capacity, albeit some studies have shown that the difference is trivial [15,16].

The aforementioned studies have dealt with the case that the winged monopiles are fully embedded in a flat soil bed, showing that setting the wings near the mudline (or ground surface) mobilizes higher soil resistance against lateral pile deflection [10,11,15,16]. However, in marine environments, waves and/or currents can induce a local scour hole around piles without scour countermeasures [17,18]. Local scour could also develop, sometimes even with countermeasures, given that the countermeasures are not deployed in time or the protections fail during severe storms. It has been proven that flow structures such as horseshoe vortices and lee-wake vortices can affect the development of scour and equilibrium scour depth [19,20]. As a consequence, local scour would reduce the effective embedment of monopiles and, in turn, the lateral bearing capacity [21,22]. For winged monopiles, local scour may still exist and expose the wings out of the seabed, since the wings embedded near the mudline have no impact on the large-scale coherent flow structures (e.g., horseshoe vortex) for initiating the scour process [23]. Li et al.'s [24] scour tests in a water flume revealed that wings near the mudline can significantly enlarge the range of the scour hole and even the scour depth, which impedes the effectiveness of the wings. Therefore, the wings need to be installed below the estimated maximum scour depth to avoid the scour caused by the wings and ensure the wings are fully functioning.

This study investigates the combined effects of local scour and wing configuration on laterally loaded winged monopiles in the sandy seabed, using a series of three-dimensional (3D) finite element (FE) analyses. A modified MMC which can describe the pre-peak hardening and post-peak softening behavior of dense sand is adopted and implemented to represent the sandy soils. The model is validated against the database of triaxial element tests and centrifuge tests of regular monopiles in Fontainebleau sand. A parametric analysis is carried out to examine the effects of scour depth, wing usage, the relative density of sand, wing load orientation, as well as aspect ratio of wings for three- and four-wing monopiles. The wing efficiency is quantitatively evaluated and the optimal arrangement of wings is proposed to aid the engineering design of winged monopile foundations for applications in offshore wind energy.

2. Numerical Modelling

2.1. Geometry and FE Mesh

The 3D FE model of laterally loaded winged monopiles embedded in the sandy seabed was established using the software Abaqus [25], as shown schematically in Figure 1. The pile diameter D is set as 6 m, while the embedded depth L_p is 48 m (8D) and the wall thickness t_p is 70 mm. Three or four wings are mounted evenly on the external surface of the pile at a vertical distance L_e from the pile head. The wings have a width equal to D, a thickness equal to t_p and a height varying between 0 and 1.75D. An inverted cone located at the center of the monopile with a slope angle of 30° is cut from the mudline to represent the local scour hole around the pile foundation in the marine environment [21]. The maximum scour depth S is conservatively assumed to be 1.9D, based on observed scour depths in field cases [5,26]. To avoid scour-induced wing outcropping, the distance from the top of the wing to the pre-scour mudline L_e is set as 2D, which exceeds the assumed maximum scour depth. It is assumed that the forces in different wing load orientations (β) apply to the center of the pile head to model the marine environment.

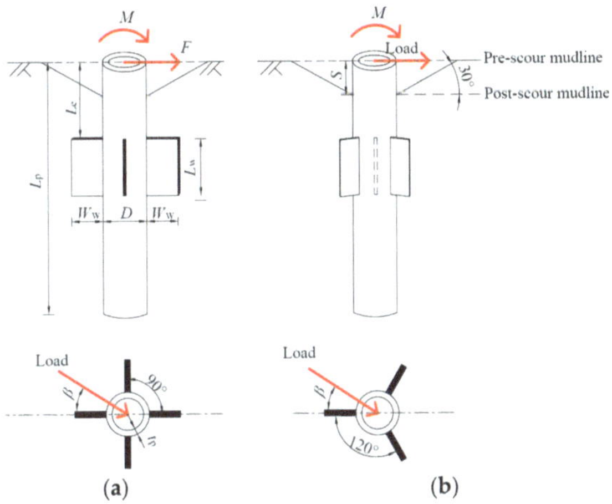

Figure 1. Schematic diagram of the winged monopile model: (**a**) a four-wing monopile; (**b**) a three-wing monopile.

Figure 2 shows a typical FE mesh used in the numerical analyses. The soil domain is divided into two zones to distinguish the scoured soil (colored brown) from the remaining soil (colored cyan). The diameter and thickness of the soil domain are set as 26D and 1.7L_p, respectively, which are sufficiently large to avoid boundary effects [13,14]. Both the soil and the pile are modelled with 3D eight-node reduced integration brick elements (C3D8R in Abaqus library). A total of more than 50,000 elements are generated in each model, with a finer mesh created in the vicinity of the pile. The present mesh size is selected based on trial mesh sensitivity analyses. The piles are assumed to be "wished-in-place" without considering the pile installation effect [27].

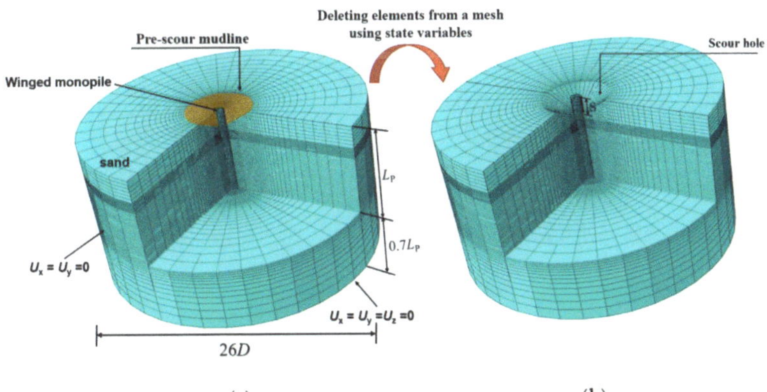

Figure 2. Meshes and boundary conditions of the FE model: (**a**) pre-scour model; (**b**) post-scour model.

2.2. Constitutive Model

The steel winged monopiles are assumed to be linearly elastic with Young's modulus E_p = 210 GPa and Poisson's ratio ν_p = 0.3, similar to previous studies on pile foundations [27–29]. The Young's modulus of sand E_s for describing the elastic behavior is stress-dependent and increases with the earth pressure at rest [30]:

$$E_s = \kappa p_a \left(\frac{\sigma_3}{p_a}\right)^{\lambda} \quad (1)$$

in which p_a is the atmospheric pressure (100 kPa), σ_3 is the confining stress, κ is the soil stiffness at the reference stress state and λ determines the stress dependence of the soil stiffness. The specific values of κ and λ for medium-dense, dense, and very dense sand are listed in Table 1 [31].

Table 1. Properties of sand used in FE analyses [31,32].

Description	Symbol	Unit	Medium-Dense	Dense		Very Dense
Relative density	D_r	%	60	70	80	90
Submerged unit weight	γ'	kN/m^3	6.28	6.58	6.90	7.24
Elastic stiffness parameters	κ	—	400	500	500	700
	λ	—	0.6	0.55	0.55	0.5
Poisson's ratio	υ	—	0.25	0.225	0.225	0.2
Critical state friction angle	φ'_c	°		31		
MMC parameters	φ'_{in}	°		29		
	A_ψ	—		3.8		
	k_ψ	—		0.6		
	Q	—		10		
	R	—		1		
	C_1	—		0.22		
	C_2	—		0.11		
	m	—		0.25		
Cohesion	c	kPa		0.1		
Constant friction angle	φ'_0	°	34.3	36.4	38.8	41.5

Though the Mohr–Coulomb (MC) model has been widely adopted to simulate the elastoplastic behavior of the sandy seabed under drained conditions due to its simplicity and acceptable accuracy [13,16,30,33–36], the plastic properties (e.g., friction angle φ', dilation angle ψ) of dense sand remain constant once the plastic state is reached. This is inconsistent with the stress–strain curves obtained from the triaxial test or the direct shear test for dense sand, which manifests evident pre-peak hardening and post-peak softening behaviors. Roy et al. [32] proposed a modified Mohr–Coulomb (MMC) model to analyze the pipe–soil interactions based on the stress–strain curves of dense sand, as indicated in Figure 3.

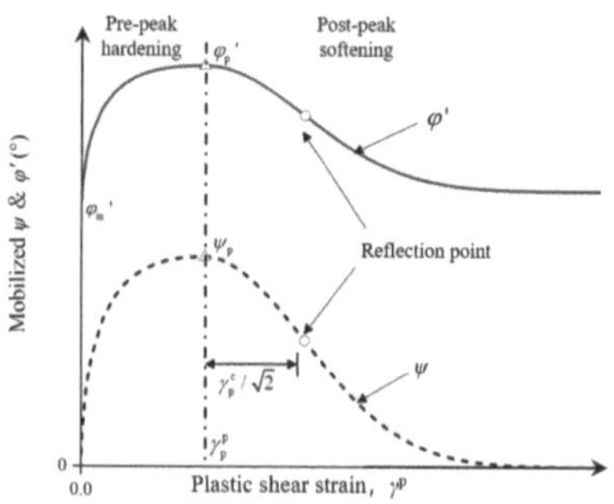

Figure 3. Typical curves of modified friction angle and modified dilation angle with plastic shear strain (after Roy et al. [32]).

In the MMC model, the peak friction angle φ'_{p} and the peak dilation angle ψ_p are governed by the relative density of sand D_{r} and the confining pressure p' [37] (see Equations (2)–(4)):

$$\varphi'_{\mathrm{p}} - \varphi'_{\mathrm{c}} = A_\psi I_{\mathrm{R}} \tag{2}$$

$$\varphi'_{\mathrm{p}} - \varphi'_{\mathrm{c}} = k_\psi \psi_{\mathrm{p}} \tag{3}$$

$$I_{\mathrm{R}} = I_{\mathrm{D}}(Q - \ln p\prime) - R \tag{4}$$

where $(A_\psi, k_\psi) = (3, 0.5)$ for the triaxial condition and $(A_\psi, k_\psi) = (5, 0.8)$ for the plane strain condition, respectively. φ'_c is the critical friction angle. Chakraborty and Salgado [38] demonstrated that $(A_\psi, k_\psi) = (3.8, 0.6)$ are applicable to both triaxial and plane strain shear conditions for Toyoura sand. I_{R} is the relative density index. $I_{\mathrm{R}} = 0 \sim 4$ is generally adopted as the acceptable range for modelling the interaction between structures and sands ([37,39]). I_{D} (=D_{r} (%)/100) and D_{r} are both the relative density with different units. Q and R are taken as 10 and 1, respectively [37].

The friction angle φ' and dilation angle ψ in the pre-peak hardening zone vary with the plastic shear strain γ^{P}, employing the sine function as the backbone function (refer to Figure 3):

$$\varphi' = \varphi_{\mathrm{in}} + \sin^{-1}\left[\left(\frac{2\sqrt{\gamma_{\mathrm{P}}\gamma_{\mathrm{P}}^{\mathrm{P}}}}{\gamma_{\mathrm{P}} + \gamma_{\mathrm{P}}^{\mathrm{P}}}\right)\sin(\varphi_{\mathrm{P}}' - \varphi_{\mathrm{in}}')\right] \tag{5}$$

$$\psi = \sin^{-1}\left[\left(\frac{2\sqrt{\gamma_{\mathrm{P}}\gamma_{\mathrm{P}}^{\mathrm{P}}}}{\gamma_{\mathrm{P}} + \gamma_{\mathrm{P}}^{\mathrm{P}}}\right)\sin(\psi_{\mathrm{P}})\right] \tag{6}$$

where $\gamma_{\mathrm{P}}^{\mathrm{P}}$ represents the γ^{P} when φ' and ψ reach the peak;

$$\gamma_{\mathrm{P}}^{\mathrm{P}} = \gamma_{\mathrm{P}}^{\mathrm{c}}(p'/p_{\mathrm{a}})^m \tag{7}$$

where $\gamma_{\mathrm{P}}^{\mathrm{c}}$ is the strain-softening factor, which determines the rate of softening with plastic shear strain. $\gamma_{\mathrm{P}}^{\mathrm{c}}$ is calculated by

$$\gamma_{\mathrm{P}}^{\mathrm{c}} = C_1 - C_2 I_{\mathrm{D}} \tag{8}$$

where the values of C_1 and C_2 are calibrated by triaxial tests or direct shear tests [32]. Based on the contributions of interparticle friction and soil fabric on the initial internal friction angle, $\varphi'_{\mathrm{in}} = 29°$ is assumed in this study [32].

The friction angle φ' and dilation angle ψ in the post-peak softening zone vary as a function of the plastic shear strain (γ^{P}), which can be represented by the exponential function as a backbone function:

$$\varphi' = \varphi_{\mathrm{c}}' + (\varphi_{\mathrm{P}}' - \varphi_{\mathrm{c}}')\exp\left[-\left(\frac{\gamma^{\mathrm{P}} - \gamma_{\mathrm{P}}^{\mathrm{P}}}{\gamma_{\mathrm{c}}^{\mathrm{P}}}\right)^2\right] \tag{9}$$

$$\psi = \psi_{\mathrm{P}}\exp\left[-\left(\frac{\gamma_{\mathrm{P}} - \gamma_{\mathrm{P}}^{\mathrm{P}}}{\gamma_{\mathrm{P}}^{\mathrm{c}}}\right)^2\right] \tag{10}$$

Figure 4 shows typical variations in the mobilized friction angle and dilation angle of dense sand with plastic shear strain, confining pressure, and relative density. It can be seen that a reduction in the confining pressure or an increase in the relative density amplifies the hardening and softening effects of dense sand. More detailed information on the MMC model can be found in Roy [40].

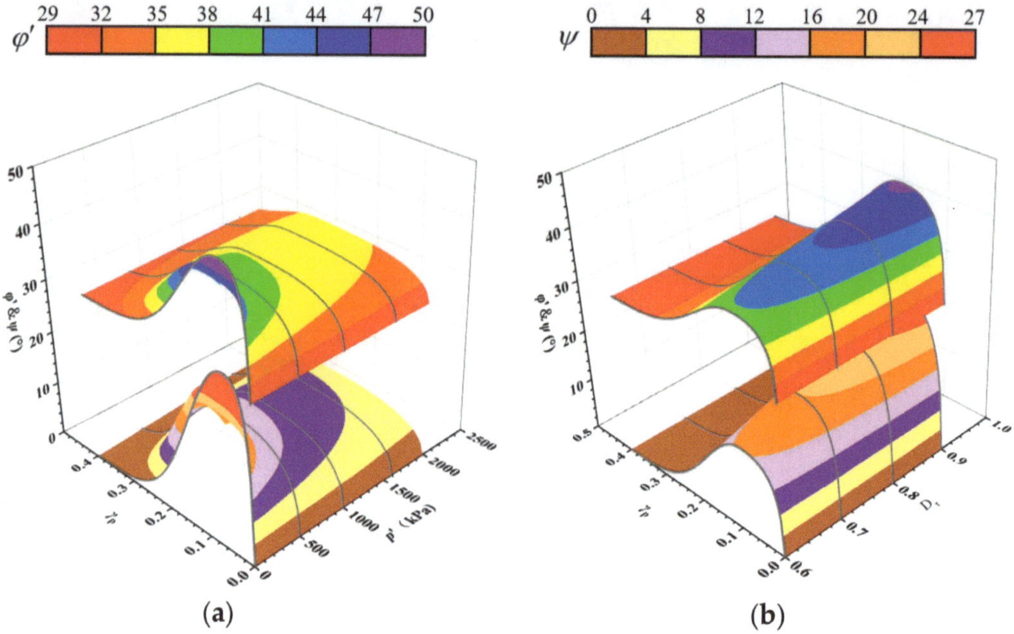

Figure 4. Variations in mobilized friction angle and dilation angle with plastic shear strain (φ'_c = 31°): (**a**) various confining pressures with D_r = 90%; (**b**) various relative densities with p'= 40 kPa.

In Abaqus, the MMC model is not built-in and requires to be implemented using subroutines VUSDFLD in FORTRAN. First, the plastic strain tensor (ε_{ij}) and the stress tensor (σ_{ij}) at each integration point are called and updated in each time increment, then stored in the state variables in the subroutine. Next, the incremental plastic shear strain $\Delta\gamma_p$ and the mean effective stress p' are calculated for each time increment. Finally, the value of γ^p (calculated as the sum of $\Delta\gamma_p$ throughout the analysis) and p' are assigned to two field variables (FV1, FV2), which are used to update the values of φ' and ψ, accordingly. Ahmed and Hawlader [27] defined the incremental plastic shear strain as

$$\Delta\gamma_p = \int \sqrt{\frac{3}{2}(\Delta\varepsilon_{ij}\Delta\varepsilon_{ji})}dt \tag{11}$$

where $\Delta\varepsilon_{ij}$ is the plastic strain increment tensor. In addition, Ahmed and Hawlader [27] verified the existence of both triaxial compression and plane strain shear conditions for the soil elements surrounding the pile under eccentric loading. Therefore, the values of $(A_\psi, k_\psi) = (3.8, 0.6)$ were used in this study. The properties of sand adopted in this study are summarized in Table 1. The maximum (γ_{max}) and minimum (γ_{min}) unit dry weight of the sand is assumed to be 17.58 kN/m^3 and 14.65 kN/m^3, respectively [41]. The unit weight of sand with different relative densities can be determined by

$$\gamma' = \gamma_s - \gamma_w = \frac{\gamma_{max}\gamma_{min}}{\gamma_{max}(1 - D_r) + \gamma_{min}D_r} - \gamma_w \tag{12}$$

A small value of cohesion c (0.1 kPa) is used in this study to avoid numerical instability.

The pile–soil interface is modelled in Abaqus using the "contact pair" algorithm. The "hard contact" model allows separation in the normal direction, while the Coulomb friction model is adopted in the tangential direction. The interface friction angle δ is taken as $2/3\varphi'_0$ to simulate the typical steel–sand contact property [29,42], where φ'_0 is the constant friction angle. The value of φ'_0 can be estimated according to the formula recommended by API [43]:

$$\varphi'_0 = 16D_r^2 + 0.17D_r + 28.4 \quad (13)$$

2.3. Loading Steps

Each FE analysis has four steps to simulate the formation of local scour and the subsequent lateral loading phase:

(1) Initial step: As shown in Figure 2, the bottom of the soil domain is fixed in all three directions ($U_x = 0$, $U_y = 0$, $U_z = 0$), while the vertical boundary is supported by rollers to restrict the lateral displacements ($U_x = 0$, $U_y = 0$). The initial stress is applied to the whole domain by means of the "geostatic stress" function built into the initial predefined field;

(2) Geostatic step: The uniform vertical body force is exerted on the whole domain to simulate the gravitational effect. During this step, the normal displacement of the pile–soil contact surface is restrained. At the end of this step, the model produces only a negligible deformation;

(3) Scour step: As shown in Figure 2, the predefined set of scour elements is deleted through state variables to simulate the scour unloading effect. The deleted elements cannot carry stresses in the subsequent analysis and, hence, have no contribution to the stiffness of the model. The remaining soil domain redistributes the stresses to reach a new stress balance after the deletion of scoured soil elements;

(4) Load step: After setting up the boundary conditions and stresses in the post-scour domain, the normal displacement restraint on the pile–soil contact surface is removed and replaced by the "contact pair" algorithm. A monotonic lateral load (F) and bending moment ($M = F{\cdot}e$, e the load eccentricity) are then applied at the geometric center of the pile head. For accuracy and efficiency, the quasi-static analysis is loaded using the smoothed step amplitude curve. During the rotation of the pile foundation, the load point couples to and moves with the pile head.

2.4. Simulation Cases

Table 2 summarizes the five series of simulation cases considered in this study. There are two aims in Series I: One is to compare the effects of scour depth on the lateral bearing capacity of winged monopiles, and the other is to compare the benefits of introducing wings and extending the embedment depth to improve the lateral bearing capacity of pile foundations with local scour. The relative density of sand, the configuration and the wing load orientation of hybrid monopiles are kept constant. Series II mainly compares the effects of load eccentricity on the lateral bearing capacity of pile foundations in the scoured sandy seabed. In Series III, the sandy seabed with various relative densities is examined to explore its effect on the lateral bearing capacity. Series IV aims to assess the effect of wing load orientation on the lateral bearing capacity of winged monopiles. The effect of aspect ratio on the lateral bearing capacity of hybrid monopiles with three or four wings is further investigated in Series V. In all simulations except Series II, the load eccentricity $e = 8D$ is generally assumed, referring to Bhattacharya's [5] report for typical wave and wind loads on OWTs, unless otherwise specified.

Table 2. Numerical simulation cases for parametric study.

Series	Constant Parameters	Variable Parameters
I	D_r = 90%, four-wing monopile with $L_p = 8D$, $L_w/W_w = 1$, $\beta = 0°$; D_r = 90%, regular monopile with $L_p = 8D$; D_r = 90%, extended regular monopile with $L_p = 9.29D$;	S = 0, 0.5D, 1D, 1.5D, 1.9D

Table 2. *Cont.*

Series	Constant Parameters	Variable Parameters
II	D_r = 90%, four-wing monopile with, $S = 1.9D$, $\beta = 0°$; D_r = 90%, four-wing monopile with, $S = 1.0D$, $\beta = 0°$; D_r = 90%, four-wing monopile with, $S = 0D$, $\beta = 0°$; D_r = 90%, regular monopile with $L_p = 8D$, $S = 1.9D$, $\beta = 0°$; D_r = 90%, regular monopile with $L_p = 8D$, $S = 1.0D$, $\beta = 0°$; D_r = 90%, regular monopile with $L_p = 8D$, $S = 0D$, $\beta = 0°$;	$e = 0, 2D, 4D, 8D$ and pure moment
III	Four-wing monopile with $L_w/W_w = 1$, $S = 1.5D$, $\beta = 0°$;	D_r = 60%, 70%, 80%, 90%
IV	D_r = 90%, four-wing monopile with, $S = 1.5D$, $L_w/W_w = 1$; D_r = 90%, three-wing monopile with, S =1.5D, $L_w/W_w = 1$;	$\beta = 0°, 15°, 30°, 45°$ $\beta = 0°, 15°, 30°, 45°, 60°$
V	D_r = 90%, four-wing monopile with, $W_w = D_P$, $S = 1.5D$, $\beta = 0°$; D_r = 90%, three-wing monopile with $W_w = D_P$, $S = 1.5D$, $\beta = 0°$;	L_w/W_w = 0, 0.5, 0.75, 1, 1.5, 1.75

3. Validation by Centrifuge Tests

The numerical model was validated based on the results of a large-diameter monopile centrifuge test in dense sand reported by Klinkvort [44]. The prototype diameter D of the monopile is 3 m and the embedment depth L is $6D$. The pile was modeled as linear elastic with Young's modulus E_p = 210 GPa and Poisson's ratio of 0.3. To simulate the offshore conditions in the North Sea, centrifuge tests were conducted in dense homogeneous Fontainebleau sand with a relative density of about 90%. Young's modulus follows Equation (1) [45]. Setting the Mohr–Coulomb (MC) model as the reference group, the constant friction angle φ'_0 was calculated by Equation (13) recommended by API [43]. The dilation angle in the MC model was calculated by Equation (14) proposed by Bolton [37]:

$$\psi = \varphi'_0 - \varphi'_c \tag{14}$$

The properties of Fontainebleau sand and the other model parameters are shown in Table 3.

Table 3. Pile–soil interaction model parameters used for validation [32,45].

Description	Symbol	Unit	Dense Fontainebleau Sand
Relative density	D_r	%	90
Submerged unit weight	γ'	kN/m^3	10.2
Stiffness parameter	κ	—	1380.7
Stiffness parameter	λ	—	0.6226
Poisson's ratio	v	—	0.3
Critical state friction angle	φ'_c	°	30
Parameter of MMC	φ'_{in}	°	29
	A_ψ	—	3.8
	k_ψ	—	0.6
	C_1	—	0.22
	C_2	—	0.11
	m	—	0.25

Figure 5a shows the stress–strain behavior of Fontainebleau sand captured by MC and MMC, respectively, against the drained triaxial compression test from Latini and Zania [46]. It shows that the MMC model can accurately capture the hardening and softening effects during the deformation of dense sand. For the MC model, the shear stress of the dense sand increases with the axial strain ε_a to the peak and then remains constant because a fixed friction angle φ'_0 is used in the analysis. As for Test 9 with a high relative density of D_r = 80%, the hardening and softening effects during the deformation of sand are salient, while the MC model is only capable of capturing the peak stress in the plastic zone. It is indicated that the MC model overestimates the load capacity of dense sand. Figure 5b compares the volumetric properties of the dense sand described by the MMC and MC

models. By adopting the MMC model, the volumetric compression occurs initially, then the sand dilates until the volume approximately remains constant. In contrast, the MC model shows that dense sand dilates almost linearly after volumetric compression due to a fixed dilation angle ψ, deviating from the test results obviously.

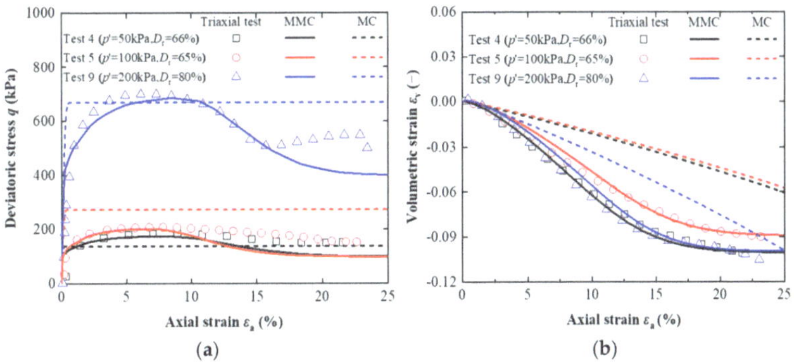

Figure 5. Comparison of FE simulation and triaxial compression soil element tests: (**a**) stress–strain behavior; (**b**) volume change behavior.

Figure 6 compares the numerical results with the centrifuge test regarding the lateral response of a regular monopile under different load eccentricities. It can be seen that the normalized load–displacement curves obtained from the MMC model match the centrifuge test results well. In contrast, the MC model tends to overestimate the lateral bearing capacity. The difference between the MC and MMC models can be further explained by the formation of shear bands (accumulated plastic shear strain concentrated zones). The evolution of the plastic shear zone with a constant load eccentricity (e = 2.5D) between the MC and MMC models is compared in Figure 7. The plastic shear strain does not develop significantly when the displacement is small (θ = 0.5°). As the displacement increases, the plastic shear band f_1 appears near the pile head first and then extends in a diagonally downward direction (θ = 2.0°). A second plastic shear band f_2 appears as the displacement continues to increase and extends obliquely upward to the surface to form a damaging wedge (θ = 8.0°). As the displacement further increases, a plastic shear band f_3 forms below the shear band f_2 and extends similarly to the surface (θ = 10°). The contours from the MC model show that the failure wedge is developed by a simple shear band from beginning to end. Such a difference causes an increasing deviation in the simulated lateral bearing capacity as the pile deflection increases. These comparisons with soil element tests and centrifugal pile tests all demonstrate the reliability of the developed MMC model in estimating the lateral response of monopile foundations embedded in dense sand.

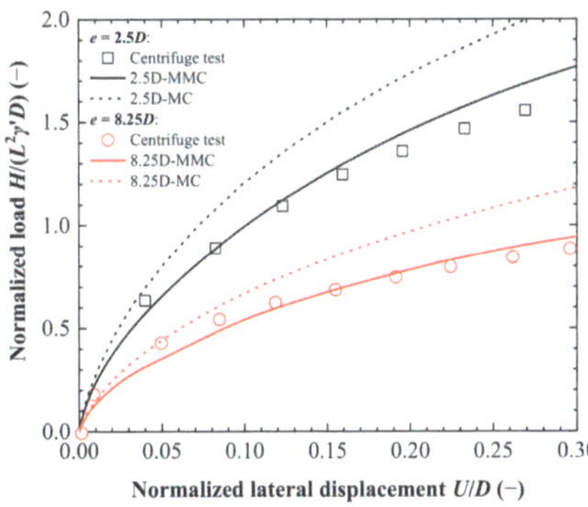

Figure 6. Comparison of finite element results with centrifuge test results [44].

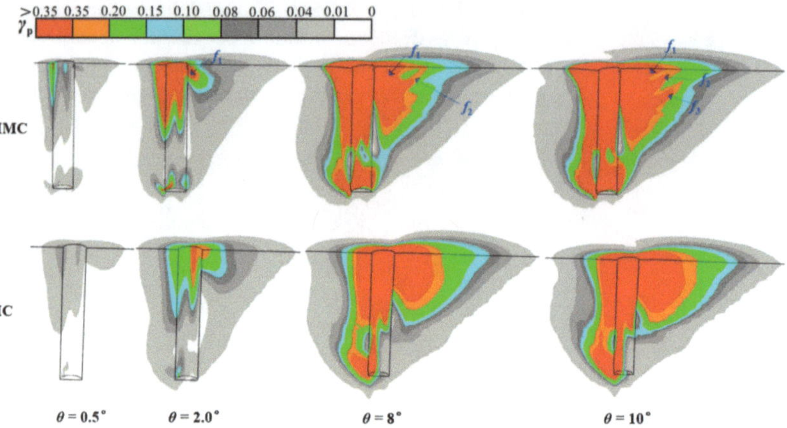

Figure 7. Comparison of the development process of the plastic zone around the pile (scaling factor: 0.1).

4. Parametric Study and Discussion

4.1. Effect of Scour Depth

4.1.1. Scour Effect on Four-Wing Monopiles

Cases with various scour depths were simulated to investigate the scour effect on the lateral bearing capacity of winged monopiles. In Series I (Table 2), four-wing monopiles (WP4) with L_W/W_W = 1 in the sand with relative density D_r = 90% were considered. Figure 8a compares the load–displacement curves at the head of the winged monopiles. The vertical axis represents the horizontal load H (expressed as the normalized load $H/L_p{}^2D\gamma'$). The horizontal axis represents the angle of rotation (θ) of the pile head at the mudline. As expected, the lateral bearing capacity of winged monopiles decreases with increasing scour depth. Two key design limits when designing pile foundations for OWTs are marked in Figure 8a, which are the serviceability limit state (SLS) and the ultimate limit state (ULS) [47]. The corresponding lateral loads are denoted as H_s and H_u, respectively. The corresponding moments are denoted as M_s and M_u, respectively. SLS failure refers to the tilt angle of OWT at the mudline exceeding the allowable value. This value is commonly set to 0.5°, which is generally small but sufficient to affect the normal operation of the turbine [48]. In contrast, the ULS criterion is characterized by the excessive rotation of piles as OWTs are typically supported by short piles with rigid behavior [49]. This larger tilt angle is set to 2° at ULS in the FE simulations performed in this study [50,51].

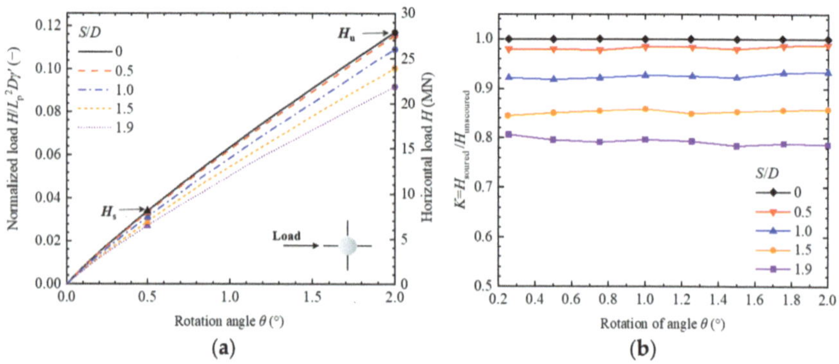

Figure 8. The effects of scour depth on bearing capacity of winged monopiles: (**a**) load–displacement curves of winged monopiles with various scour depths; (**b**) variation in lateral bearing capacity ratio with displacement at different scour depths.

It can be seen from Figure 8a that the lateral bearing capacity at SLS and ULS of winged monopiles with maximum scour depth $S = 1.9D$ are reduced by 20% compared to the case without local scour. Figure 8b compares the variation in the lateral bearing capacity ratio (K) with the rotation of winged monopiles for different scour depths. The lateral bearing capacity ratio K represents the ratio of the lateral load under the scoured condition to that of the unscoured condition. The value of K remains almost constant with the rotation for all the examined scour depths. This indicates that the effect of pile deformation on the lateral bearing capacity ratio of pile foundations under different scour depths is generally constant.

As shown in Figure 9, the discrepancy in the previous lateral responses can be attributed to the difference in failure mechanism due to scour. In Figure 9, profile AA′ is 12 m from the top of the pile (top of the wings); profile BB′ is 15 m from the head of the pile (middle of the wings); and profile CC′ is 18m from the top of the pile (bottom of the wings). When the pile is rotated clockwise under the load, the soil near the mudline follows the winged pile and the pile toe kicks back the sand. Profile AA′ shows that the scour hole reduces the width and length of mobilized soil at the top of the wings by 6%. This demonstrates that winged monopiles without local scour (Figure 9a) can mobilize a larger amount of soil in the shallow layer to obtain a higher lateral bearing capacity than winged monopiles with local scour (Figure 9b).

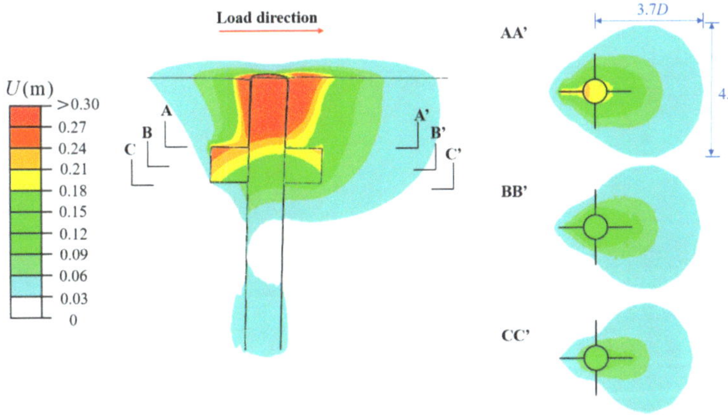

(a)

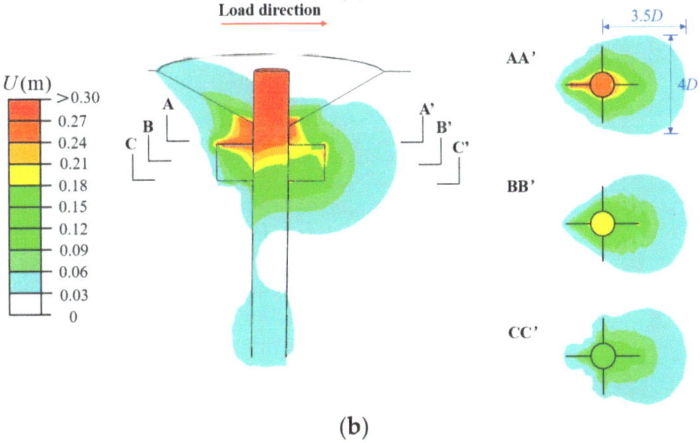

(b)

Figure 9. Soil displacement in the plane of symmetry and the planform at $\theta = 2°$: (**a**) the winged monopile without local scour; (**b**) the winged monopile with local scour ($S/D = 1.5$).

The effect of scour on soil–pile interaction can be further explained by the evolution of the plastic zone in dense sand. In Figure 10, $\gamma_p/\gamma_p^P > 1$ indicates that the soil is in the post-peak region; conversely, the soil is in the pre-peak region when this condition is not met. As shown in Figure 10a, the soil in front of the pile head first enters the post-peak softening zone (θ = 0.5°) and gradually evolves downward with increasing rotation (θ = 2°). The plastic zone around the wing is mainly gathered above the rear wing and ahead of the front wing (θ = 2°). However, in Figure 10b, a large volume of soil is in the post-peak softening stage when the pile head displacement is small (θ = 0.5°). Nearly all of the soil around the wings enters the plastic zone after the formation of the local scour hole (θ = 2°), suggesting that the soil resistance mobilized by the wings acts as the main force to resist pile deformation. The γ_p above the front wing develops to the post-peak softening stage with pile rotation.

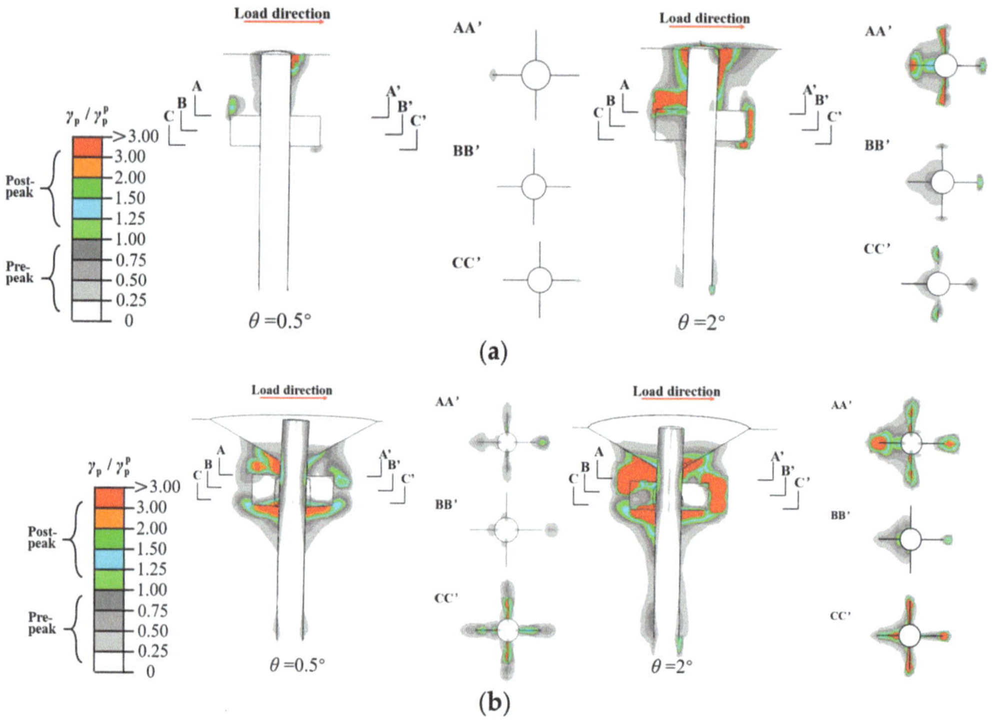

Figure 10. Distribution of plastic shear strain in the plane of symmetry of the winged monopile (WP4): (**a**) without local scour; (**b**) with local scour (*S*/*D* = 1.5).

Figure 11 shows the modified friction angle φ' and the modified dilation angle ψ of the sand for pile head rotation θ = 2° under unscoured and scoured (*S* = 1.5*D*) conditions, respectively. The soil elements that reach the critical state are highlighted in the figure. It is evident that more soil near the wings of the hybrid monopile is at the critical state for the scoured case, with a large accumulated plastic shear strain. The profiles show the wings perpendicular to the direction of the load dividing the soil around the wings into two parts: "front" and "rear". The friction and dilation angles of the front area are generally smaller than those of the rear area. This is attributed to the relaxation of the restraint (lower stress level) in the rear zone due to the displacement of wings, which allows the soil to obtain a high friction angle (dilation angle) with only a small plastic strain.

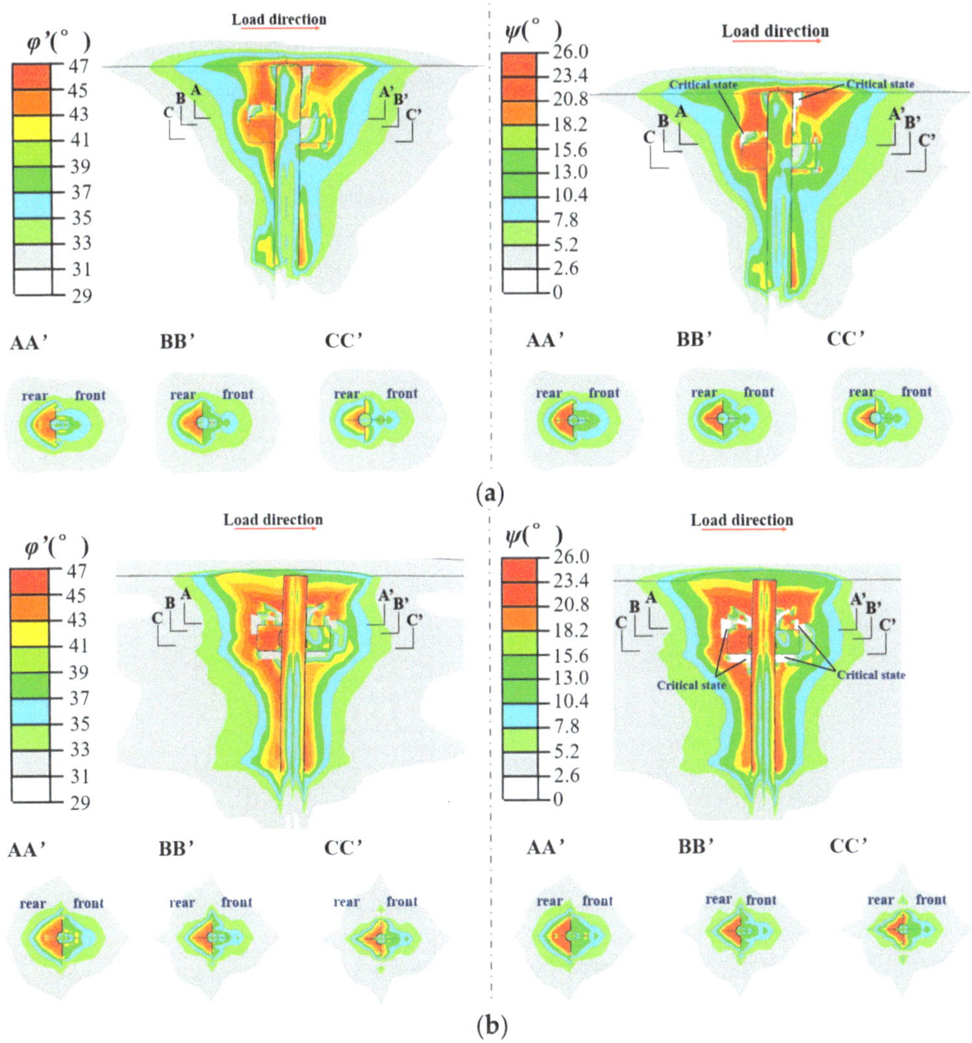

Figure 11. Modified friction angle φ' and dilation angle ψ of the sand in the plane of symmetry at $\theta = 2°$: (**a**) without local scour; (**b**) with local scour ($S/D = 1.5$).

4.1.2. Comparison between Four-Wing Monopiles and Regular Monopiles

The lateral bearing capacity is usually improved by increasing the embedded depth of the pile foundation. A group of regular monopiles (RP) with the same embedment depth and a group of extended monopiles (RP-E) consuming the same material as WP4 were established as reference groups to examine the sensitivity of wings to the variation in scour depth. Figure 12 shows the load–displacement curves of regular monopiles (RP) and extended regular monopiles (RP-E) with different scour depths. Figure 13 shows the ratio of the lateral bearing capacity of the winged monopiles (WP4) and the extended regular monopiles (RP-E) to the regular monopiles ($RP_{S=0}$) without local scour at SLS and ULS. It can be seen that the lateral bearing capacity of the pile foundation can be enhanced by increasing the embedment depth and introducing wings, with the latter being more effective. The presence of wings can increase the lateral bearing capacity of the pile by up to 9.1% at SLS and 8.6% at ULS. It is favorable to have a slightly higher increase in the lateral bearing capacity at SLS as OWTs are more sensitive to deformation.

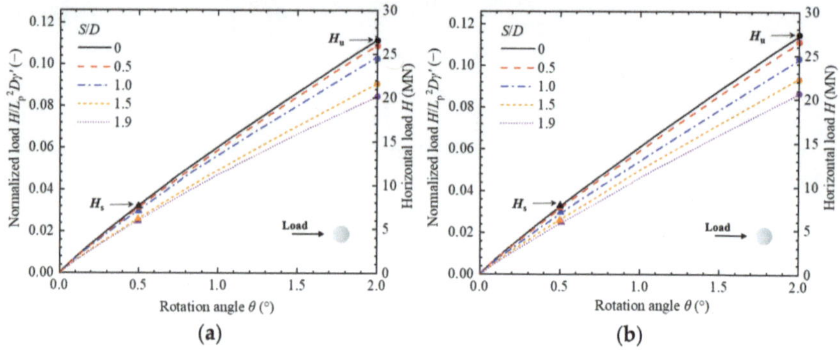

Figure 12. Load–displacement curves with different scour depths: (**a**) regular monopiles (RP); (**b**) extended regular monopiles (RP-E).

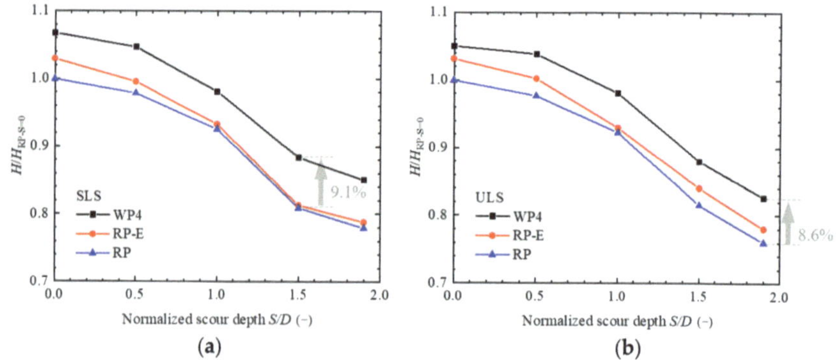

Figure 13. Variation in the ratio of lateral loads of WP4 and RP-E with scour depths at (**a**) SLS and (**b**) ULS.

Figure 14 shows the soil displacement of regular monopiles with different embedment depths at ULS. It is found that extending the embedded length of a regular monopile hardly produces any displacement at the bottom of the pile. Nevertheless, profile AA′ reveals that slightly more soil at shallow depths can be mobilized to resist pile deformation with the increase in embedment depth. In addition, comparing Figures 14b and 9b, the presence of the wings changes the pile circumferential displacement profile from "gourd-shaped" to "teardrop-shaped", leading to higher lateral bearing capacity than extending the embedment depth of regular monopiles.

The lateral bearing capacity of piles can be better described by using moment–force interaction diagrams (Figure 15), which show an almost linear trend. As expected, the lateral bearing capacities (both H_s and M_s) of piles decrease with an increase in scour depth. In addition, the weakening of the lateral bearing capacity is more pronounced when the scour depth develops from 1D to 1.9D, compared with that from 0 to 1D (i.e., $\Delta H_1 < \Delta H_2$, $\Delta M_2 < \Delta M_3$). This implies that the detrimental effect of scour on the lateral bearing capacity of monopiles accelerates with the development of scour. Another observation from the interaction diagram is that at a small eccentricity (e = 1D), the lateral force weakened by scour is equivalent to four times as much as that with e = 8D (i.e., $\Delta H_3 \approx 4\Delta H_4$), while the bending moment variation is relatively small (i.e., $\Delta M_1 \approx 0.5\Delta M_4$). This indicates that the effect of scour on the lateral bearing capacity of piles shifts from lateral force to a bending moment with the increase in load eccentricity e. When the scour depth S = 1D, the envelope of the winged monopile (WP4) almost coincides with the regular monopile under the unscoured condition, demonstrating the effectiveness of the wings in improving the lateral bearing capacity of pile foundations.

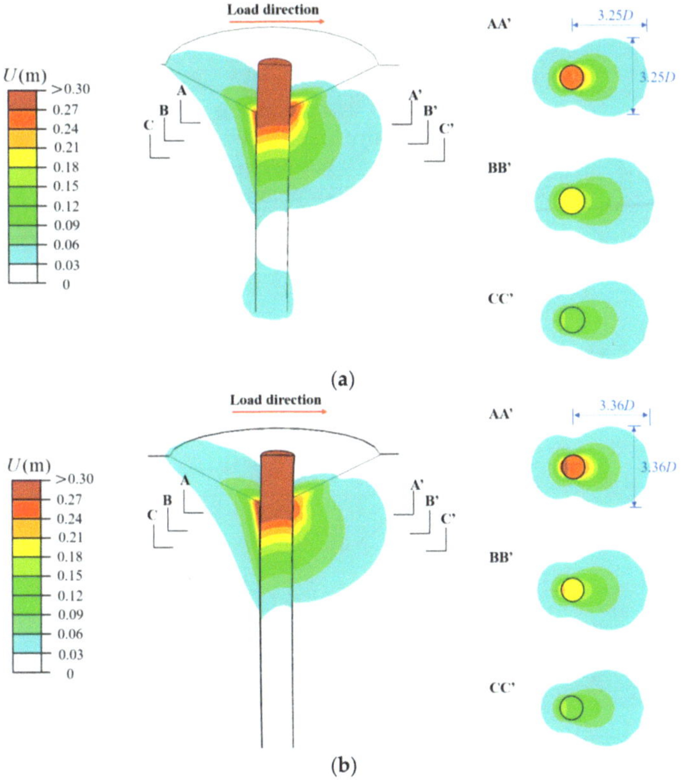

Figure 14. Soil displacement in the plane of symmetry and the planform at $\theta = 2°$: (**a**) the regular monopile with scour (S/D = 1.5); (**b**) the extended monopile with scour (S/D = 1.5).

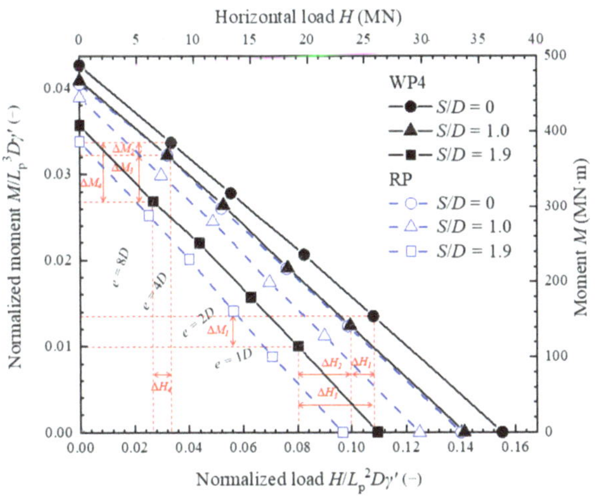

Figure 15. Moment–lateral load interaction diagram at SLS.

4.2. Effect of Relative Density of Sand

This section modelled three groups of sand with different relative densities (i.e., medium-dense sand: $D_r = 60\%$; dense sand: $D_r = 70\%$, 80%; and very dense: sand $D_r = 90\%$) to analyze their effect on the lateral bearing capacity of pile foundations under scoured conditions ($S = 1.5D$). Note that the MMC model is not recommended to capture the mechanical behavior of loose sand [40]. It is also worth noting that the relative density influences both the elastic and plastic behaviors of the sand, as reflected in the model parameters (Table 1). The lateral bearing capacity curves of winged monopiles ($L_w/W_w = 1$) with different relative densities of sand are shown in Figure 16. Both the lateral load at SLS (H_s) and ULS (H_u) of winged monopiles increase with increasing relative density. Compared with medium-dense sand ($D_r = 60\%$), the values of H_s and H_u of winged monopiles embedded in very dense sand ($D_r = 90\%$) are increased by 12.4% and 16.7%, respectively. On the one hand, this is because the increased relative density magnifies the peak friction angle φ'_p and the peak dilation angle ψ_p (Figure 4b), leading to a higher soil resistance. On the other hand, more soil reaches the critical state in the sand with lower relative density, with a lower shear strength level maintained at a large accumulated plastic shear strain (marked as dashed circles in Figure 17).

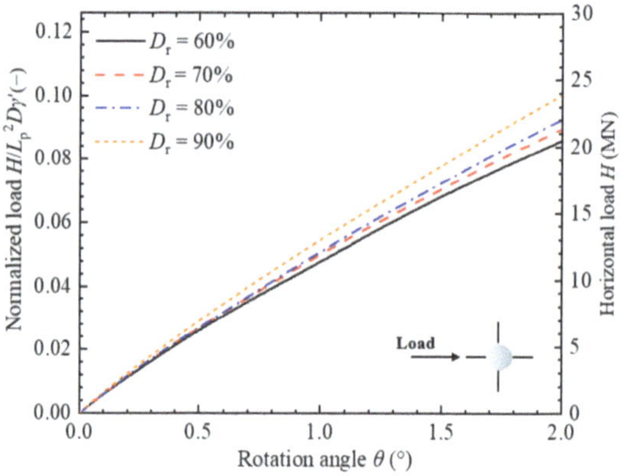

Figure 16. Load–displacement curves of winged monopiles in sand with different relative densities (D_r).

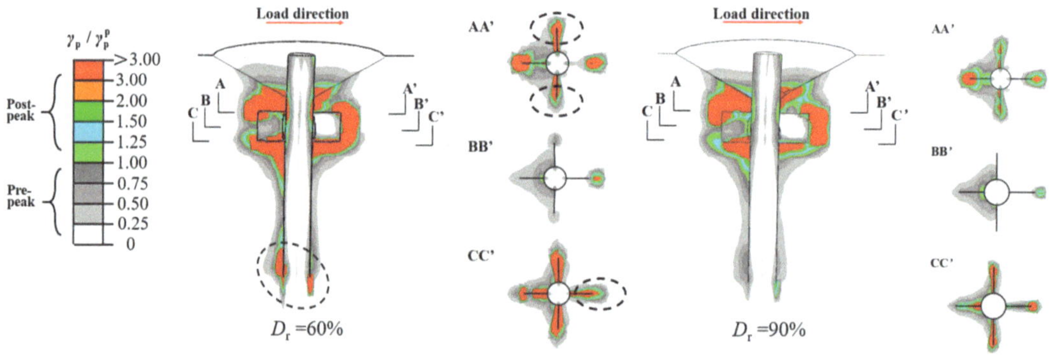

Figure 17. Distribution of plastic shear strain in the plane of symmetry of the winged monopile with different relative densities (D_r) at $\theta = 2°$.

4.3. Evaluation of Wing Efficiency

OWT foundations in service bear loads in multiple directions. The effect of wing load orientation on the wings' ability to mobilize soil resistance needs to be considered if the winged monopiles are to be applied in marine environments. Figure 18a,b shows the load–displacement curves for different wing load orientations (β). Considering the symmetry of the winged monopile itself, the load direction β acting on the pile head of the four-wing monopile (WP4) is set in the range of 0° to 45°. The load direction β on the head of the three-wing monopile (WP3) is set in the range of 0~60°. The results show that the effect of load direction on the load capacity of winged monopiles is marginal, implying the excellent adaptability of three- or four-wing monopiles to multi-directional loads in marine environments.

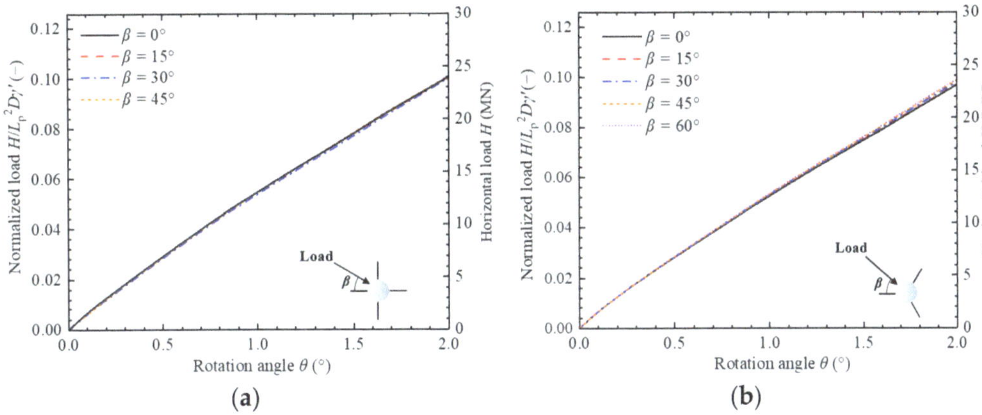

Figure 18. Effect of wing load orientation on the load capacity of winged monopiles: (**a**) a four-wing monopile; and (**b**) a three-wing monopile.

In terms of the wing geometry effect, Figure 19 shows the enhancement of lateral capacity with the larger wing height for both four-wing monopiles and three-wing monopiles with a fixed width of wings, with four wings providing a higher capacity. This is because less soil is mobilized by three wings at shallow soil depths, as shown in Figure 20 (compared with Figure 9b for four wings). Similarly, less soil is in the post-peak zone for three-wing monopiles (Figure 21), which can also explain the capacity difference. It should be noted that no solid conclusion can be drawn at this stage that four-wing monopiles perform better than three-wing monopiles. The compared results only apply to the conditions investigated in this study, including the scour depth and the geometry of wings.

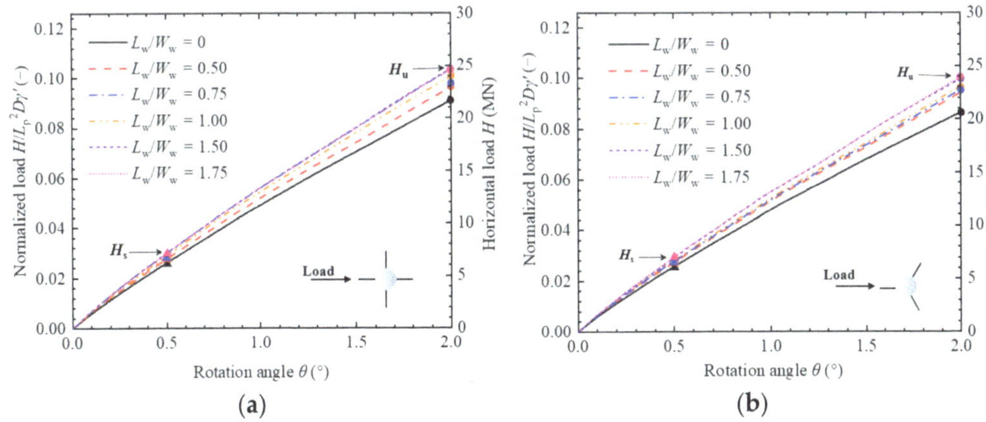

Figure 19. Variation in the lateral bearing capacity of winged monopiles under different wing heights: (**a**) four-wing monopiles; (**b**) three-wing monopiles.

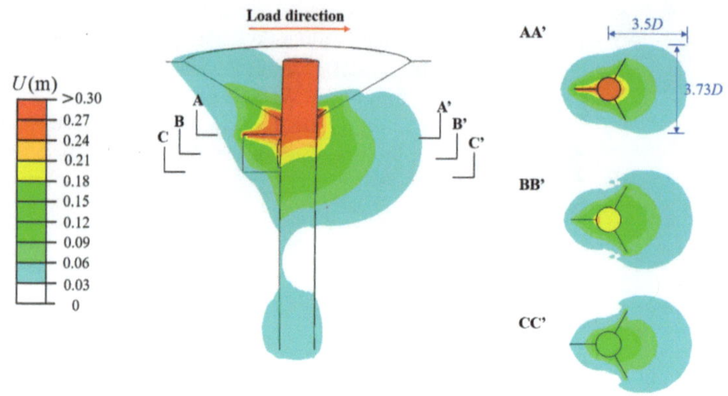

Figure 20. Sand displacement in the plane of symmetry and the planform of the three-wing monopile under scour ($S/D = 1.5$) at $\theta = 2°$.

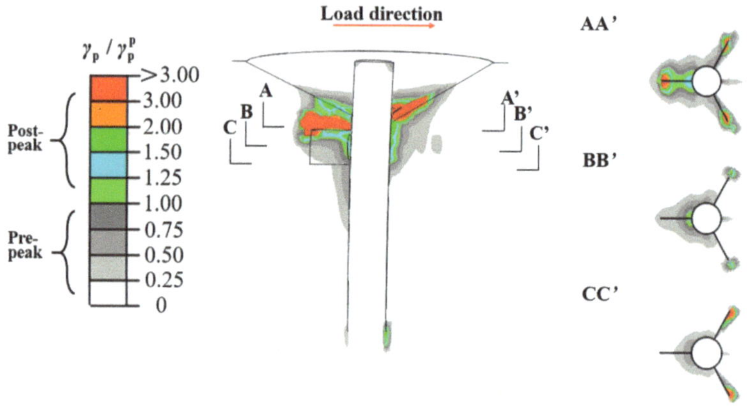

Figure 21. Distribution of plastic shear strain in the plane of symmetry of the winged monopile (WP3) with $S/D = 1.5$ at $\theta = 2°$.

The lateral capacity of winged monopiles can be quantified by the wing efficiency, which is defined as follows:

$$\eta = \frac{H_W - H_{RP}}{A_W} / \frac{H_{RP}}{A_{pr}} \tag{15}$$

where H_W and H_{RP} are loads of winged monopiles and regular monopiles at the same rotation, respectively. A_W is the sum of the areas of the wings of the winged monopile ($A_W = n \times L_W \times W_W$). n is the number of wings mounted on the monopile. A_{pr} is the external surface area of the regular monopile ($A_{pr} = \pi \times D \times L_P$). η in Equation (15) is the ratio of the incremental magnitude of the lateral load of the hybrid monopile caused by the unit area of wings to the lateral load of the unit area of the monopile. This is a different definition regarding the wing efficiency compared with those in previous studies [13,14,33], which only considered load efficiency. Rather, the newly defined η is capable of evaluating both load and economic efficiency. As the wing thickness is identical to the pile wall thickness in this study, the areas A_W and A_{pr} are used in Equation (15). If the difference between the wing and pile wall thicknesses needs to be considered, the volume of the consumed material can be used instead.

Figure 22 shows the trend of wing efficiency with various wing heights at SLS and ULS. The wing efficiency calculated by Equation (15) can take different wing numbers and wing heights into account. In the range of the aspect ratio (W_L/W_W) considered, the wing efficiencies for four-wing and three-wing monopiles decrease with increasing L_W/W_W, which is valid for both SLS and ULS. For four-wing monopiles with $L_W/W_W = 1$, the wing efficiency is nearly equal to 1, indicating that the soil resistance mobilized per unit area of the wings is the same as that of a regular monopile. Therefore, a wing efficiency $\eta > 1$ is recommended to fully utilize the potential of the wings for the enhancement of the lateral capacity of monopiles.

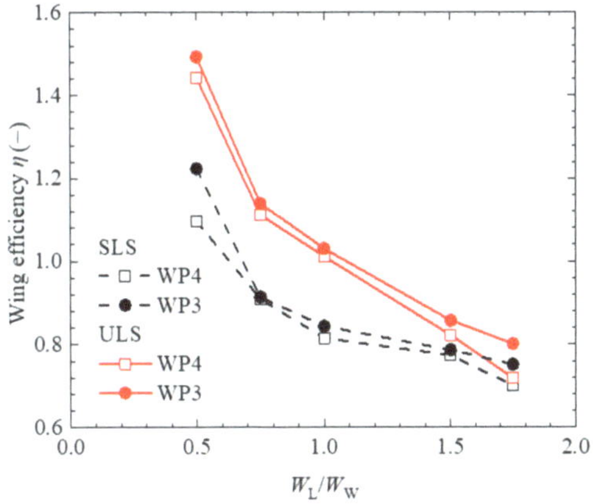

Figure 22. Variation in wing efficiency for winged monopiles with various wing heights.

5. Conclusions

In this paper, the combined effects of wing configuration and local scour on the lateral capacity of a hybrid monopile (monopile with "wings") are investigated through finite element analyses. A modified Mohr–Coulomb model (MMC) for dense sand is used to capture the load–displacement behavior and the evolution of failure mechanisms for the pile–soil interaction. The following conclusions can be drawn from this study:

(1) The increase in local scour depth reduces the lateral capacity of winged monopiles in the same wing load orientation. The capacity reduction can be as high as 20% at the serviceability limit state (SLS) and the ultimate limit state (ULS) for four-wing monopiles (WP4) with a scour depth of $S/D = 1.9$. More soil around the wings will reach the critical state caused by post-peak softening if a scour hole is developed around the pile, consequently undermining the lateral capacity;

(2) Extending the embedment length of the pile foundation contributes marginally to the lateral capacity, while introducing wings can significantly enhance the capacity. The lateral capacity of the winged monopile is increased by 9.1% compared to the regular monopile when the scour depth $S/D = 1.5$. Due to the presence of wings, the circumferential failure pattern around the pile shifts from a "gourd" to a "teardrop" shape;

(3) Higher lateral capacity can be expected for winged monopiles in denser sand. The lateral capacity at SLS and ULS can be raised by 12.4% and 16.7%, respectively, with the relative density increasing from 60% to 90%;

(4) A wing efficiency number is proposed to determine the optimal design of wing configuration, accounting for both load and economic efficiency. The wing efficiency at SLS reaches its maximum value when the ratio of wing height and wing width reaches its minimum ($L_W/W_W = 0.5$) and only decreases slightly for the wing height $L_W/W_W > 1$. A wing efficiency larger than unity is recommended to maximize the effectiveness of wings for enhancing the lateral capacity of monopiles.

This paper mainly aims to reveal the combined effects of local scour and wing configuration on a winged monopile under horizontal loading. Future research is required to quantify the effect of wings on monopile capacity with various pile diameters and embedment depths, which is essential to taking full advantage of the wings in the design. The modified p–y curve accounting for the influence of wings could be a feasible approach to accomplish the optimal design.

Author Contributions: Conceptualization, W.Q.; numerical simulations and data analyses, B.L., W.Q. and Y.W.; writing—original draft, B.L.; writing—review and editing, W.Q., Y.W., S.W. and F.G.; supervision, W.Q., F.G. and Y.W.; funding acquisition, W.Q. and F.G. All authors have read and agreed to the published version of the manuscript.

Funding: This research was funded by the National Natural Science Foundation of China (11972036; 11825205) and the Youth Innovation Promotion Association CAS (Grant No. 2021018).

Institutional Review Board Statement: Not applicable.

Informed Consent Statement: Not applicable.

Data Availability Statement: The data presented in this study are available upon request from the corresponding author.

Conflicts of Interest: The authors declare no conflict of interest.

References

1. GWEC. *Global Wind Report 2021*; GWEC: Brussels, Belgium, 2022.
2. BNEF. *Offshore Wind Market Outlook*; BNEF: Paris, France, 2020.
3. Oh, K.Y.; Nam, W.; Ryu, M.S.; Kim, J.Y.; Epureanu, B.I. A review of foundations of offshore wind energy convertors: Current status and future perspectives. *Renew. Sustain. Energy Rev.* **2018**, *88*, 16–36. [CrossRef]
4. Gupta, B.K.; Basu, D. Offshore wind turbine monopile foundations: Design perspectives. *Ocean Eng.* **2020**, *213*, 107514. [CrossRef]
5. Bhattacharya, S. *Design of Foundations for Offshore Wind Turbines*; John Wiley & Sons: Hoboken, NJ, USA, 2019.
6. Qi, W.G.; Gao, F.P. Local scour around a monopile foundation for offshore wind turbines and scour effects on structural responses. In *Geotechnical Engineering-Advances in Soil Mechanics and Foundation Engineering*; IntechOpen: London, UK, 2019.
7. Bienen, B.; Dührkop, J.; Grabe, J.; Randolph, M.F.; White, D.J. Response of piles with wings to monotonic and cyclic lateral loading in sand. *J. Geotech. Geoenviron. Eng.* **2012**, *138*, 364–375. [CrossRef]
8. Murphy, G.; Cadogan, D.; Doherty, P.; Gavin, K.; Caprani, C.; O'Connor, A. Experimental investigation of novel foundation solutions for offshore wind turbines. *Proc. BCRI* **2012**, *169*, 227–239.
9. Murphy, G.; Doherty, P.; Cadogan, D.; Gavin, K. Field experiments on instrumented winged monopiles. *Proc. Inst. Civ. Eng. Geotech. Eng.* **2016**, *169*, 227–239. [CrossRef]
10. Babu, K.; Viswanadham, B. Numerical investigations on lateral load response of fin piles. In *International Congress and Exhibition Sustainable Civil Infrastructures: Innovative Infrastructure Geotechnology*; Springer: Berlin/Heidelberg, Germany, 2017; pp. 317–329.
11. Abongo, K.O. *Model Study of the Static and Cyclic Lateral Capacity of Finned Piles*; Lehigh University: Bethlehem, PA, USA, 2019.
12. Dührkop, J.; Grabe, J. Laterally loaded piles with bulge. *J. Offshore Mech. Arct. Eng.* **2008**, *130*, 041602. [CrossRef]
13. Nasr, A.M.A. Experimental and theoretical studies of laterally loaded finned piles in sand. *Can. Geotech. J.* **2014**, *51*, 381–393. [CrossRef]
14. Pei, T.; Qiu, T. A Numerical investigation of laterally loaded steel fin pile foundation in sand. *Int. J. Geomech.* **2022**, *22*, 04022102. [CrossRef]
15. Peng, J.R. *Behaviour of Finned Piles in Sand under Lateral Loading*; Newcastle University: Newcastle upon Tyne, UK, 2006.
16. Peng, J.R.; Rouainia, M.; Clarke, B.G. Finite element analysis of laterally loaded fin piles. *Comput. Struct.* **2010**, *88*, 1239–1247. [CrossRef]
17. Qi, W.G.; Gao, F.P. Equilibrium scour depth at offshore monopile foundation in combined waves and current. *Sci. China-Technol. Sci.* **2014**, *57*, 1030–1039. [CrossRef]
18. Qi, W.G.; Gao, F.P. Physical modeling of local scour development around a large-diameter monopile in combined waves and current. *Coast. Eng.* **2014**, *83*, 72–81. [CrossRef]
19. Sumer, B.; Fredsøe, J.; Christiansen, N. Scour around vertical pile in waves. *J. Waterw. Port Coast. Ocean Eng.* **1992**, *118*, 15–31. [CrossRef]
20. Sumer, B.M.; Christiansen, N.; Fredsøe, J. The horseshoe vortex and vortex shedding around a vertical wall-mounted cylinder exposed to waves. *J. Fluid Mech.* **1997**, *332*, 41–70. [CrossRef]
21. Qi, W.G.; Gao, F.P.; Randolph, M.F.; Lehane, B.M. Scour effects on *p–y* curves for shallowly embedded piles in sand. *Géotechnique* **2016**, *66*, 648–660. [CrossRef]
22. Li, C.; Xiao, Y.; Liu, J.; Lin, Q.; Zhang, T.; Liu, J. The impact of scour on laterally loaded piles bored and socketed in marine clay. *J. Mar. Sci. Eng.* **2022**, *10*, 1636. [CrossRef]
23. Qi, W.G.; Liu, J.; Gao, F.P.; Li, B.; Chen, Q.G. Quantifying the spatiotemporal evolution of the turbulent horseshoe vortex in front of a vertical cylinder. *Phys. Fluids* **2022**, *34*, 015110. [CrossRef]

24. Li, B.; Wang, S.-Y.; Qi, W.-G.; Wang, Y.; Gao, F.-P. Lateral bearing capacity of a winged monopile in the scoured Sand-Bed. In Proceedings of the 32nd International Ocean and Polar Engineering Conference, Shanghai, China, 6–10 June 2022.
25. Hibbitt, D.; Karlsson, B.; Sorensen, P. *Abaqus: Analysis User's Manual, Version 6.11*; Simulia DCS: Pawducket, RI, USA, 2011.
26. Whitehouse, R.J.S.; Harris, J.M.; Sutherland, J.; Rees, J. The nature of scour development and scour protection at offshore windfarm foundations. *Mar. Pollut. Bull.* **2011**, *62*, 73–88. [CrossRef]
27. Ahmed, S.S.; Hawlader, B. Numerical analysis of large-diameter monopiles in dense sand supporting offshore wind turbines. *Int. J. Geomech.* **2016**, *16*, 04016018. [CrossRef]
28. Qi, W.; Tian, J.; Zheng, H.; Wang, H.; Yang, J.; He, G.; Gao, F. Bearing capacity of the high-rise pile cap foundation for offshore wind turbines. *Sustain. Dev. Crit. Infrastruct.* **2014**, 413–420.
29. Wang, H.; Wang, L.; Hong, Y.; Askarinejad, A.; He, B.; Pan, H. Influence of pile diameter and aspect ratio on the lateral response of monopiles in sand with different relative densities. *J. Mar. Sci. Eng.* **2021**, *9*, 618. [CrossRef]
30. Achmus, M.; Kuo, Y.S.; Abdel Rahman, K. Behavior of monopile foundations under cyclic lateral load. *Comput. Geotech.* **2009**, *36*, 725–735. [CrossRef]
31. Thieken, K.; Achmus, M.; Lemke, K. A new static p-y approach for piles with arbitrary dimensions in sand. *Geotechnik* **2015**, *38*, 267–288. [CrossRef]
32. Roy, K.; Hawlader, B.; Kenny, S.; Moore, I. Finite element modeling of lateral pipeline–soil interactions in dense sand. *Can. Geotech. J.* **2016**, *53*, 490–504. [CrossRef]
33. Abongo, K.; Gu, J.; Pervizpour, M.; Quiel, S.; Pamukcu, S. Parametric lab-scale testing and numerical simulation of finned monopiles under directional loading. *Ocean Eng.* **2022**, *259*, 111871. [CrossRef]
34. Jung, S.; Kim, S.-R.; Patil, A.; Hung, L.C. Effect of monopile foundation modeling on the structural response of a 5-MW offshore wind turbine tower. *Ocean Eng.* **2015**, *109*, 479–488. [CrossRef]
35. Wang, H.; Wang, L.Z.; Hong, Y.; He, B.; Zhu, R.H. Quantifying the influence of pile diameter on the load transfer curves of laterally loaded monopile in sand. *Appl. Ocean Res.* **2020**, *101*, 102196. [CrossRef]
36. Wang, J.; Sun, G.; Chen, G.; Yang, X. Finite element analyses of improved lateral performance of monopile when combined with bucket foundation for offshore wind turbines. *Appl. Ocean Res.* **2021**, *111*, 102647. [CrossRef]
37. Bolton, M. The strength and dilatancy of sands. *Geotechnique* **1986**, *36*, 65–78. [CrossRef]
38. Chakraborty, T.; Salgado, R. Dilatancy and shear strength of sand at low confining pressures. *J. Geotech. Geoenviron. Eng.* **2010**, *136*, 527–532. [CrossRef]
39. White, D.; Cheuk, C.; Bolton, M. The uplift resistance of pipes and plate anchors buried in sand. *Géotechnique* **2008**, *58*, 771–779. [CrossRef]
40. Roy, K.S. *Numerical Modeling of Pipe-Soil and Anchor-Soil Interactions in Dense Sand*; Memorial University of Newfoundland: St. John's, NL, Canada, 2018.
41. Leblanc, C.; Houlsby, G.T.; Byrne, B.W. Response of stiff piles in sand to long-term cyclic lateral loading. *Géotechnique* **2010**, *60*, 79–90. [CrossRef]
42. Wang, X.; Li, J. Parametric study of hybrid monopile foundation for offshore wind turbines in cohesionless soil. *Ocean Eng.* **2020**, *218*, 108172. [CrossRef]
43. API. *Recommended Practice for Planning, Designing and Constructing Fixed Offshore Platforms*, 17th ed.; API Recomm. Pract. 2a (RP 2a); API: Washington, DC, USA, 1987.
44. Klinkvort, R.T. Centrifuge Modelling of Drained Lateral Pile-Soil Response. Ph.D. Thesis, The Technical University of Denmark, DTU Civil Engineering, Lyngby, Denmark, 2012.
45. Klinkvort, R.T.; Hededal, O. Lateral response of monopile supporting an offshore wind turbine. *Proc. Inst. Civ. Eng.-Geotech. Eng.* **2013**, *166*, 147–158. [CrossRef]
46. Latini, C.; Zania, V. Triaxial Tests in Fontainebleau Sand. 2016. Available online: https://orbit.dtu.dk/en/publications/triaxial-tests-in-fontainebleau-sand (accessed on 1 August 2022).
47. Arany, L.; Bhattacharya, S.; Macdonald, J.; Hogan, S.J. Design of monopiles for offshore wind turbines in 10 steps. *Soil Dyn. Earthq. Eng.* **2017**, *92*, 126–152. [CrossRef]
48. DNV. *DNVGL-ST-0126: Support Structures for Wind Turbines*; DNV GL: Oslo, Norway, 2016.
49. Doherty, P.; Gavin, K. Laterally loaded monopile design for offshore wind farms. *Proc. Inst. Civ. Eng. Energy* **2012**, *165*, 7–17. [CrossRef]
50. Hu, Q.; Han, F.; Prezzi, M.; Salgado, R.; Zhao, M. Lateral load response of large-diameter monopiles in sand. *Géotechnique* **2022**, *72*, 1035–1050. [CrossRef]
51. Luo, R.; Yang, M.; Li, W. Numerical study of diameter effect on accumulated deformation of laterally loaded monopiles in sand. *Eur. J. Environ. Civ. Eng.* **2020**, *24*, 2440–2452. [CrossRef]

Journal of
*Marine Science
and Engineering*

MDPI

Article

Experimental Investigation on Behavior of Single-Helix Anchor in Sand Subjected to Uplift Cyclic Loading

Dongxue Hao [1,2,*], Jianyi Che [2], Rong Chen [1,2,*], Xin Zhang [3], Chi Yuan [4] and Xichao Chen [2]

[1] Key Lab of Electric Power Infrastructure Safety Assessment and Disaster Prevention of Jilin Province, Northeast Electric Power University, Jilin 132012, China
[2] School of Civil Engineering and Architecture, Northeast Electric Power University, Jilin 132012, China
[3] Northeast Electric Power Design Institute Co., Ltd. of China Power Engineering Consulting, Changchun 130021, China
[4] College of Architecture and Civil Engineering, Beijing University of Technology, Beijing 110124, China
* Correspondence: 20102291@neepu.edu.cn (D.H.); 20112384@neepu.edu.cn (R.C.); Tel.: +86-432-6480-6481 (D.H.)

Abstract: Helical anchors have been widely used in geotechnical engineering due to their large uplift resistance. However, the current knowledge of the cyclic performance of helical anchors is still insufficient. Consequently, a series of small-scale model tests are carried out in sand to investigate the influences of embedment ratio, sand compactness and the cyclic parameters on the monotonic, cyclic and post-cyclic performance of single-helix anchors. The tests results indicate that the single-helix anchors with optimal embedment ratio still exhibit a relatively high uplift capacity after suffering cyclic load. The cyclic frequency has the greatest influence on the accumulated displacement, and the influence of amplitude is relatively greater than that of the mean cyclic load. The anchors in dense sand exhibit better performance to resist pullout than those in medium–dense sand under the same cyclic parameter ratios. Moreover, the correlation of post-cyclic uplift capacity and displacement after cyclic loading as well as the possible influence of the upward displacement on the sand flow above the helix are discussed.

Keywords: helical anchor; sand compactness; embedment ratio; cyclic uplift response; post-cyclic monotonic uplift capacity

Citation: Hao, D.; Che, J.; Chen, R.; Zhang, X.; Yuan, C.; Chen, X. Experimental Investigation on Behavior of Single-Helix Anchor in Sand Subjected to Uplift Cyclic Loading. *J. Mar. Sci. Eng.* **2022**, *10*, 1338. https://doi.org/10.3390/jmse10101338

Academic Editor: Erkan Oterkus

Received: 15 August 2022
Accepted: 18 September 2022
Published: 21 September 2022

1. Introduction

Helical anchors or helical piles are widely used as foundations for various structures, such as transmission towers, onshore wind turbine foundations and floating offshore installations, to resist tension forces and vertical cyclic loads from wind, waves or current loads. They can provide large uplift capacity and good cyclic performance due to the anchor effect of the helix [1–5]. In recent years, this type of foundation has been suggested as a potential alternative to driven piles in offshore renewable energy structures due to the rapid installation, lesser disturbance, low noise and convenience for recycling [6,7].

The environmental cyclic loads are predominant in many applications of helical anchors, which need to be properly considered during design. The cyclic and post-cyclic responses of helical anchors with different geometries and the variations of loads the helix and shaft can resist during cyclic loading under different cyclic loading parameters and different loading sequences with different amplitudes in different soils have been investigated by 1 g model tests, centrifuge and field tests [8–20].

Clemence and Smithling [8] carried out the axial cyclic loading model test of single helical anchor in medium–dense sand to study the effects of displacement amplitude and pre-stressed load on the anchor response. The results show that the application of prestressed load can prolong the fatigue life of the anchor and the post-cyclic capacity decreases. Cerato and Buhler [9,10] and Buhler and Cerato [11] conducted field tests of multi-helical anchors in layered soil to investigate the influence of the number of anchor plates, dynamic load application sequence, load characteristics and groundwater level fluctuations on the uplift behavior of helical anchors under long-term wind load, and validated the existing uplift bearing capacity prediction methods. They found that the triple-helical anchor has the best cyclic performance under long-term dynamic loading, while cyclic loads at 25–40% of the static uplift capacity may visibly increase the post-dynamic uplift capacity and minimize the long-term creep, while the amplitude of cyclic load has a greater influence on the dynamic response of the helical anchor than the maximum load. Sharnouby and Naggar [12,13] conducted field tests on steel fiber–reinforced helical pulldown micropiles under axial compressive cyclic loading to study the cyclic bearing characteristics and load transfer mechanism of piles under different loading sequences. Newgard et al. [14] performed static and cyclic loading tests on a helical anchor model at shallow embedment in saturated medium–dense sand and observed the rapid increase in the rate of accumulation of displacements when the anchor reaches the displacement near the peak static load and the degradation of post-cyclic capacity. Wada et al. [15] investigated the bearing and pullout capacities of steel piles with continuous helix wings under two-way stepwise cyclic loading with an increment of 1/6 of ultimate static capacity for every three cycles by laboratory and field tests. It was found that the bearing and pullout capacities of continuous helix piles under cyclic reversal loading decreased to approximately 60–80% of those under monotonic loading, and the decrease in resistance was mainly due to the reduction in shaft friction. Schiavon et al. [16,17] carried out centrifugal tests for cyclic and post-cyclic monotonic loading of single-helix anchors in very dense sand. The results indicated that a rapid degradation of shaft resistance occurred during cyclic loading, no or slight reduction of post-cyclic uplift capacity occurred for stable anchors, cumulative permanent displacements developed rapidly in the approximately first 100 cycles and previous large cyclic amplitude improved the anchor cyclic performance. Schiavon et al. [18] also performed field tests for single-helix anchors in residual soil of sandstone. The results of cyclic loading tests show no significant degradation of helix bearing resistance and reduced displacement accumulation with increasing load cycles during the first stage cyclic loading. Thorel et al. [19] investigated the effect of the installation rotation rate on the tension resistance and the behavior of helical pile under cyclic loading based on centrifuge models. Hao et al. [20] investigated the influence of embedment ratio of single-helix anchor and number of helices on the cyclic uplift capacity in dense sand by centrifugal stepwise cyclic uplift tests. They found that the ultimate cyclic loading level increases gradually with embedment ratio to the maximum value at the embedment ratio of 6, and then keeps almost constant for greater embedment ratios. The uplift displacements at the beginning of ultimate cyclic loading level are very close to the failure displacements under monotonic loading for single-helix anchors. In addition, the double-helix anchor accumulates less uplift displacement than the single-helix anchor during cyclic loading.

Although these previous studies provided important information, the current knowledge of the cyclic performance of helical piles is still insufficient to enable the development of an appropriate design procedure [16]. Additionally, the influences of sand compactness and frequency on the performance of helical anchors under cyclic loading, as well as the possible change rule of sand compactness with anchor displacement after cyclic loading are rarely reported in the previous literature. Therefore, more investigations on the cyclic behavior of helical piles are necessary to confirm the findings of the previous studies, and to provide new results for onshore and offshore applications [1].

This paper carried out monotonic and cyclic loading model tests of single-helix anchors in medium–dense sand and dense sand to investigate the development of accumulated displacement, axial cyclic stiffness and post-cyclic uplift capacity. It focused on the comprehensive effects of embedment ratio on accumulated displacement, static and post-cyclic ultimate uplift capacity, and the differences in the development of accumulated displacement, axial stiffness and the post-cyclic uplift capacity ratio in different sand compactness under various cyclic loading conditions.

2. Materials and Methods

2.1. Sand Preparation and Installation of Anchors

The silica sand sample was prepared by the air pluviation method in a rectangular strongbox with internal dimensions of 1000 × 500 × 1000 mm (length × width × depth). The particle size of the sand is in between 0.1 and 1 mm, and the particle size in the range of 0.25–0.5 mm accounts for 89.5% of the total mass. The properties of the silica sand are given in Table 1.

Table 1. Properties of silica sand.

Property	Value
Specific gravity, G_s	2.63
Average grain size d_{50}: mm	0.35
Coefficient of uniformity, C_u	1.57
Curvature coefficient, C_c	0.96
Maximum dry density, ρ_{dmax} (g/cm^3)	1.75
Minimum dry density, ρ_{dmin} (g/cm^3)	1.55
Critical friction angle, ϕ_{cv} (°)	28

Dry sand samples with relative density D_r = 62% (medium–dense sand) and D_r = 92% (dense sand) were obtained, respectively, by controlling the height, travel speed and opening width of the sand rain. The helical anchors are made of stainless steel. The thickness of the anchor plate is 1 mm and the diameter of the anchor plate D is 50 mm. The pitch and rod diameter are determined by referring to the size proportion of the helical anchor RS series products of A.B Chance [21]. The ratio of rod diameter to plate diameter d/D = 0.25, and the ratio of pitch to plate diameter t/D = 0.3. The helical plate and each rod part can be connected by screws and the base of the anchor is conical. The single-helix anchor models were installed by manually turning the handle at the speed of 20 rpm (80 mm/min). Anchor models and torsion mounting bracket are shown in Figure 1. Two anchors are installed in one strongbox. The distance between the two anchors is 8D and the minimum distance between the anchor and the wall of the box is 5D, which meets the requirements of boundary conditions [22]. Figure 2 shows the process of sand preparation and installation of anchors, as well as the connection with MTS actuator.

Figure 1. Model of helical anchor and torsion mounting frame.

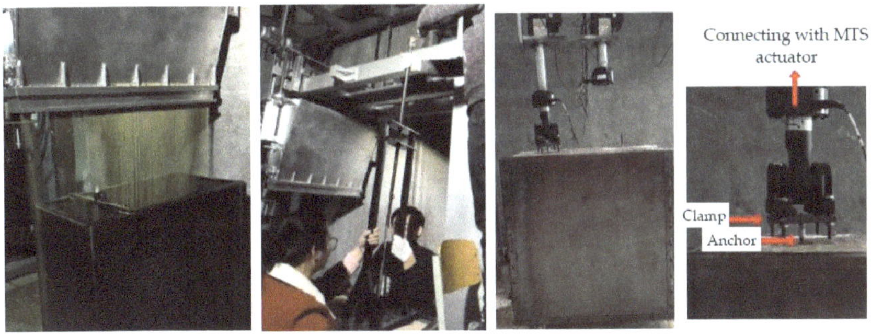

Figure 2. Sand preparation, installation and connection with actuator of anchors.

2.2. Test Program and Loading Mode

In order to analyze the influence of anchor depth, cyclic loading parameters and sand compactness on the cyclic performance of helical anchors, four groups of comparative tests are designed, as shown in Table 2. The cyclic loading pattern is shown in Figure 3. The anchors were pulled monotonically to the mean cyclic load Q_{mean} first, and then vibrated for 1200 cycles in a one-way sinusoidal cyclic manner. The mean cyclic load Q_{mean} and amplitude Q_{cyc} are both set according to the ratio of static ultimate uplift capacity Q_t that is obtained from the monotonic pull-out test of another anchor in parallel with the cyclic loading test. The maximum mean cyclic load Q_{mean} is taken as $0.5Q_t$ considering the design safety factor of 2 for static loading, while the amplitude Q_{cyc} varies from 0.1–0.3 Q_t. Cyclic frequency f is set 0.5 Hz, and when comparing the effect of frequency, f is set 1 Hz and 2 Hz. The anchors that are still stable after vibration are subjected to monotonic pull-out load to determine the post-cyclic uplift capacity Q_{pt}. The testing program encompasses 28 monotonic and cyclic tests across 15 sand samples. The first letter of the test name represents the loading mode, i.e., M represents monotonic loading and C stands for cyclic loading; the second letter represents the compactness of sand samples, i.e., D for dense sand and M for medium–dense sand; the third letter H and the number close to H represents the embedment ratio, and the number after "-" is the number of parallel tests; the figures in brackets are the relative mean cyclic load Q_{mean}/Q_t and the relative amplitude Q_{cyc}/Q_t in turn. For the test on the effect of frequency, the frequency is listed last, and the ones unlisted are 0.5 Hz.

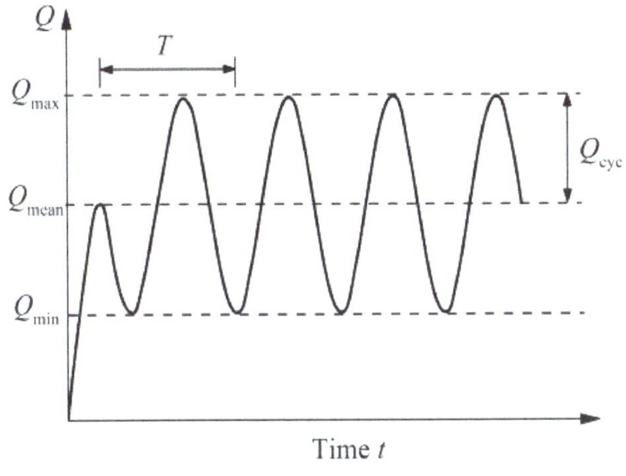

Figure 3. Patten of cyclic loading.

Table 2. Test program.

Name	Sand Sample	Influence Factor	Frequency f/Hz	H/D	$\frac{Q_{mean}}{Q_t}$	$\frac{Q_{cyc}}{Q_t}$	Number of Cycles	Q_t/N	Uplift Capacity Ratio, β
MDH8	1	depth	—	8	—	—	—	316.3	—
CDH8 (0.4, 0.2)	1		0.5	8	0.4	0.2	1200	318.5	1.007
MDH10	2		—	10	—	—	—	580.1	—
CDH10 (0.4, 0.2)	2		0.5	10	0.4	0.2	1200	439.6	0.758
MDH12-1	3		—	12	—	—	—	863.3	—
CDH12 (0.4, 0.2, 0.5 Hz)	3		0.5	12	0.4	0.2	1200	850.6	0.985
MDH14	4		—	14	—	—	—	890.3	—
CDH14 (0.4, 0.2)	4		0.5	14	0.4	0.2	1200	733.9	0.824
MDH12-2	5	cyclic amplitude and mean load	—	12	—	—	—	822.3	—
CDH12 (0.4, 0.3)	5		0.5	12	0.4	0.3	14	—	—
MDH12-3	6		—	12	—	—	—	802.2	—
CDH12 (0.3, 0.3)	6		0.5	12	0.3	0.3	1200	876.7	1.093
CDH12 (0.4, 0.1)	7		0.5	12	0.4	0.1	1200	679.7	0.847
CDH12 (0.3, 0.2)	7		0.5	12	0.3	0.2	1200	780.5	0.973
CDH12 (0.2, 0.2, 0.5 Hz)	8		0.5	12	0.2	0.2	1200	738.3	0.920
CDH12 (0.5, 0.1)	8		0.5	12	0.5	0.1	1200	805.8	1.004
MDH12-4	9	frequency	—	12	—	—	—	813.2	—
CDH12 (0.4, 0.2, 1 Hz)	9		1	12	0.4	0.2	12	—	—
MDH12-5	10		-	12	-	-	-	803.8	—
CDH12 (0.2, 0.2, 2 Hz)	10		2	12	0.2	0.2	526	808.5	1.006
MMH12-1	11	compactness	—	12	—	—	—	658.8	—
CMH12 (0.3, 0.1)	11		0.5	12	0.3	0.1	1200	570.7	0.866
MMH12-2	12		—	12	—	—	—	638.4	
CMH12 (0.3, 0.2)	13		0.5	12	0.3	0.2	1200	655.0	1.026
MMH12-3	14		—	12	—	—	—	630.4	
CMH12 (0.3, 0.3)	14		0.5	12	0.3	0.3	1200	521.3	0.827
CMH12 (0.2, 0.2)	15		0.5	12	0.2	0.2	1200	620.5	0.984
CMH12 (0.4, 0.2)	15		0.5	12	0.4	0.2	1200	623.6	0.989

The monotonic and cyclic loading of the helical anchor is realized by MTS hydraulic actuator. The anchor head is rigidly connected with the actuator, as shown in Figure 2. Monotonic loading is controlled by displacement at the rate of 0.125 mm/s, and the pullout displacement is $1D$. Cyclic loading is applied according to the set load parameters. After cyclic loading, displacement control is used for monotonic pullout of the helical anchors. The data acquisition frequency is 12.8 Hz.

3. Results

3.1. Test Results in Dense Sand

3.1.1. Results of Anchors with Different Embedment Ratios

- Ultimate uplift capacity of monotonic loading

In order to analyze the influence of the embedment ratio on the monotonic and cyclic uplift behavior of helical anchors in dense sand, the tests of anchors with embedment ratio H/D varying from 8 to 14 were carried out with a fixed vibration frequency of f = 0.5 Hz, a mean cyclic load ratio Q_{mean}/Q_t = 0.4 and amplitude ratio Q_{cyc}/Q_t = 0.2.

Figure 4 shows the ultimate uplift capacity Q_t and the breakout factor N_γ of helical anchors with various embedment ratios under monotonic loading, where $N_\gamma = Q_t/\gamma AH$, among which Q_t is the peak values before the load displacement curve exhibits an obvious oscillation, as listed in Table 2; γ is the unit weight of soil mass; A is the cross-sectional area of anchor plate and H is the embedment depth of anchor plate. It can be seen that the static ultimate uplift capacity Q_t increases with the increase in embedment ratio H/D, and when $H/D \geq 12$, the growth rate of Q_t slows down. The N_γ reaches the maximum at

$H/D = 12$, beyond which the anchor can be considered as deeply buried [22,23]. The smaller the compactness of the sand sample, the smaller the critical embedment ratio corresponding to deep anchor [23,24]. Therefore, the embedment ratio of 12 can ensure that the anchor is deeply buried for medium–dense sand.

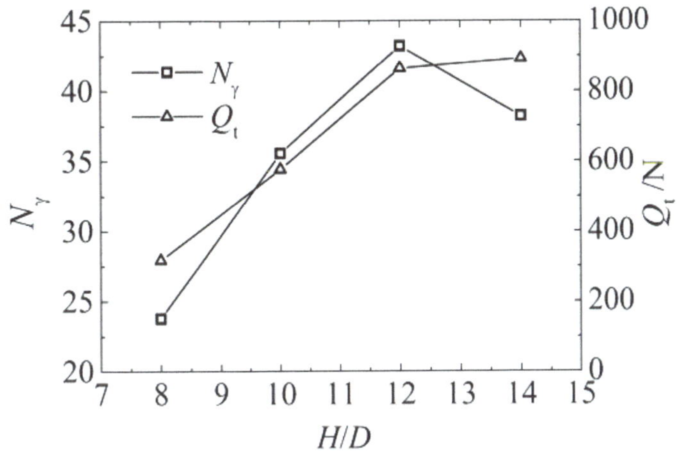

Figure 4. N_γ and Q_t of helical anchors with different embedment ratios in dense sand.

Figure 5 shows the standardized load and displacement curve u/D-Q/Q_t under cyclic loading with different embedment ratios. The results of the first six cycles are shown in Figure 5b to demonstrate the loading process. There is an error between the actually applied load and the target load during the first three cycles. The applied load is slightly smaller than the target load when the embedment ratio H/D equals 8 due to the relative small cyclic load and the loading accuracy of the actuator.

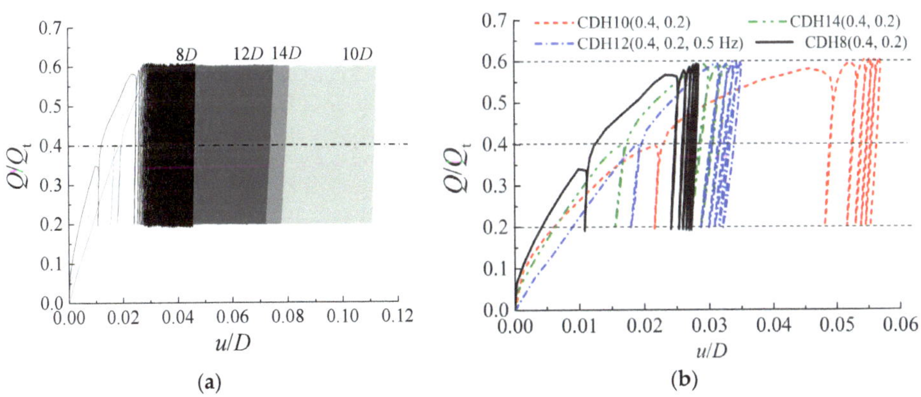

Figure 5. Standardized load–displacement curves at different embedment ratios (**a**) $N = 1200$; (**b**) $N = 6$.

Figure 6 shows the relationship between the number of cycles and standardized displacement, including the initial monotonic uplift displacement u_0 at $Q = 0.4Q_t$. The values of u_0 of the anchors with the embedment ratio $H/D = 8$, 10, 12 and 14 are $0.01D$, $0.022D$, $0.02D$ and $0.016D$, respectively. The value of u_0 for anchor CDH8 (0.4, 0.2) is smaller than that of the other cases, which may be related to the fact that the initial monotonic loading $Q = 0.35Q_t$ does not reach the standard of $0.4Q_t$. Additionally, the value of u_0 for anchor CDH10 (0.4, 0.2) is the largest, which may be due to the initial difference of sand samples or installation disturbance. The standardized pullout displacements u/D for the anchors with various embedment ratios after cyclic loading are less than 0.1, with the exception of CDH10 (0.4, 0.2).

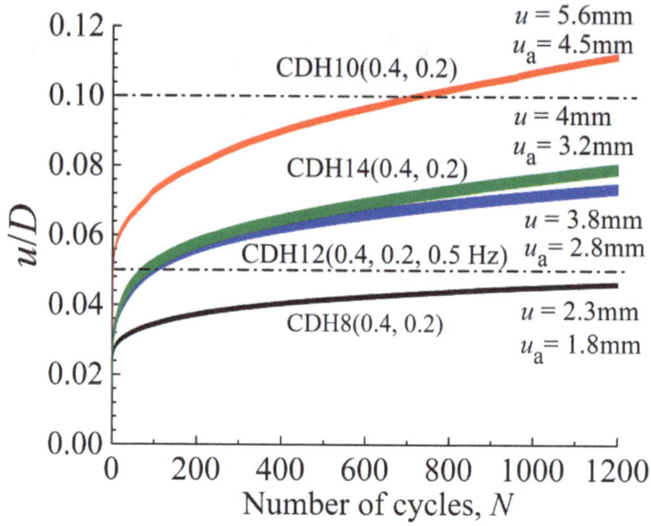

Figure 6. Developments of uplift displacement at different embedment ratios.

- Accumulated displacement

The developments of accumulated displacement u_a and displacement accumulation rate with number of cycles N for the anchors with different embedment ratios are plotted in Figure 7. The displacement accumulation rate is the accumulated displacement in each cycle. It can be seen from Figure 7a that u_a increases nonlinearly when $N < 200$ and then develops in a linear and slow way. The accumulated displacement of anchor CDH8 (0.4, 0.2) is the smallest, which may be related to the relatively small amplitude of the anchor and the first few cyclic loads being less than the target value compared to other anchors. Anchor CDH12 (0.4, 0.2, 0.5 Hz) shows better cyclic performance than anchor CDH10 (0.4, 0.2) and CDH14 (0.4, 0.2). Figure 7b displays the semi-log relationships of u_a and N, which are linear before exhibiting hollow circles and then develops nonlinearly in different extents. The values of u_a at the hollow circles are close to 0.02D, except for CDH10 (0.4, 0.2), and the numbers of cycles corresponding to hollow circles are less than 50 for all anchors with different embedment ratios. The values of accumulated displacement at $N = 50$ are all more than half of the total accumulated displacement, which indicates that the accumulated displacement develops significantly before $N = 50$.

It can be seen from Figure 7c that the accumulated displacement in the first cycle for the anchor CDH10 (0.4, 0.2) is 1.9–2.6 times that of other anchors with different embedment ratios, which may be due to the effect of installation disturbance. The initial displacement accumulation rate is the largest, and then decreases rapidly within three cycles, until reaching a small and stable value after 10 cycles for the anchors with different embedment ratios. The displacement accumulation rate decreases from 0.91–1.82 mm/10 cycles at the first 10 cycles to 0.14–0.3 mm/100 cycles after 100 cycles.

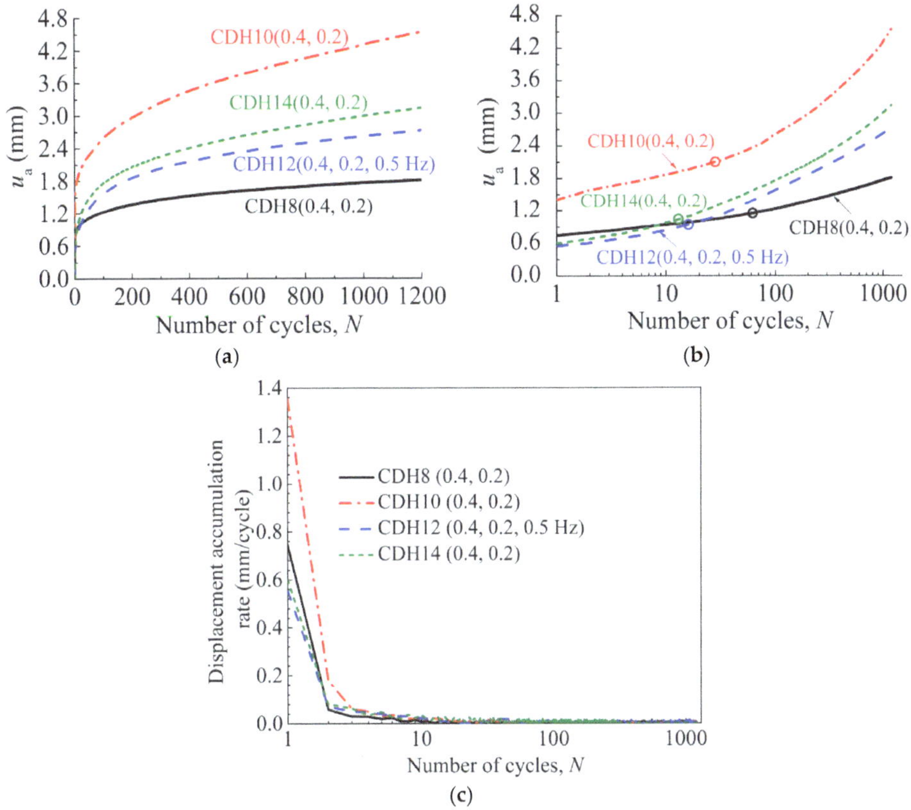

Figure 7. Developments of accumulated displacement and displacement accumulation rate at different embedment ratios: (**a**) linear scale of u_a-N; (**b**) semi-log scale of u_a-N; (**c**) semi-log scale of displacement accumulation rate and cycle number.

- Axial stiffness

The axial stiffness k of the helical anchor is defined by the ratio of the load increment in each loading cycle to its deformation increment, as shown in Figure 8a. The relationships of axial stiffness k and normalized axial stiffness k_N/k_1 with number of cycles N for different embedment ratios are plotted in the semi logarithmic coordinate system, as shown in Figure 8b,c, where k_N and k_1 are the axial stiffness of the Nth cycle and the first cycle, respectively.

It can be seen from the figure that the axial stiffness k increases rapidly and nonlinearly within 10 cycles for different embedment ratios. The developments of axial stiffness for anchor CDH8 (0.4, 0.2) and CDH10 (0.4, 0.2) are generally stable after 10 cycles, although there are slight fluctuations compared to the other two anchors. The stiffness of the first two cycles is relatively low for CDH10 (0.4, 0.2) due to the sand loosening, which is caused by installation disturbance, and then the stiffness increases rapidly when the vibration densifies the sand above the anchor. The stable values of axial stiffness for anchor CDH8 (0.4, 0.2), CDH10 (0.4, 0.2), CDH12 (0.4, 0.2, 0.5 Hz) and CDH14 (0.4, 0.2) are 10.5 times, 13.9 times, 4.4 times and 4.3 times of the initial axial stiffness, respectively.

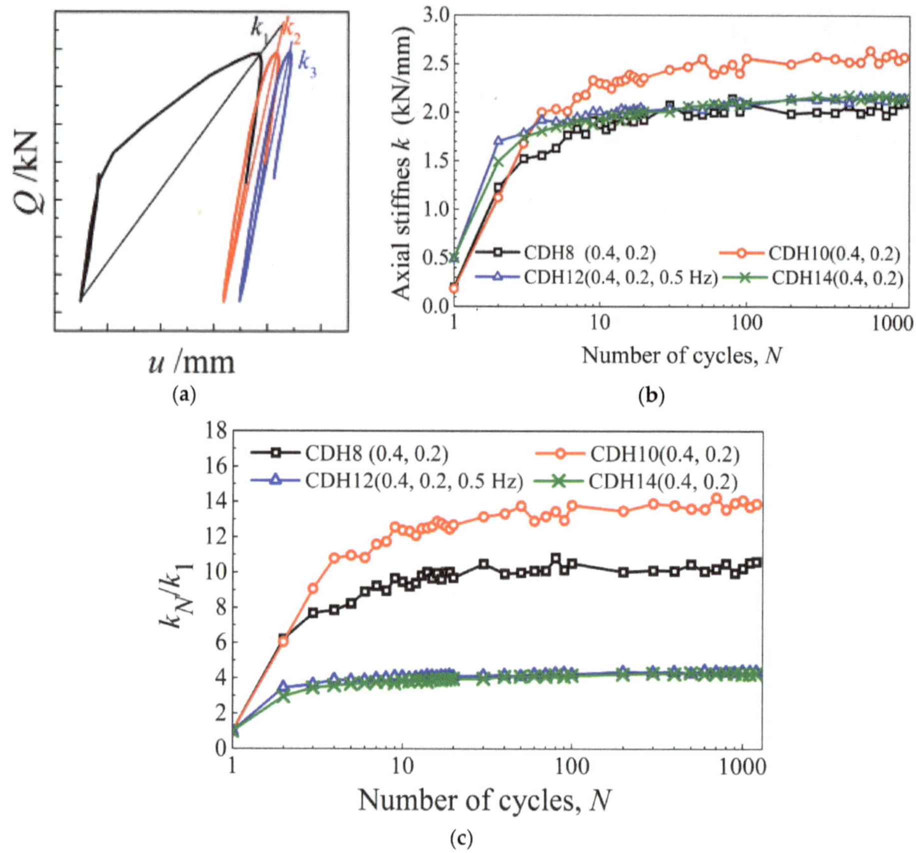

Figure 8. Axial stiffness for different embedment ratios: (**a**) definition of axial stiffness; (**b**) relationship of axial stiffness k and number of cycles N; (**c**) relationship of normalized axial stiffness k_N/k_1 and number of cycles N.

- Post-cyclic monotonic response and uplift capacity

The monotonic pull-out tests were carried out after cyclic loading to obtain the load displacement responses, which were compared with the results of monotonic tests without cyclic loading, as shown in Figure 9, where the peak points are marked in the grey circles.

It can be seen from the figure that the monotonic responses after cyclic loading for the helical anchors with different embedment ratios are more rigid than that without cyclic loading. When monotonic pullout is carried out directly, the peak points of each anchor appear after the uplift displacement of 0.1D, while the peak points after cyclic loading are reached before 0.1D. Schiavon et al. [17] also found the stiffer initial response in post-cyclic monotonic uplift test by centrifuge and the peak points of direct monotonic and post-cyclic monotonic pullout also occur after and before uplift displacement of 0.1D, respectively.

The uplift capacity ratio after cyclic loading β is defined as the ratio of ultimate uplift capacity after cyclic loading Q_{pt} and static ultimate uplift capacity Q_t. The values of β for anchor CDH8 (0.4, 0.2), CDH10 (0.4, 0.2), CDH12 (0.4, 0.2, 0.5 Hz) and CDH14 (0.4, 0.2) are 1.007, 0.756, 0.985 and 0.824, respectively, which indicates that the ultimate uplift capacities of helical anchors with different embedment ratios in dense sand after cyclic loading with cyclic parameters (0.4, 0.2, 0.5 Hz) are not improved, and in fact some of them significantly reduced. The influence of the embedment ratio on the post-cyclic uplift capacity of anchors is irregular.

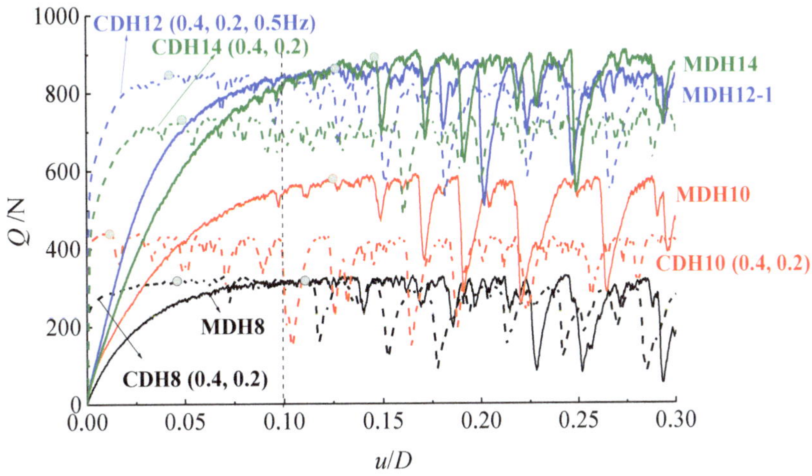

Figure 9. Post-cyclic monotonic and monotonic responses for different embedment ratios.

The single-helix anchor test with the embedment ratio of 12 exhibits the good static and cyclic behavior based on post-cyclic uplift capacity and the development of cyclic accumulated displacement and axial stiffness. Therefore, the helical anchor with $H/D = 12$ is selected for the later tests.

3.1.2. Results for Different Amplitudes and Mean Cyclic Loads

- Accumulated displacement

Figure 10 shows the accumulated displacement developments of the anchors under different amplitudes ($f = 0.5$ Hz and $Q_{mean}/Q_t = 0.4$ and 0.3). It is evident that the greater the amplitude, the greater the accumulated displacement u_a when other cyclic parameters are the same, which is similar with the observation of Hanna et al. [25] and Petereit [26] for plate anchors and Schiavon et al. [16,17] for helical anchors. The u_a increases non-linearly and rapidly for N less than 200 and then develops slowly under the situation of $Q_{max} = Q_{mean} + Q_{cyc} < 0.7Q_t$, while u_a develops very fast with cycle number and has reached $0.8D$ after 14 cycles when Q_{mean}/Q_t equals 0.4 and Q_{cyc}/Q_t equals 0.3 ($Q_{max} = 0.7Q_t$), which is likely to represent a continuously pulled out case for which the test is interrupted.

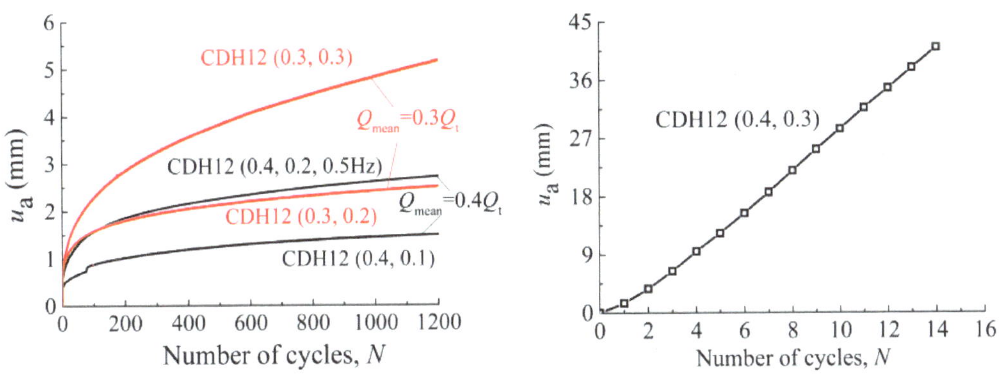

Figure 10. Developments of accumulated displacement under different amplitudes.

The curves of accumulated displacement u_a with cycle number N for the anchors subjected to different mean cyclic load ratios Q_{mean}/Q_t and the same amplitude ratios Q_{cyc}/Q_t of 0.1 or 0.2 are plotted in Figure 11. When Q_{cyc}/Q_t equals 0.2 and Q_{mean}/Q_t varies from 0.2 to 0.4, the rule according to which the greater the mean cyclic load, the larger accumulated displacement is observed. While when Q_{cyc}/Q_t equals 0.1, the accumulated displacement of anchor CDH12 (0.5, 0.1) subjected to a mean cyclic load ratio of 0.5 is obviously smaller than that of anchor CDH12 (0.4, 0.1) with a mean cyclic load ratio of 0.4, which is similar to the results of the sequence cyclic loading tests for single-helix anchors from Schiavon et al. [17] where a subsequent low-level cyclic loading produces very low permanent displacements due to the anchor performance being improved by previous large amplitude cyclic-loading. The observation indicates that the influence of the mean cyclic load on the accumulated displacement may be affected by the amplitude. When the amplitude is medium the accumulated displacement of anchors subjected to a greater mean cyclic load will be larger. Additionally, when the amplitude is small, the influence of the mean cyclic load on the accumulated displacement shows the opposite trend. This phenomenon may be due to the fact that the sand densification above the anchor caused by preloading will remain dense during small amplitude cyclic loading, and the sand densification by preloading cannot be sustained and eventually becomes loose due to the backflow of sand above the anchor under the larger maximum cyclic load (corresponding to medium amplitude). The relationship of the accumulated displacement between anchor CDH12 (0.3, 0.2) > CDH12 (0.4, 0.1) and anchor CDH12 (0.3, 0.3) > CDH12 (0.4, 0.2, 0.5 Hz) > CDH12 (0.5, 0.1) can be observed by comparing the accumulated displacement of the anchors with the same maximum cyclic load ratio Q_{max}/Q_t. This indicates that for the anchors under the same maximum cyclic load ratio Q_{max}/Q_t, the greater accumulated displacement will be obtained when the anchors are subjected to the larger value of Q_{cyc}/Q_t and the smaller value of Q_{mean}/Q_t. The same conclusion for single-helix anchors in very dense sand is obtained by Schiavon et al. [17].

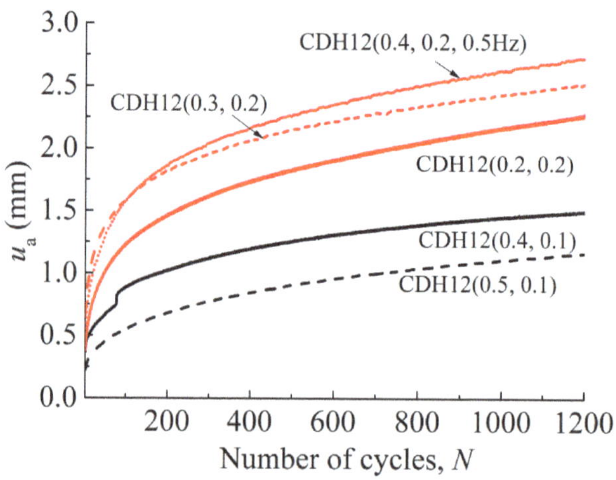

Figure 11. Developments of accumulated displacement under different mean cyclic loads.

The developments of displacement accumulation rate of the anchors under different amplitudes and mean cyclic loads are shown in Figure 12. The rules of development for all the anchors are similar. The displacement accumulation rate decreases rapidly within three cycles, and then decreases slowly when the cycle number is more than 3 and less than 10 However, there is still an oscillated displacement accumulation rate after 100 cycles, which decreases to 0.1–0.23 mm per 100 cycles.

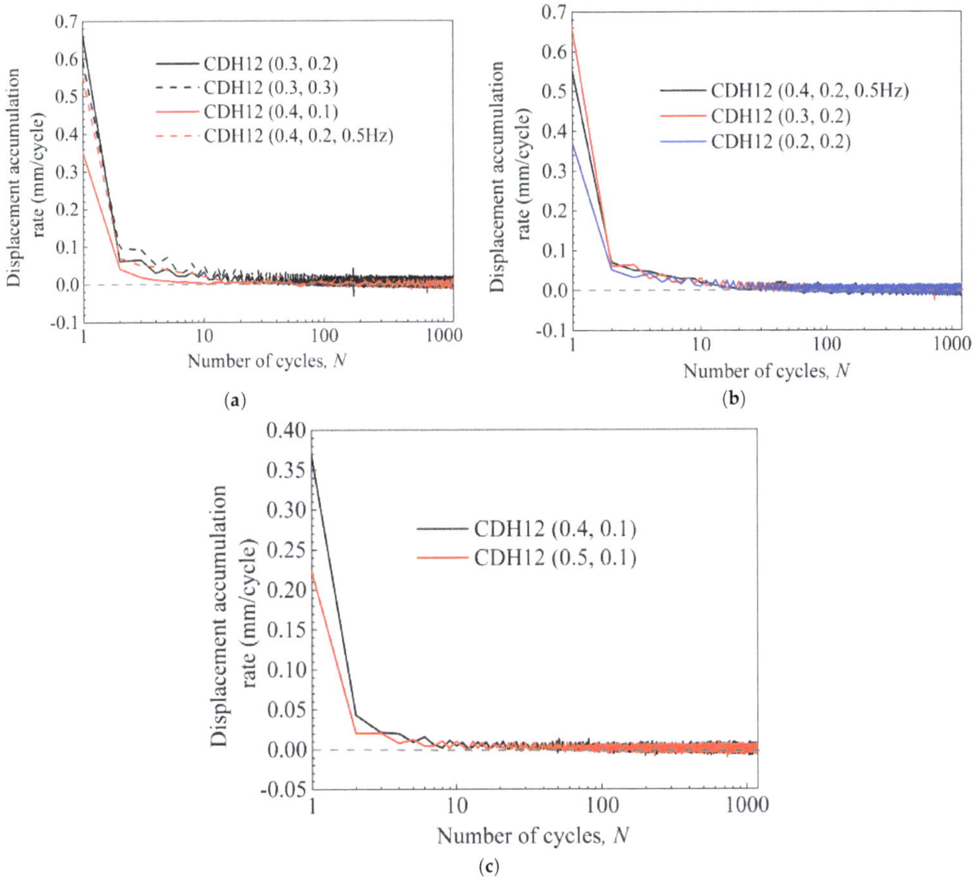

Figure 12. Developments of displacement accumulation rate under different amplitudes and mean loads: (**a**) the influence of amplitude; (**b**) the influence of mean cyclic load Q_{cyc}/Q_t = 0.2; (**c**) the influence of mean cyclic load Q_{cyc}/Q_t = 0.1.

The anchor under the greater amplitude has the larger displacement accumulation rate, and the phenomenon is more obvious for larger Q_{mean}. The displacement accumulation rate of anchors under small mean cyclic loads is relatively small in the cases of the same medium amplitude, and the displacement accumulation rate of the anchor under a smaller mean cyclic load is larger in the cases of the same small amplitude.

- Axial stiffness

The relationships between axial stiffness and the number of cycles for the anchors in dense sand subjected to different Q_{mean}/Q_t and Q_{cyc}/Q_t are shown in Figure 13, which shows similar behaviors. The axial stiffness increases with the cycle number rapidly and nonlinearly within ten cycles and reaches stability at cycle numbers between 10 and 50. Stable axial stiffness is about three to six times the initial value for all the cases of different Q_{mean}/Q_t and Q_{cyc}/Q_t, except for CDH12 (0.4, 0.3). It is observed that the axial stiffness for anchors under a larger amplitude is relatively small and has no significant correlation with the mean cyclic load.

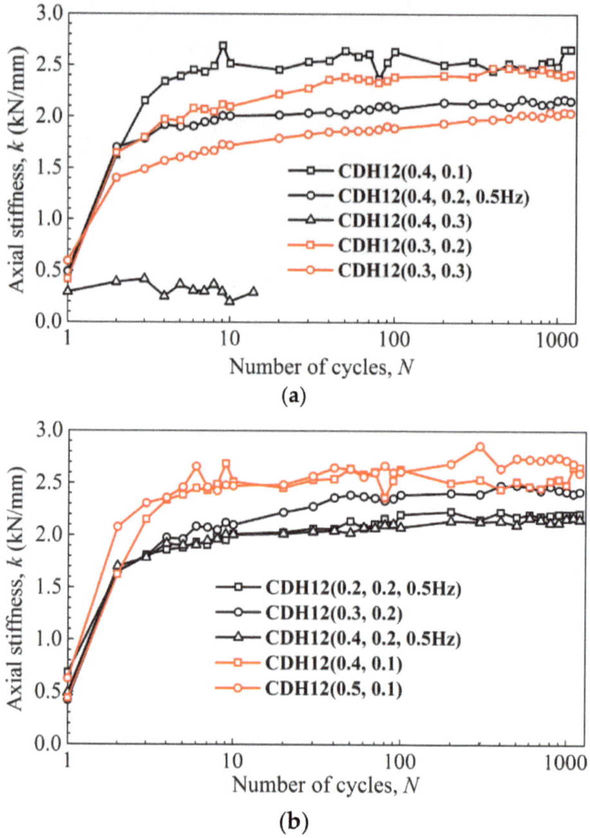

Figure 13. Developments of axial stiffness for different amplitudes and mean cyclic loads: (**a**) different amplitudes; (**b**) different mean cyclic loads.

- Post-cyclic monotonic response and uplift capacity

Figure 14 shows the monotonic responses of anchors without cyclic loading and post-cyclic monotonic responses in dense sand. The uplift capacity ratios after cyclic loading β for anchor CDH12 (0.4, 0.2, 0.5 Hz) and CDH12 (0.4, 0.1) are 0.985 and 0.847, respectively. Additionally, the values of β for anchor CDH12 (0.3, 0.2) and CDH12 (0.3, 0.3) are 0.973 and 1.093, respectively. The anchors subjected to a greater amplitude have a higher post-cyclic uplift capacity when the same mean cyclic loads are applied. The values of β vary from 0.847 to 1.004 when the amplitude ratio Q_{cyc}/Q_t = 0.1 and 0.2, and the mean cyclic load ratio Q_{mean}/Q_t = 0.2–0.5. Among these anchors, CDH12 (0.3, 0.3) has the largest displacement after cyclic loading, CDH12 (0.5, 0.1) has the smallest displacement and both uplift capacity ratios are greater than 1. Additionally, the centrifuge test results from Schiavon et al. [17] show the slight post-cyclic capacity degradation of 1–7% for the cases of Q_{max}/Q_t = 0.42–0.69, while the centrifuge test results from Schiavon et al. [16] show no reduction or a slight increase in post-cyclic capacity for the Q_{max}/Q_t = 0.61–0.93. The ratio of Q_{max}/Q_t used in this study is similar to that in Schiavon et al. [17]; however, both a decrease and increase in the post-cyclic capacity occur, and the variation of post-cyclic capacity in comparison with monotonic value range from −15% to 10%.

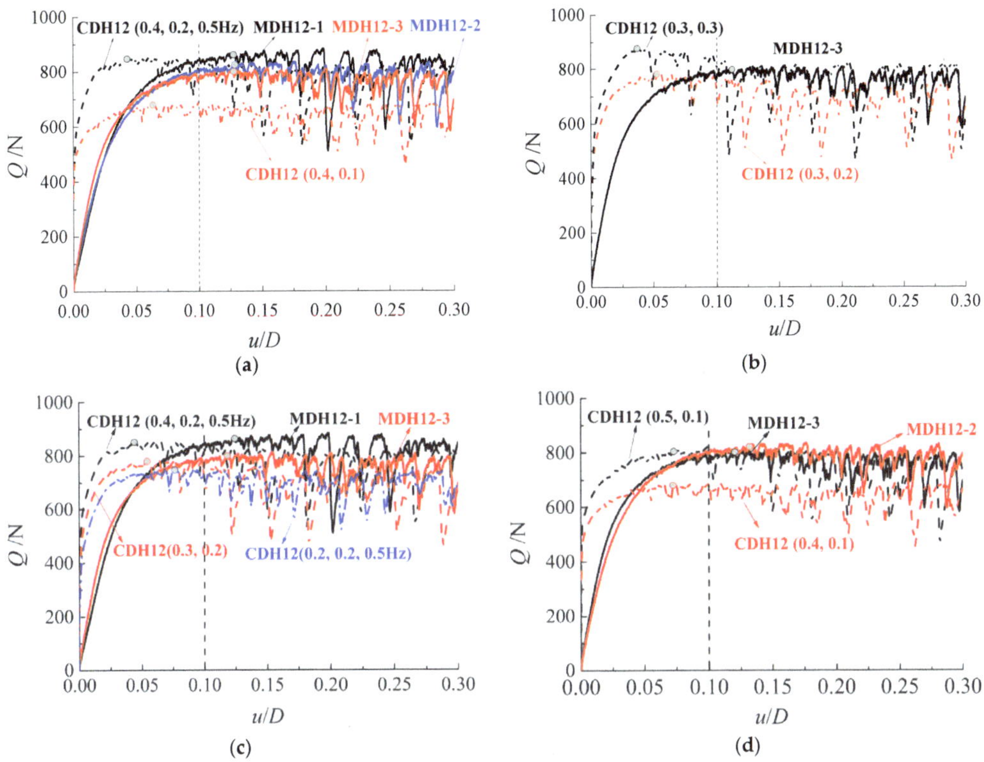

Figure 14. Monotonic and post-cyclic monotonic responses for different amplitudes and mean cyclic loads: (**a**) Q_{mean}/Q_t = 0.4; (**b**) Q_{mean}/Q_t = 0.3; (**c**) Q_{cyc}/Q_t = 0.2; (**d**) Q_{cyc}/Q_t = 0.1.

3.1.3. Results for Different Cyclic Loading Frequencies

The developments of accumulated displacement under different cyclic frequencies are shown in Figure 15. Anchor CDH12 (0.4, 0.2, 1 Hz) under a frequency of 1 Hz is pulled up sharply at the initial stage of cyclic loading, indicating that the vibration frequency is close to the natural frequency of the anchor–soil system, and resonance occurs. The test is stopped when cyclic loading is applied for 12 cycles. Comparing the accumulated displacement development of CDH12 (0.2, 0.2, 0.5 Hz) and CDH12 (0.2, 0.2, 2 Hz), it can be seen that when f = 2 Hz, the anchor displacement increases rapidly and linearly with the cycle number. The test for anchor CDH12 (0.2, 0.2, 2 Hz) is stopped at the 526th cycle, corresponding to the accumulated displacement of 23.7 mm, which is close to 0.5D. Additionally, the accumulated displacement of anchor CDH12 (0.2, 0.2, 0.5 Hz) is of only is 2.6 mm after 1200 cycles. The results show that the influence of frequency on accumulated displacement is much higher than that of embedment ratio, cyclic amplitude and mean load. Therefore, the safety margin should be increased considering the impact of frequency variations.

Figure 16 indicates that the displacement accumulation rate of anchor CDH12 (0.4, 0.2, 1 Hz) rises rapidly after the third cycle, and reaches 3.77 mm/cycle after 11 cycles. The development for anchor CDH12 (0.2, 0.2, 2 Hz) has the same trend as that for CDH12 (0.2, 0.2, 0.5 Hz), but the displacement accumulation rate in the stable stage is more than six times that of anchor CDH12 (0.2, 0.2, 0.5 Hz).

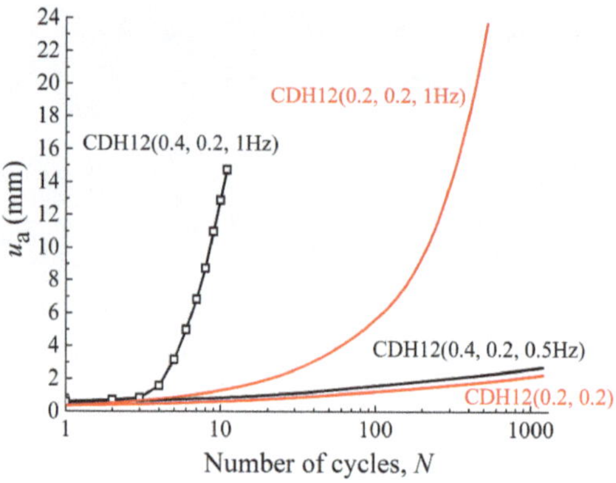

Figure 15. Developments of accumulated displacement under different frequencies.

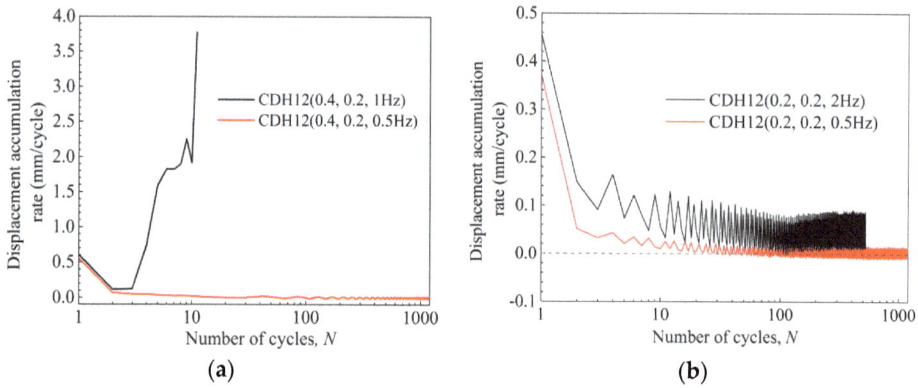

Figure 16. Developments of displacement accumulation rate under different frequencies: (**a**) $Q_{mean}/Q_t = 0.4$ and $Q_{cyc}/Q_t = 0.2$; (**b**) $Q_{mean}/Q_t = 0.2$ and $Q_{cyc}/Q_t = 0.2$.

It can be seen from Figure 17 that the development trend of axial stiffness at two frequencies is similar, and the axial stiffness at a high frequency is lower. The stable axial stiffness at a frequency of 0.5 Hz is approximately three times that at frequency of 2 Hz, and the values of stable axial stiffness are about three times those of the initial stiffness.

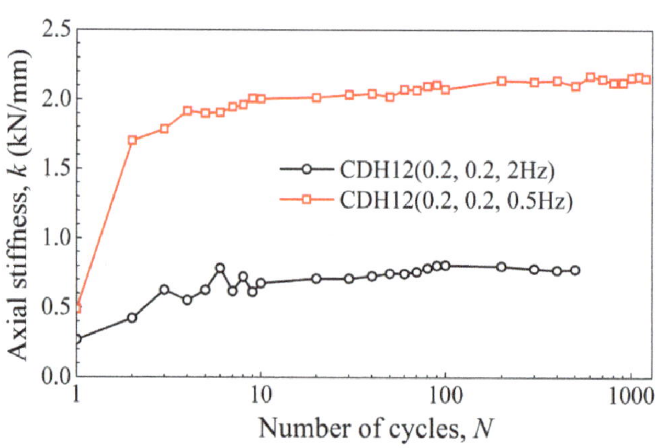

Figure 17. Developments of axial stiffness for different frequencies.

Figure 18 shows the comparisons between the post-cyclic monotonic responses of anchor CDH12 (0.2, 0.2, 0.5 Hz) and CDH12 (0.2, 0.2, 2 Hz) and the monotonic responses without cyclic loading. Their post-cyclic uplift capacity ratios are 0.920 and 1.006, respectively. After cyclic loading, the anchor CDH12 (0.2, 0.2, 2 Hz) produces a large uplift displacement, reaching 0.47*D*. Although there is a certain loss of embedment depth, the combined effect of vibration frequency and pullout displacement finally densifies the soil above the helix of anchor CDH12 (0.2, 0.2, 2 Hz), and the post-cyclic uplift capacity is not reduced compared with MDH12-5.

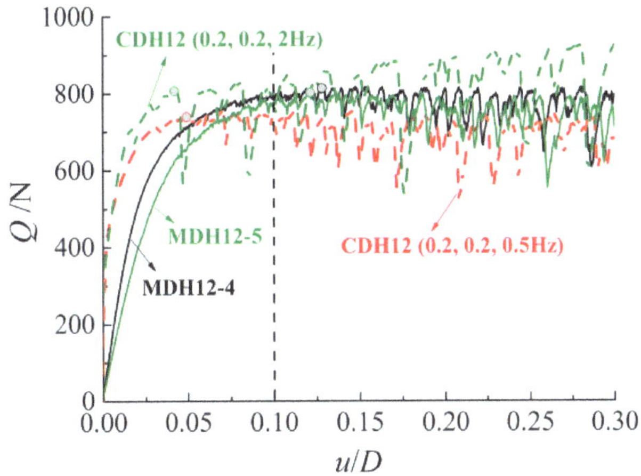

Figure 18. Post-cyclic monotonic and monotonic responses for different frequencies.

3.2. Test Results in Medium–Dense Sand

- Accumulated displacement and axial stiffness

Figure 19 shows the developments of accumulated displacement and displacement rate for the anchors in medium–dense sand under different cyclic parameters. The anchors subjected to a greater amplitude at the same of mean cyclic load produce larger accumulated displacement, which is the same as that in dense sand. The accumulated displacement of anchor CMH12 (0.3, 0.3) is significantly greater than those of anchor CMH12 (0.3, 0.2) and CMH12 (0.3, 0.1) and its pullout displacement after cyclic loading is of 14.1 mm, up to 0.29*D*, which is far more than 0.1*D*.

By comparing the accumulated displacement of anchor CMH12 (0.4, 0.2), CMH12 (0.3, 0.2) and CMH12 (0.2, 0.2) with the same amplitude, it is observed that the anchor CMH12 (0.4, 0.2) subjected to the largest mean cyclic load has the largest accumulation displacement within 150 cycles and the largest displacement accumulation rate for the first two cycles, and the accumulated displacement of anchor CMH12 (0.2, 0.2) exceeds that of anchor CMH12 (0.4, 0.2) after 150 cycles, which may be an anomaly caused by the disturbance of sand above the helix or by a large amount of sand back-flowing around the edge of the helix into the void of the helix bottom after being tightly compacted. The accumulated displacement and displacement accumulation rate of anchor CMH12 (0.3, 0.2) are always at the minimum. Therefore, it is discerned that the anchors in medium–dense sand with a greater mean cyclic load may produce larger accumulated displacement when they are subjected to the same medium amplitude. The comparisons among the accumulated displacement of the anchors with the same maximum cyclic load ratio shows that the anchor subjected to a greater amplitude and a smaller mean cyclic load will produce larger accumulated displacement, which is the same as that observed in dense sand.

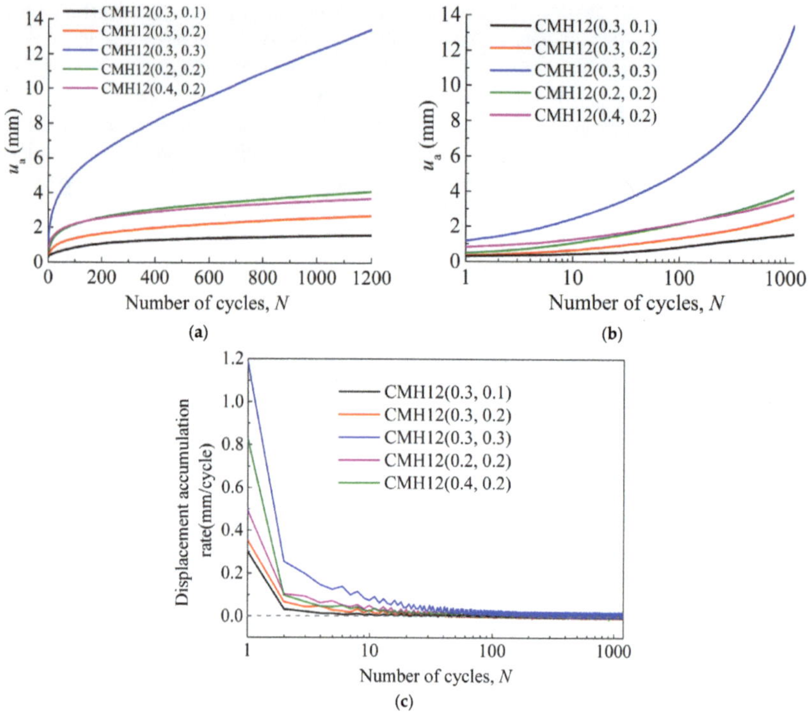

Figure 19. Developments of accumulated displacement and displacement accumulation rates under different cyclic parameters: (**a**) linear scale of u_a-N; (**b**) semi-log scale of u_a-N; (**c**) semi-log scale of displacement accumulation rate-N.

Figure 20 shows the development of axial stiffness k under different amplitudes and mean cyclic loads. It can be seen that when the cycle number is not above 3, the k value of each anchor increases rapidly. The axial stiffness of anchor CMH12 (0.3, 0.1) with Q_{cyc}/Q_t of 0.1 becomes stable after 10 cycles, which is the same as that of dense sand. However, for the anchors with Q_{cyc}/Q_t not less than 0.2, it still rises slowly after 10 cycles and becomes steady after 400 cycles. The stable axial stiffness of the anchors with different amplitudes and mean cyclic loads is 3.5–9 times the initial axial stiffness.

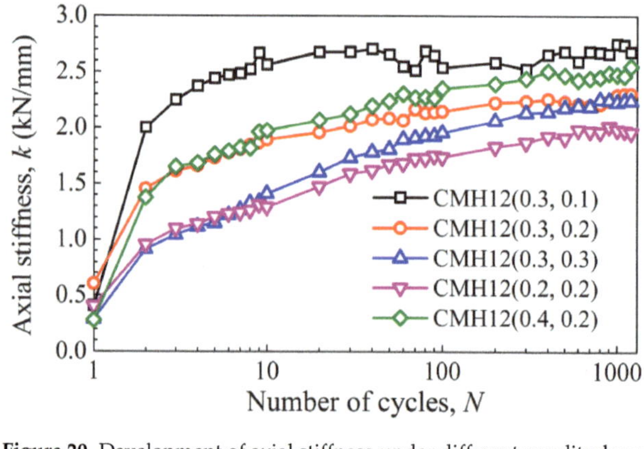

Figure 20. Development of axial stiffness under different amplitudes and mean cyclic loads.

- Post-cyclic monotonic response and uplift capacity

Figure 21 is post-cyclic monotonic and monotonic responses for the anchors in medium–dense sand under different cyclic loading parameters. The post-cyclic monotonic responses for the anchors in medium–dense sand are also stiffer than the monotonic responses without cyclic loading, which is the same as the responses in dense sand. The post-cyclic uplift capacity Q_{pt} of anchor CMH12 (0.3, 0.3) decreases the most, and the ratio β equals 0.827, which may be caused by the flow of sand above the helix to the bottom of the helix loosening the sand above the helix at high amplitudes. The value of β for anchor CMH12 (0.3, 0.1) is 0.866, which is also relatively small compared to other anchors. It indicates that a small amplitude vibration also loosens the sand above the helix. The post-cyclic uplift capacities for the anchors with an amplitude ratio of 0.2 are basically close to the static uplift capacities.

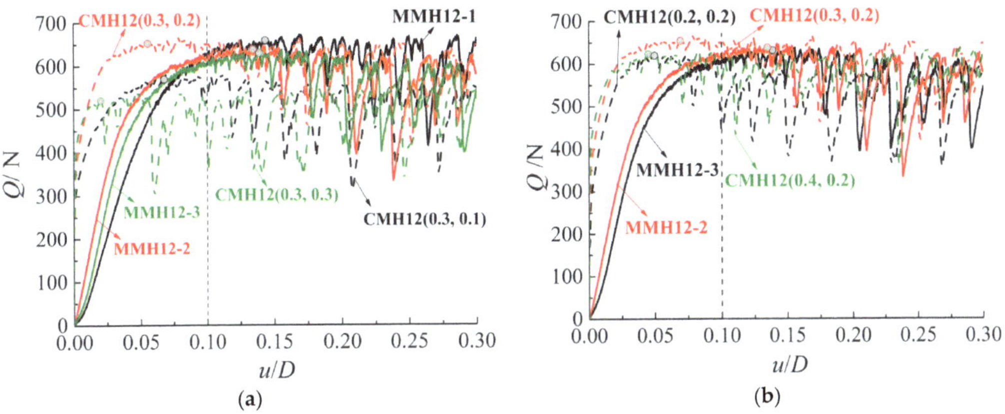

Figure 21. Post-cyclic monotonic and monotonic responses for different cyclic parameters: (**a**) Q_{mean}/Q_t = 0.3; (**b**) Q_{cyc}/Q_t = 0.2.

4. Discussion

The accumulated displacements and post-cyclic uplift capacities for the anchors in medium–dense and dense sands subjected to the same cyclic loading parameters are listed in Table 3. It can be seen that the accumulated displacements of anchors in dense sand are smaller than those in medium–dense sand at the same cyclic loading parameters, and the difference is more obvious when the anchors are pulled under a higher amplitude ratio. The data shown in Table 3 indicate that the smaller the accumulated displacement, the larger the post-cyclic uplift capacity ratio in medium–dense sand, and the trend is opposite in dense sand. Schiavon et al. [17] found that the anchor with the significant capacity degradation after cyclic loading is the one that has a stable cyclic response, and that the anchor with the slight or negligible capacity degradation may be the one that has a meta-stable cyclic response by centrifuge tests of helical anchor in dense sand. This is similar with the observation in dense sand according to which the minimum accumulated displacement does not necessarily lead to high post-cyclic capacity. The post-cyclic uplift capacity ratio of the anchor in dense sand is higher or lower than that in medium–dense sand, which may be related to the flow state of the sand above the helix caused by their accumulated displacements.

The post-cyclic uplift capacities of anchors are affected by the change of sand compactness above the helix and the loss of embedment depth caused by uplift displacement after cyclic loading. Additionally, the change of sand compactness above the helix is controlled by both the uplift displacement after cyclic loading and the amount of sand back-flowing. Lumay et al. [27] considered that the flowability of powders can be evaluated by the measurements of the angle of repose of sand. Schiavon et al. [16] exhibited the gap formation below the helix during one cycle and Schiavon et al. [17] recognized the upward accumulated displacement being smaller than 5%D at the start of sand flow for the cases of helical anchors in dry dense sand and with the ratio D/d_{50} greater than 275.

Table 3. Accumulated displacement and post-cyclic uplift capacity ratio for anchors in medium–dense and dense sands.

Cyclic Loading $Q_{mean} \pm Q_{cyc}$	Sand Compactness					
	Medium–Dense Sand			Dense Sand		
	$u_a(N = 1)$ /mm	$u_a(N = 1200)$ /mm	β	$u_a(N = 1)$ /mm	$u_a(N = 1200)$ /mm	β
(0.3 ± 0.3) Q_t	0.65	13.46	0.827	0.60	5.19	1.094
(0.4 ± 0.2) Q_t	0.83	3.73	0.989	0.55	2.72	0.985
(0.3 ± 0.2) Q_t	0.36	2.72	1.026	0.66	2.52	0.973
(0.2 ± 0.2) Q_t	0.49	4.13	0.984	0.37	2.28	0.920

Figure 22a is the relationship between post-cyclic uplift capacity ratio β and standardized displacement u/D of each anchor after cyclic loading at a frequency of 0.5 Hz. The loss ratios of embedment depth (ratio of uplift displacement to initial embedment depth) of these anchors are all less than 1%. The influence of embedment depth loss on post-cyclic uplift capacity is relatively small, and the influence of soil compactness change may be the control factor. The compactness of the sand above the anchor changes little when the uplift displacement of the anchor is small after cyclic loading, which basically does not affect the subsequent static uplift capacity, such as the test CDH12 (0.5, 0.1). It can be inferred that sand flow starts when the anchor's upward movement reaches 0.04D, corresponding to the accumulated displacement 0.0235D. This is consistent with the conclusion obtained by centrifuge test from Schiavon et al. 2019 [17].

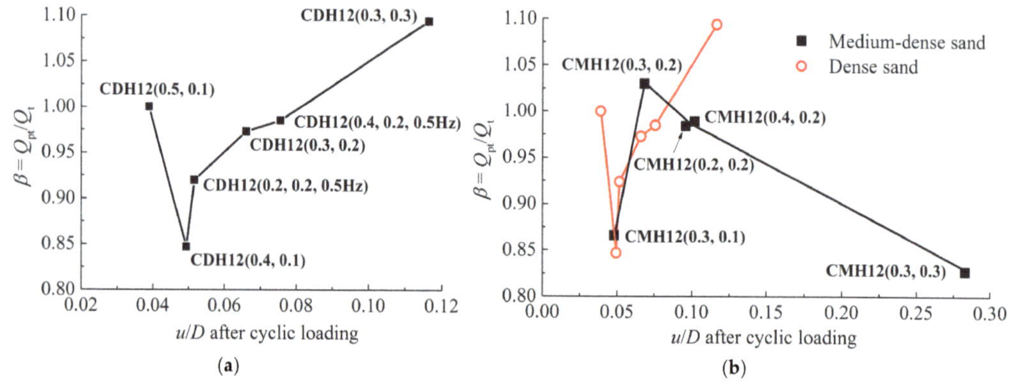

Figure 22. Relationship between u/D after cyclic loading and uplift capacity ratio β. (**a**) Anchors in dense sand; (**b**) comparison of anchors in medium–dense sand with those in dense sand.

The sand above the helix gradually becomes loose, which is caused by the backflow of sand with the development of accumulated displacement, and the value of β begins to decrease. When the anchor displacement u reaches 0.05D after cyclic loading, the subsequent monotonic uplift capacity Q_{pt} is more than 15% lower than the static uplift capacity Q_t. However, when the accumulated displacement continues to increase, the sand above the helix will be compacted again, and the uplift capacity ratio β will gradually increase. The value of Q_{pt} is close to Q_t when the uplift displacement u exceeds 0.08D, and even Q_{pt} will be higher than Q_t when uplift displacement after cyclic loading exceeds 0.1D, for example, when the post-cyclic uplift capacity of anchor CDH12 (0.3, 0.3) has increased by nearly 10%.

Figure 22b shows the relationship between β and u/D for the anchors in medium–dense sand and the comparison with dense sand. Anchor CMH12 (0.3, 0.1) produces an uplift displacement of 0.05D after cyclic loading, corresponding to an embedment depth loss rate of 0.4%. The small embedment depth loss should have a slight influence on the post-cyclic uplift capacity Q_{pt}; however, the value of Q_{pt} reduces by more than 13%, which indicates that cyclic vibration loosens the sand above the helix of anchor CMH12 (0.3, 0.1). Therefore, it is inferred that the sand above the helix may also flow into the gap at the bottom of the helix in the case that the anchor in medium–dense sand produces a small displacement under cyclic loading. Additionally, when the uplift displacement after cyclic loading of anchor CMH12 (0.3, 0.2) is greater than 0.07D, the sand above the anchor recovers to the pre-vibration state or is densified, and thus Q_{pt} reaches the static uplift capacity Q_t, or an even higher magnitude. However, when the uplift displacement continues to increase, the sand on the helix is pulled tight enough to cause more sand flowing back to the helix's bottom gap, and the sand above the helix becomes loose again, such as for the anchors CMH12 (0.4, 0.2), CMH12 (0.2, 0.2) and CMH12 (0.3, 0.2).

It can be seen from Figure 22b that the variation rule of the post-cyclic uplift capacity ratio in medium–dense sand with the standardized uplift displacement is different from that in dense sand. When the uplift displacement caused by vibration is about 0.05D, the compactness of the sand above the helix in medium–dense sand and dense sand will be looser than that before cyclic loading. With the continuous increase in uplift displacement after cyclic loading, the sand above the helix returns to the initial state before vibration, and the uplift displacement required for medium–dense sand to return to the initial state is smaller than that of dense sand. However, the post-cyclic uplift capacity ratio of anchor in dense sand continues to increase with the increase in uplift displacement after cyclic loading. After the uplift displacement exceeds 0.1D, Q_{pt} still improves, indicating that the sand above the helix in dense sand is less likely to flow back to the bottom of the helix than that in medium–dense sand.

5. Conclusions

A series of reduced-scale cyclic loading model tests of a single-helix anchor in sand with different compactness have been carried out, and the effects of anchor embedment ratio and cyclic loading parameters on the accumulated displacement, the development of axial stiffness and the post-cyclic monotonic uplift capacity have been analyzed. The correlation of post-cyclic monotonic uplift capacity ratio and post-cyclic displacement and possible influence of the cyclic uplift displacement of the anchors on the sand flow above the helix were discussed. The main conclusions are as follows:

- The single-helix anchor in dense sand with an optimal embedment ratio that is determined according to the relationship between breakout factors and embedment ratio under monotonic loading still exhibits a relatively high uplift capacity after cyclic loading. The post-cyclic monotonic responses of all the anchors exhibit stiffer behaviors.
- The anchors in both dense sand and medium–dense sand subjected to greater amplitude ratios will produce greater accumulated displacement when the same frequency and mean cyclic load ratio are applied. The influence of the mean cyclic load ratio on the accumulated displacement of anchors in dense sand may be affected by the amplitude. Under the same medium amplitude ratio, the anchor in dense sand subjected to a greater mean cyclic load ratio will produce a larger accumulated displacement, which is similar to the anchor in medium–dense sand. Additionally, for the case of anchors with the same small amplitude in dense sand, the influence of the mean cyclic load ratio on the accumulated displacement is opposite. The cyclic frequency has the greatest influence on the accumulated displacement, and the influence of cyclic amplitude is relatively greater than that of the mean cyclic load ratio. For the anchors in both dense sand and medium–dense sand under the same maximum cyclic load

ratio Q_{max}/Q_t, the greater accumulated displacement is obtained when the anchors are subjected to a larger value of Q_{mean}/Q_t and a smaller value of Q_{cyc}/Q_t.

- There are some differences in the development of axial cyclic stiffness of single-helix anchors in medium–dense sand and dense sand. When the anchors in medium–dense sand are subjected to a small amplitude, the development between axial stiffness and cycle number is the same as that of dense sand, and axial stiffness tends to stabilize at 10 to 50 cycles. However, the axial stiffness of the anchors in medium–dense sand with a medium amplitude still rises slowly after 10 cycles, and becomes stable after 400 cycles. The ratio of stable axial stiffness and initial value of anchors in dense sand is between 3 and 6, and the ratio varies between 3.5 and 9 in medium–dense sand.
- The accumulated displacements at the first cycle and after cyclic loading in dense sand for the anchors under the same standardized cyclic load parameters are basically smaller than those in medium–dense dense sand, indicating that the anchor in dense sand has higher pullout resistance.
- The backflow of sand above the helix has been inferred by analyzing the relationship of post-cyclic uplift capacity ratio and uplift displacement after cyclic loading. The variation of the post-cyclic uplift capacity ratio in medium–dense sand with the standardized uplift displacement is different from that in dense sand. However, the compactness of sand above the helix both in medium–dense sand and dense sand will be much looser than that before cyclic loading at the uplift displacement of $0.05D$ after cyclic loading. This relationship of post-cyclic capacity with anchor upward movement in dense sand and medium–dense sand can provide the reference of evaluation of stability and bearing capacity of helical anchors.

This investigation on the cyclic behavior of single-helix anchors in dense sand and medium–dense sand was carried out based on small scale model tests; therefore, the results may be different from those from field tests and centrifuge tests. In addition, the cyclic frequency and cyclic load parameters are selected based on the wind load on the transmission tower. Therefore, further studies are required to provide more comprehensive and conclusive observations.

Author Contributions: Conceptualization, formal analysis, and writing—review and editing, D.H. and R.C.; investigation and writing—original draft preparation, J.C. and X.Z.; investigation and data curation, C.Y. and X.C. All authors have read and agreed to the published version of the manuscript.

Funding: This research was funded by National Natural Science Foundation of China grant number 52078108, Jilin Province Youth Science and Technology Innovation Leader and Team Project of Provincial Department of Science and Technology Grant number 20210509058RQ and Scientific Research Project of Jilin Provincial Department of Education Grant number JJKH20210103KJ.

Institutional Review Board Statement: Not applicable.

Informed Consent Statement: Not applicable.

Conflicts of Interest: The authors declare no conflict of interest.

References

1. Spagnoli, G.; Tsuha, C.D.C. A review on the behavior of helical piles as a potential offshore foundation system. *Mar. Georesources Geotechnol.* **2020**, *38*, 1013–1036. [CrossRef]
2. Tsuha, C.H.C.; Aoki, N.; Rault, G.; Thorel, L.; Garnier, J. Evaluation of the efficiencies of helical anchor plates in sand by centrifuge model tests. *Can. Geotech. J.* **2012**, *49*, 1102–1114. [CrossRef]
3. Merifield, R.S. Ultimate uplift capacity of multiplate helical type anchors in clay. *J. Geotech. Geoenviron. Eng.* **2011**, *137*, 704–716. [CrossRef]
4. Lutenegger, A.J. Behavior of multi-helix screw anchors in sand. In Proceedings of the 2011 Pan-Am CGS Geotechnical Conference, Toronto, ON, Canada, 2–6 October 2011.
5. Feng, S.J.; Fu, W.D.; Chen, H.X.; Li, H.X.; Li, J. Field tests of micro screw anchor piles under different loading conditions at three soil sites. *Bull. Eng. Geol. Environ.* **2021**, *80*, 127–144. [CrossRef]
6. Cerfontaine, B.; Knappett, J.; Brown, M.; Davidson, C.; Sharif, Y. Screw Pile Design Optimisation under Tension in Sand. In Proceedings of the ECSMGE 2019, Reykjavik, Iceland, 1–6 September 2019.

7. Ding, H.; Wang, L.; Zhang, P.; Liang, Y.; Tian, Y.; Qi, X. The recycling torque of a single-plate helical pile for offshore wind turbines in dense sand. *Appl. Sci.* **2019**, *19*, 4105. [CrossRef]
8. Clemence, S.P.; Smithling, A.P. Dynamic uplift capacity of helical anchors in sand. In Proceedings of the 4th Australia–New Zealand Conference on Geomechanics, Perth, WA, Australia, 14–18 May 1984.
9. Cerato, A.B.; Victor, R. Effects of helical anchor geometry on long-term performance of small wind tower foundations subject to dynamic loads. *J. Deep. Found. Inst.* **2008**, *2*, 30–41. [CrossRef]
10. Cerato, A.B.; Victor, R. Effects of long-term dynamic loading and fluctuating water table on helical anchor performance for small wind tower foundations. *J. Perform. Constr. Facil.* **2009**, *23*, 251–261. [CrossRef]
11. Buhler, R.; Cerato, A.B. Design of dynamically wind-loaded helical piers for small wind turbines. *J. Perform. Constr. Facil.* **2010**, *24*, 417–426. [CrossRef]
12. Sharnouby, M.M.E.; Naggar, M.H.E. Field investigation of axial monotonic and cyclic performance of reinforced helical pulldown micropiles. *Can. Geotech. J.* **2012**, *49*, 560–573. [CrossRef]
13. Sharnouby, M.M.E.; Naggar, M.H.E. Axial monotonic and cyclic performance of fibre-reinforced polymer (FRP)—Steel fibre-einforced helical pulldown micropiles (FRP-RHPM). *Can. Geotech. J.* **2012**, *49*, 1378–1392. [CrossRef]
14. Newgard, J.T.; Schneider, J.A.; Thompson, D. Cyclic response of shallow helical anchors in a medium dense sand. In Proceedings of the 3rd Conference on Frontiers in Offshore Geotechnics, Oslo, Norway, 10–12 June 2015.
15. Wada, M.; Tokimatsu, K.; Maruyama, S.; Sawaishi, M. Effects of cyclic vertical loading on bearing and pullout capacities of piles with continuous helix wing. *Soils Found.* **2017**, *57*, 141–153. [CrossRef]
16. Schiavon, J.A.; Tsuha, C.H.C.; Thorel, L. Cyclic and post-cyclic monotonic response of a single-helix anchor in sand. *Géotechnique Lett.* **2017**, *7*, 11–17. [CrossRef]
17. Schiavon, J.A.; Tsuha, C.H.C.; Neel, A.; Thorel, L. Centrifuge modelling of a helical anchor under different cyclic loading conditions in sand. *Int. J. Phys. Model. Geotech.* **2019**, *19*, 72–88. [CrossRef]
18. Schiavon, J.A.; Tsuha, C.; Thorel, L. Monotonic, cyclic and post-cyclic performances of single-helix anchor in residual soil of sandstone. *J. Rock Mech. Geotech. Eng.* **2019**, *11*, 824–836. [CrossRef]
19. Thorel, L.; Haffar, I.E.; Maatouk, S.; Schiavon, J.A.; Tsuha, C. Cyclic loading of helical pile as anchor for floating windturbines: Centrifuge tests. In Proceedings of the ISFOG2020, Austin, TX, USA, 28 August 2022.
20. Hao, D.X.; Chen, R.; Yuan, C.; Kong, G.Q.; Shi, D.D. Centrifugal model tests on cyclic uplift performance of wished-in-place helical anchors in dense sand. *Chin. J. Rock Mech. Eng.* **2021**, *40*, 2896–2904. (In Chinese)
21. Product drawings and ratings. In *Technical Design Manual*, 4th ed.; Hubbell Power Systems Inc.: Columbia, SC, USA, 2018; pp. 7-31–7-62.
22. Hao, D.X.; Wang, D.; O'Loughlin, C.D.; Gaudin, C. Tensile monotonic capacity of helical anchors in sand: Interaction between helices. *Can. Geotech. J.* **2019**, *56*, 1534–1543. [CrossRef]
23. Ilamparuthi, K.; Dickin, E.A.; Muthukrisnaiah, K. Experimental investigation on the uplift capacity of circular anchors in sand. *Can. Geotech. J.* **2002**, *39*, 648–664. [CrossRef]
24. Hao, D.X.; Chen, R.; Fu, S.N. Experimental study on uplift capacity of multi-helix anchors in sand. *Chin. J. Geotech. Eng.* **2015**, *37*, 126–132. (In Chinese)
25. Hanna, T.; Sivapalan, E.; Senturk, A. The behaviour of dead anchors subjected to repeated and alternating loads. *Ground Eng.* **1978**, *11*, 28–34. [CrossRef]
26. Petereit, R. The Static and Cyclic Pullout Behavior of Plate Anchors in Fine Saturated Sand. Master's Thesis, Oregon State University, Corvallis, OR, USA, 1987.
27. Lumay, G.; Boschini, F.; Traina, K.; Bontempi, S.; Remy, J.C.; Cloots, R.; Vandewalle, N. Measuring the flowing properties of powders and grains. *Powder Technol.* **2012**, *224*, 19–27. [CrossRef]

Journal of
Marine Science and Engineering

MDPI

Article

Dynamic Penetration Process of Torpedo Anchors into Sand Foundation

Gang Li [1], Jinli Zhang [2,*], Jinglin Niu [2], Jia Liu [3] and Yiran Yang [1]

1 Shaanxi Key Laboratory of Safety and Durability of Concrete Structures, Xijing University, Xi'an 710123, China
2 State Key Laboratory of Coastal and Offshore Engineering, Dalian University of Technology, Dalian 116024, China
3 School of Geological Engineering and Geomatics, Chang'an University, Xi'an 710054, China
* Correspondence: jlzhang@dlut.edu.cn; Tel.: +86-139-4094-9667

Abstract: Torpedo anchors are a new type of anchoring system in deepwater that is much more economical than conventional anchoring methods. The dynamic penetration process is vitally important to the installation of torpedo anchors. Based on the spherical cavity expansion theory, the dynamic response characteristics of pressure-locked soils are analyzed using the Mohr-Coulomb criterion. The equations for the penetration of torpedo anchors with bullet-shaped heads are established considering rod friction. Subsequently, the analytical solutions for velocity, acceleration, and final penetration depth and the approximate analytical solution for penetration depth vs. time are obtained. The established penetration equation is solved using MATLAB software to obtain a semi-analytical solution, and the model tests on the penetration of a torpedo anchor with different initial velocities into saturated sand with different densities were conducted. A comparison of the test results shows that the analytical solution and the semi-analytical solution can well predict the model test results, indicating that the established analytical method can be used to analyze the penetration process of torpedo anchors. The research results can provide a guideline to the installation of torpedo anchors into the seabed in actual engineering.

Keywords: spherical cavity expansion; side friction; torpedo anchor; penetration; sand

Citation: Li, G.; Zhang, J.; Niu, J.; Liu, J.; Yang, Y. Dynamic Penetration Process of Torpedo Anchors into Sand Foundation. *J. Mar. Sci. Eng.* **2022**, *10*, 1097. https://doi.org/10.3390/jmse10081097

Academic Editor: José A.F.O. Correia

Received: 28 June 2022
Accepted: 8 August 2022
Published: 11 August 2022

1. Introduction

Gravity anchors are used in the mooring system of deep-sea oil and gas exploitation platforms, and their pull-out capacity is very important for engineering design. Gravity anchors such as torpedo anchors are freely released at a height of 30 to 50 m above the seafloor and penetrate into the seafloor through their own weight and velocity [1]. In 2000, Petrobras used torpedo anchors in the mooring system of a floating production, storage, and offloading (FPSO) platform. The process of the dynamic penetration of torpedo anchors into seafloor soils can be considered a penetration problem. In the military field, the penetration depth of an earth penetrator is a primary concern, for which the penetration equation has been proposed through a large number of tests. The torpedo anchor is composed of an anchor head, an anchor rod, and anchor wings, with a length of 12 to 15 m, a diameter of 762 to 1077 mm, and a weight of 240 to 950 kN [2]. The factors affecting the bearing capacity of the torpedo anchor include its own weight, the penetration depth, the mechanical properties of the seafloor soils, and the geometry of the torpedo anchor.

Chen et al. [3] conducted laboratory tests to investigate the penetration behavior and pullout capacity of a torpedo anchor under vibration, and found that the anchoring force and penetration depth of a self-penetration torpedo anchor are not restrained by water depth and drop height. Ads et al. [4] reported that the fin length of a torpedo anchor negatively correlated with penetration depth and positively correlated with maximum extraction resistance, and soil displacement increased with increasing penetration depth till full embedment. Based on the discrete element method, Zhang et al. [5] investigated the effect of anchor mass, impact velocity, and interparticle friction on the penetration process of a torpedo anchor, and concluded that the penetration increased with increasing impact velocity, and decreased with increasing interparticle friction. Based on large deformation finite element analysis, Kim et al. [6,7] reported that under the condition of the demarcation point that lies within the top stiff layer, the penetration depth of the torpedo anchor decreased with increasing of strength ratio, and the penetration is stopped between two layers when the strength ratio is higher than 15. Furthermore, an improved analytical embedment model was proposed based on strain rate dependent shearing resistance and drag resistance. Hossain et al. [8] found that the embedment depth of a torpedo anchor increased with increasing impact velocity, and decreased with increasing soil shear strength. The geometries of fin and tip had a significant effect on the bearing capacity, and the rectangular fin and conical tip proved to be more effective. Coupled with the advantages of the torpedo anchor and plate anchor, Lai et al. [9] concluded that the penetration depth of the new hybrid anchor increased with increasing impact velocity, and the new hybrid anchor could penetrate through the stiff layers. Fernandes et al. [10] concluded that the shape and mass distribution of the torpedo anchor have a remarkable effect on the directional stability, and the rear line influenced the terminal velocity and directional stability. Liu et al. [11] noted that the factors affecting the penetration of a gravity-installed anchor followed the order of undrained shear strength, impact velocity, strain rate dependency, friction coefficient, and strain-softening of soil. Based on a computational fluid dynamics approach, Raie and Tassoulas [12] developed a procedure to predict the embedment depth of the torpedo anchor, the pressure and shear distribution on the interface and in the soil. The research results provide a guideline to the installation of a torpedo anchor in a deep-sea project. Based on finite element analysis, Sabetamal et al. [13,14] revealed that a smooth discretization of the contact interface between soil and structure is a crucial factor to avoid oscillations in the prediction of a dynamic response of soil during penetration of a torpedo anchor. Based on small-scale model tests, True [15–17] measured the acceleration time-history curves of torpedo anchors with different anchor tip shapes during their penetration into soft clay, silt, and cement soils and established the torpedo anchor penetration equation based on Newton's second law of motion and the theory of the ultimate bearing capacity of foundations. Considering the nonmonotonicity of the measured acceleration time-history curves and assuming that the net resistance is a composite function of the velocity squared and depth, Boguslavskii et al. [18] obtained analytical solutions for velocity, acceleration, and final penetration depth, determined the parameters in the analytical solutions based on dimensional analysis, and compared the analytical solutions with the test results. O'Loughlin et al. [19,20] studied the penetration process of torpedo anchors with a centrifuge model test and combined the existing test results to establish an empirical formula of penetration depth based on the energy balance. O'Beirne et al. [21] analyzed the entire process from release to rest of a penetrating torpedo anchor, established the calculation model for the entire release-to-rest process, and compared it with the field model test. Nazem et al. [22] used the ALE method to perform a numerical analysis on the process of the dynamic penetration of a torpedo anchor into a uniform clay layer and provided a dynamic penetration factor. Bishop et al. [23] applied the cylindrical cavity and spherical cavity expansion theory in a quasistatic state to calculate the pressure on the conical surface when a conical object is slowly pressed into a metal. Forrestal and Luk [24] applied the spherical cavity expansion theory to analyze the dynamic response of the soil at a constant cavity expansion velocity, established the equation for the vertical penetration of a long rod with a bullet-shaped tip into the soil, and provided the analytical solutions for velocity, acceleration, and penetration depth. The reliability of the calculation results was verified by actual penetration tests in the field. Shi et al. [25] used the p-α state equation and the Mohr-Coulomb-Tresca criterion to describe the constitutive relationship of dry sand, obtained the formula for calculating the depth of a projectile vertically penetrating dry sand based on the spherical cavity expansion theory, and compared the analytical results with the test results. Chian et al. [26] studied the influences of the projectile nose shape, the relative density of sand, and the projectile mass on the penetration process and found that the shape and mass of the projectile nose significantly affect the total absorption energy of sand.

The penetration of torpedo anchors into soil was mainly studied using torpedo anchors as the research object. The force was analyzed, Newton's second law was used to establish the equation of motion, and the resistance of soil to the torpedo anchor was expressed by the ultimate tip resistance. However, the penetration process of a torpedo anchor into soil is a dynamic problem, thereby it is necessary to consider the dynamic response of soil. In addition, because the torpedo anchor has a low installation speed (20 to 30 m/s) and a large length-to-diameter ratio, the cavity expansion phenomenon is not remarkable, so the frictional resistance [27] between the anchor rod and the cavity wall cannot be ignored. In this study, the locked hydrostat model and the Mohr-Coulomb criterion are used to describe the constitutive relationship of soil. Based on the spherical cavity expansion theory and assuming that the cavity expansion velocity is constant, the governing equations for the elastic and plastic regions of soil and the normal stress of the cavity wall are established. On this basis, assuming a plane strain condition and considering the friction between the anchor rod and the cavity wall, Newton's second law is applied to establish the equation of motion of torpedo anchors, from which the analytical solution for the torpedo anchor penetration equation is obtained. Furthermore, a semi-analytical method by MATLAB software was used to solve the penetration equation. A self-made penetration device was used to carry out a small-scale model test of torpedo anchor penetration into saturated sand. Finally, a comparison between analytical, semi-analytical, and test results were conducted to verify the feasibility of the analytical method.

2. Torpedo Anchor Penetration Equations

2.1. Calculation for the Axial Force of the Anchor Tip

It is assumed that the torpedo anchor vertically penetrates the seafloor soil with an initial velocity of V_0 and proceeds to penetrate at velocity V_z. The anchor tip squeezes the soil around it to form a cavity, assuming that the anchor and surrounding soil makecontact. The diameter of the cavity gradually increases from zero to the rod diameter during the penetration process. When the velocity reaches zero, the penetration process ends. The radial stress and tangential stress acting at the anchor tip are denoted as σ_n and σ_t, respectively, as shown in Figure 1. For the torpedo anchor, the motion and final depth can be calculated when forces on the anchor tip are known. Therefore, we first model the anchor tip resistance and then calculate velocity, deceleration, and penetration depth.

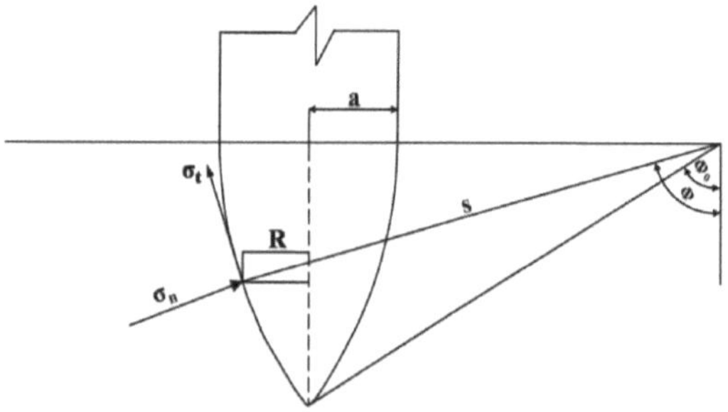

Figure 1. Stress analysis of the bullet-shaped anchor tip.

Let the anchor tip be bullet-shaped, with a shape factor of $CRH = \psi = s/2a$, where a is the radius of the anchor rod, and s is the radius of the anchor tip arc. Let the friction coefficient between the anchor tip and the soil be η_1 and the penetration velocity be V_z. The normal force acting on the segment $sd\phi$ of the anchor tip is

$$dF_n = 2\pi R s \sigma_n d\phi \tag{1}$$

where $R = s\sin\phi - (s-a)$.

The component in the axial direction of the anchor tip is

$$dF_{z_1} = 2\pi s^2\sigma_n(V_z,\phi)\cos\phi\left[\sin\phi - \frac{s-a}{s}\right]d\phi \tag{2}$$

The upward component of tangential friction in the axial direction of the anchor tip is

$$dF_{z_2} = 2\pi s^2\sigma_n(V_z,\phi)\eta_1\sin\phi\left[\sin\phi - \frac{s-a}{s}\right]d\phi \tag{3}$$

Then, the resultant force in the axial direction of the anchor tip is

$$dF_z = 2\pi s^2\sigma_n(V_z,\phi)\left[\sin\phi - \frac{s-a}{s}\right](\cos\phi + \eta_1\sin\phi)d\phi \tag{4}$$

The total resultant force in the axial direction of the anchor tip is obtained by integrating Equation (4)

$$F_z = 2\pi s^2\int_{\varphi_0}^{\frac{\pi}{2}}\left\{\left[\sin\phi - \frac{s-a}{s}\right](\cos\phi + \eta_1\sin\phi)\right\}\sigma_n(V_z,\phi)d\phi \tag{5}$$

where $\varphi_0 = \arcsin\left(\frac{s-a}{s}\right) = \arcsin\left(\frac{2\psi-1}{2\psi}\right)$.

2.2. Spherical Cavity Expansion Theory

2.2.1. Plastic Region Response

Let the velocity of the spherical cavity be a constant, and the radius of the spherical cavity r be increased from zero to a. It is assumed that the soil is a homogeneous, non-viscous and isotropic elastic-plastic material. The soil is saturated and incompressible, and the expansion of spherical cavity is regarded as an undrained process. As the soil cavity is expanded, it is assumed that a plastic response region and an elastic response region are formed in the soil surrounding the cavity (Figure 2). The plastic region is bounded by the radii V_t and c_t, where t is the time, V is the cavity expansion velocity and c is the elastic-plastic interface velocity. It is assumed that the constitutive relationship of the soil in the plastic region is expressed by a locked hydrostat (pressure-volumetric strain) and follows the Mohr-Coulomb yield criterion. The elastic region is taken as an incompressible elastic material and the constitutive relationship follows Hooke law. In addition, the stress and strain are defined as positive when subject to compression.

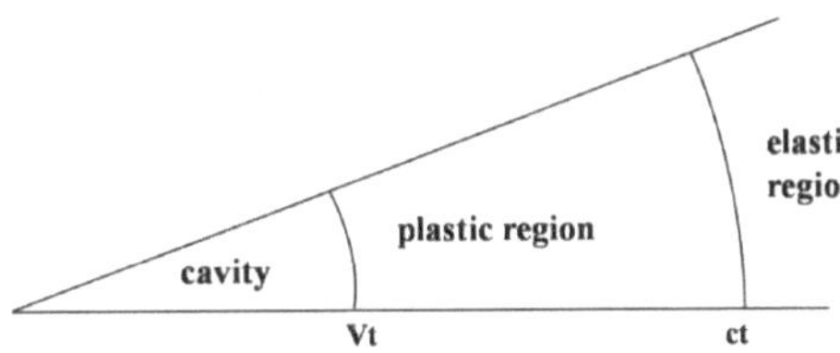

Figure 2. Soil response regions.

For a locked hydrostat, the Mohr-Coulomb yield criterion and spherical symmetry can be expressed [24]:

$$\eta^* = 1 - \frac{\rho_0}{\rho^*},\ \sigma_r - \sigma_\theta = \tau_0 + \lambda p,\ p = \frac{(\sigma_r + 2\sigma_\theta)}{3}$$

where ρ_0 and ρ^* are the initial and locked (compacted) densities of soil, respectively, η^* is the locked volumetric strain, σ_r and σ_θ are the radial and tangential components of Cauchy stress, respectively, τ_0 and λ are yield parameters, and p is the average principal stress.

In the Lagrangian coordinate system, the momentum and mass conservation equations can be expressed as

$$(r+u)^2\frac{\partial\sigma_r}{\partial r}+2\left(1+\frac{\partial u}{\partial r}\right)(r+u)(\sigma_r-\sigma_\theta)+\rho_0 r^2\frac{\partial^2 u}{\partial t^2}=0 \tag{6}$$

$$\frac{1}{3}\frac{\partial}{\partial r}\left[(r+u)^3\right]=\frac{\rho_0}{\rho}r^2 \tag{7}$$

where r is the radial coordinate, u is the radial displacement, and ρ is the current density.

The boundary condition at the cavity wall is

$$u(r=0,t)=Vt \tag{8}$$

2.2.2. Elastic-Plastic Contact Surface

The Hugoniot jump condition, i.e., the momentum and mass conservation condition, is satisfied on the elastic-plastic interface

$$\begin{aligned}\rho_2(v_2-c)&=\rho_1(v_1-c)\\ \sigma_2+\rho_2 v_2(v_2-c)&=\sigma_1+\rho_1 v_1(v_1-c)\end{aligned} \tag{9}$$

where the subscripts 1 and 2 represent the plastic and elastic regions, respectively.

2.2.3. Incompressible Elastic Region Response

The mass conservation equation is

$$\frac{\mathrm{d}u}{\mathrm{d}r}+2\frac{u}{r}=0 \tag{10}$$

If the convection term is ignored, the momentum equation can be expressed as

$$\frac{\partial\sigma_r}{\partial r}+\frac{2(\sigma_r-\sigma_\theta)}{r}=-\rho\frac{\partial^2 u}{\partial t^2} \tag{11}$$

By solving Equations (6)–(11) using the similarity transformation, we can obtain the solutions corresponding to the plastic and elastic regions. In particular, the normal stress at the cavity wall can be expressed as

$$\sigma_n(V_z,\varphi)=\sigma_r=A\tau_0+B\rho_0 V^2 \tag{12}$$

where $A=\frac{1}{\alpha}\left(\frac{1+\tau_0/2E}{\gamma}\right)-\frac{1}{\lambda}$, $B=\frac{3}{(1-\eta^*)(1-2\alpha)(2-\alpha)}+\frac{1}{\gamma^2}\left(\frac{1+\tau_0/2E}{\gamma}\right)^{2\alpha}(C_1+C_2)$, $C_1=\frac{3\tau_0}{E}+\eta^*\left(1-\frac{3\tau_0}{2E}\right)^2$, $C_2=-\frac{\gamma^3\left[2(1-\eta^*)(2-\alpha)+3\gamma^3\right]}{(1-\eta^*)(1-2\alpha)(2-\alpha)(1+\tau_0/2E)^4}$, $\gamma=\frac{V}{c}$, $\alpha=3\lambda/(3+2\lambda)$.

2.3. *Penetration Equations*

Let the penetration velocity of the torpedo anchor at any time be V_z; then the cavity expansion velocity is $V=V_z\cos\varphi$.

2.3.1. Calculation of the Axial Force of the Anchor Tip

Substituting Equation (12) into Equation (5) gives the upward resultant force in the axial direction of the anchor tip as

$$F_z=\alpha_s+\beta_s V_z^2 \tag{13}$$

where $\alpha_s=\pi a^2\tau_0 A\left[1+4\eta_1\psi^2(\pi/2-\varphi_0)-\eta_1(2\psi-1)(4\psi-1)^{1/2}\right]$, $\beta_s=\pi a^2\rho_0 B\left[\frac{8\psi-1}{24\psi^2}+\eta_1\psi^2(\pi/2-\varphi_0)-\frac{\eta_1(2\psi-1)\left(6\psi^2+4\psi-1\right)(4\psi-1)^{1/2}}{24\psi^2}\right]$.

2.3.2. Motion Equation of the Torpedo Anchor

Let the torpedo anchor vertically penetrate the soil with an initial velocity of V_0. The seafloor surface is taken as the coordinate origin, and the positive z coordinate direction is vertically downward. Let the time be t and the penetration depth of the anchor tip be z. The equation of motion can be expressed as

$$m\frac{\mathrm{d}^2 z}{\mathrm{d}t^2} = mg - F_z - F_f \tag{14}$$

where F_z is the axial force of the anchor tip and can be calculated by Equation (13). F_f is the friction between the anchor rod and the soil. Under the assumption that the anchor rod is not separated from the soil, the anchor shank bears the earth pressure in the normal direction. Let the rod-soil friction coefficient be η_2 and the anchor length be L. Then, the friction F_f acting on the anchor rod can be expressed as

When $z \leq L$

$$F_f = \int_0^z 2\pi a\eta_2 K\rho g h dh = \pi a\eta_2 K\gamma z^2 \tag{15}$$

When $z > L$

$$F_f = \int_{z-L}^z 2\pi a\eta_2 K\rho g h dh = \pi a\eta_2 K\gamma(2z - L)L \tag{16}$$

where K is the earth pressure coefficient.

By substituting Equations (13) and (15) into Equation (14), we have

When $z \leq L$

$$m\frac{\mathrm{d}^2 z}{\mathrm{d}t^2} = mg - \alpha_s - \beta_s\left(\frac{\mathrm{d}z}{\mathrm{d}t}\right)^2 - \pi\eta_2 K a\gamma z^2 \tag{17}$$

When $z > L$

$$m\frac{\mathrm{d}^2 z}{\mathrm{d}t^2} = mg - \alpha_s - \beta_s\left(\frac{\mathrm{d}z}{\mathrm{d}t}\right)^2 - \pi\eta_2 K a L\gamma(2z - L) \tag{18}$$

Equations (17) and (18) are the equations of motion of the torpedo anchor.

To solve Equation (17), let $y = \frac{\mathrm{d}z}{\mathrm{d}t}$, and $\frac{\mathrm{d}^2 z}{\mathrm{d}t^2} = \frac{\mathrm{d}y}{\mathrm{d}z}\frac{\mathrm{d}z}{\mathrm{d}t} = y\frac{\mathrm{d}y}{\mathrm{d}z}$.

Then, Equation (17) is reduced to a first-order ordinary differential equation

$$y\frac{\mathrm{d}y}{\mathrm{d}z} = g - \frac{\alpha_s}{m} - \frac{\beta_s}{m}y^2 - \frac{\pi a\eta_2 K\gamma}{m}z^2 \tag{19}$$

Let $u = y^2$; then, Equation (19) can be expressed as

$$\frac{\mathrm{d}u}{\mathrm{d}z} + \frac{2\beta_s}{m}u = 2g - \frac{2\alpha_s}{m} - \frac{2\pi a\eta_2 K\gamma}{m}z^2 \tag{20}$$

The initial conditions are $t = 0$, $z = 0$, and $u = {V_0}^2$.

The general solution for Equation (20) is

$$u = -\frac{2\pi a\eta_2 K\gamma}{m}\left(\frac{m}{2\beta_s}\right)^3\left[\left(\frac{2\beta_s z}{m}\right)^2 - \frac{4\beta_s z}{m} + 2\right] + C_3 e^{-\frac{2\beta_s z}{m}} + \frac{mg - \alpha_s}{\beta_s} \tag{21}$$

The constant C_3 in Equation (21) is determined by the initial conditions:

$$C_3 = \frac{4\pi a\eta_2 K\gamma}{m}\left(\frac{m}{2\beta_s}\right)^3 - \frac{mg - \alpha_s}{\beta_s} + {V_0}^2$$

Then, the solution of Equation (19) is

$$u = a_1 e^{-\frac{2\beta_s z}{m}} - a_2\left(\frac{2\beta_s z}{m}\right)^2 + 2a_2\left(\frac{2\beta_s z}{m}\right) + a_3 \tag{22}$$

where $a_1 = \frac{4\pi a\eta_2 K\gamma}{m}\left(\frac{m}{2\beta_s}\right)^3 - \frac{mg-\alpha_s}{\beta_s} + V_0{}^2$, $a_2 = \frac{2\pi a\eta_2 K\gamma}{m}\left(\frac{m}{2\beta_s}\right)^3$, $a_3 = \frac{mg-\alpha_s}{\beta_s} - 2a_2$.

By substituting $u = y^2 = V_z{}^2$ into Equation (22), we have

$$V_z{}^2 = a_1 e^{-\frac{2\beta_s z}{m}} - a_2\left(\frac{2\beta_s z}{m}\right)^2 + 2a_2\left(\frac{2\beta_s z}{m}\right) + a_3 \tag{23}$$

When the penetration depth of the torpedo anchor exceeds one anchor length, the solution for Equation (18) can be obtained using the same method and expressed as

$$V_z{}^2 = a_4 e^{\frac{-2\beta_s z}{m}} + \frac{mR}{2\beta_s} - W\left(\frac{m}{2\beta_s}\right)^2\left(\frac{2\beta_s z}{m} - 1\right) \tag{24}$$

where $a_4 = e^{\frac{2\beta_s L}{m}}\left[a_1 e^{\frac{-2\beta_s L}{m}} - a_2\left(\frac{2\beta_s L}{m}\right)^2 + 2a_2\left(\frac{2\beta_s L}{m}\right) + a_3 + W\left(\frac{m}{2\beta_s}\right)^2\left(\frac{2\beta_s L}{m} - 1\right) - \frac{mR}{2\beta_s}\right]$, $R = 2g - \frac{2\alpha_s}{m} + \frac{2\pi a K\eta_2\gamma L^2}{m}$, $W = \frac{4\pi a K\eta_2\gamma L}{m}$.

Equations (23) and (24) show the relationships between the penetration velocity and the penetration depth when the penetration depth of the torpedo anchor is less than and greater than the anchor length, respectively.

By taking the derivative of both sides of Equations (23) and (24) with respect to t, we can obtain the relationship between the acceleration and penetration depth of the torpedo anchor as

$$\frac{dV_z}{dt} = -\frac{a_1\beta_s}{m}e^{-\frac{2\beta_s z}{m}} - a_2\left(\frac{2\beta_s}{m}\right)^2 z + \frac{2\beta_s a_2}{m} \tag{25}$$

$$\frac{dV_z}{dt} = -\frac{a_4\beta_s}{m}e^{\frac{-2\beta_s z}{m}} - \frac{Wm}{4\beta_s} \tag{26}$$

To obtain the variation in the penetration depth with time, let $x = \frac{2\beta_s z}{m}$; then Equation (23) is simplified as

$$\frac{m}{2\beta_s}\frac{dx}{dt} = \sqrt{a_1 e^{-x} - a_2 x^2 + 2a_2 x + a_3} \tag{27}$$

By integrating Equation (27), we have

$$\frac{2\beta_s}{m}\int dt = \int \frac{dx}{\sqrt{a_1 e^{-x} - a_2 x^2 + 2a_2 x + a_3}} \tag{28}$$

There is no analytical solution for the integral on the right side of Equation (28). To obtain an approximate solution, we can approximately express e^{-x} as $e^{-x} = A_0 x^2 + B_0 x + C_0$.

Then, Equation (28) is approximately expressed as

$$\frac{2\beta_s}{m}\int dt = \int \frac{dx}{\sqrt{(a_1 A_0 - a_2)x^2 + (a_1 B_0 + 2a_2)x + a_1 C_0 + a_3}} \tag{29}$$

When $a_1 A_0 - a_2 > 0$,

$$\frac{2\beta_s}{m}t = C_4 + \frac{1}{\sqrt{a_1 A_0 - a_2}}\ln\left|2(a_1 A_0 - a_2)x + a_1 B_0 + 2a_2 + 2\sqrt{a_1 A_0 - a_2}\sqrt{(a_1 A_0 - a_2)x^2 + (a_1 B_0 + 2a_2)x + a_1 C_0 + a_3}\right. \tag{30}$$

The integral constant C_4 in Equation (30) can be determined by the initial conditions, and the approximate solution for Equation (27) is obtained as

$$\frac{2\beta_s}{m}t = \frac{1}{\sqrt{a_1A_0 - a_2}}\ln\left|\frac{2(a_1A_0 - a_2)x + a_1B_0 + 2a_2 + 2\sqrt{a_1A_0 - a_2}\sqrt{(a_1A_0 - a_2)x^2 + (a_1B_0 + 2a_2)x + a_1C_0 + a_3}}{a_1B_0 + 2a_2 + 2\sqrt{a_1A_0 - a_2}\sqrt{a_1C_0 + a_3}}\right| \tag{31}$$

By substituting $x = \frac{2\beta_s z}{m}$ into Equation (31), we have

$$t = \frac{m}{2\beta_s\sqrt{a_1A_0 - a_2}}\ln\left|\frac{2(a_1A_0 - a_2)\frac{2\beta_s}{m}z + a_1B_0 + 2a_2 + 2\sqrt{a_1A_0 - a_2}\sqrt{(a_1A_0 - a_2)(\frac{2\beta_s}{m})^2 z^2 + (a_1B_0 + 2a_2)\frac{2\beta_s}{m}z + a_1C_0 + a_3}}{a_1B_0 + 2a_2 + 2\sqrt{a_1A_0 - a_2}\sqrt{a_1C_0 + a_3}}\right| \tag{32}$$

When the penetration depth $z = L$, the required time t_1 is determined by Equation (32):

$$t_1 = \frac{m}{2\beta_s\sqrt{a_1A_0 - a_2}}\ln\left|\frac{2(a_1A_0 - a_2)\frac{2\beta_s}{m}L + a_1B_0 + 2a_2 + 2\sqrt{a_1A_0 - a_2}\sqrt{(a_1A_0 - a_2)(\frac{2\beta_s}{m})^2 L^2 + (a_1B_0 + 2a_2)\frac{2\beta_s}{m}L + a_1C_0 + a_3}}{a_1B_0 + 2a_2 + 2\sqrt{a_1A_0 - a_2}\sqrt{a_1C_0 + a_3}}\right| \tag{33}$$

When the penetration depth $z > L$, Equation (24) is solved using the same method to obtain

$$t = t_1 + \frac{m}{2\beta_s\sqrt{a_4A_0}}\ln\left|\frac{2a_4A_0\frac{2\beta_s}{m}z + a_4B_0 - W\left(\frac{m}{2\beta_s}\right)^2 + 2\sqrt{a_4A_0}\sqrt{a_4A_0\left(\frac{2\beta_s}{m}\right)^2 z^2 + \left[a_4B_0 - W\left(\frac{m}{2\beta_s}\right)^2\right]\frac{2\beta_s}{m}z + a_4C_0 + \frac{mR}{2\beta_s} + W\left(\frac{m}{2\beta_s}\right)^2}}{2a_4A_0\frac{2\beta_s}{m}L + a_4B_0 - W\left(\frac{m}{2\beta_s}\right)^2 + 2\sqrt{a_4A_0}\sqrt{a_4A_0\left(\frac{2\beta_s}{m}\right)^2 L^2 + \left[a_4B_0 - W\left(\frac{m}{2\beta_s}\right)^2\right]\frac{2\beta_s}{m}L + a_4C_0 + \frac{mR}{2\beta_s} + W\left(\frac{m}{2\beta_s}\right)^2}}\right| \tag{34}$$

Equations (32) and (34) represent the time needed for the penetration depth of the torpedo anchor when the penetration depth is less than and greater than the anchor length, respectively.

The final penetration depth of the torpedo anchor Z_{max} can be determined by Equation (23) or Equation (24):

(1) $Z_{max} \leq L$

$$a_1e^{-\frac{2\beta_s Z_{max}}{m}} - a_2\left(\frac{2\beta_s Z_{max}}{m}\right)^2 + 2a_2\left(\frac{2\beta_s Z_{max}}{m}\right) + a_3 = 0 \tag{35}$$

(2) $Z_{max} > L$

$$\frac{2a_4\beta_s}{m}e^{\frac{-2\beta_s Z_{max}}{m}} - WZ_{max} + R + \frac{Wm}{2\beta_s} = 0 \tag{36}$$

If the friction on the rod is not considered, the final penetration depth Z_{max} is

$$Z_{max} = \frac{m}{2\beta_s}\ln\left(\frac{\beta_s {V_0}^2}{mg - \alpha_s} - 1\right) \tag{37}$$

Therefore, the analytical solution for penetration depth vs. the velocity, penetration depth vs. acceleration, final penetration depth, and penetration depth vs. time of the torpedo anchor are obtained. From the engineering application viewpoint, Equations (23) and (24) can be used to calculate penetration depth vs. velocity, Equations (25) and (26) can be used to calculate penetration depth vs. acceleration, Equations (32)–(34) can be used to calculate penetration depth vs. time, and Equations (35)–(37) can be used to calculate the final penetration depth.

3. Model Test of Torpedo Anchor Penetration into Saturated Sand

3.1. Test Materials and Device

The soil samples used in the test were Fujian standard sand with a specific gravity of 2.66, and the maximum and minimum dry densities were 1.93 g/cm^3 and 1.53 g/cm^3, respectively. Given the particle-size distribution curve of the test sand, as shown in Figure 3, it can be concluded that the test sand was poorly graded sand. Figure 4 shows the schematic of the model test setup. The model torpedo has an anchor length of 175 mm, a diameter of 25 mm, and a mass of 235 g. The anchor tip is bullet shaped with $CRH = \psi = 1.69$, the anchor head is made of stainless steel, and the anchor rod is made of an aluminum alloy. The anchor rod is hollow with a built-in micro electro mechanical system (MEMS) accelerometer, which has a measurement range of ± 500 g.

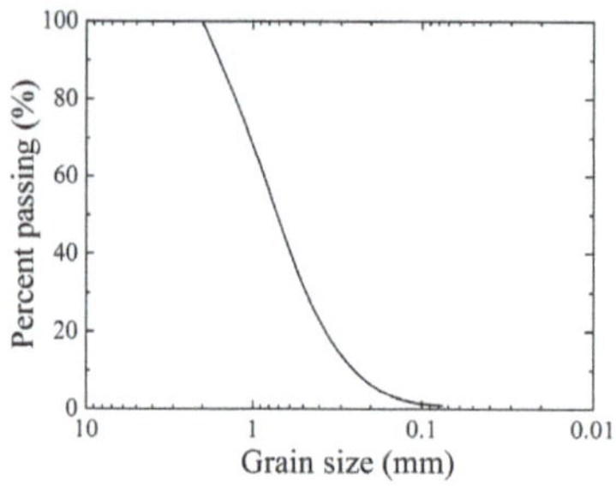

Figure 3. Particle-size distribution curve of the test sand.

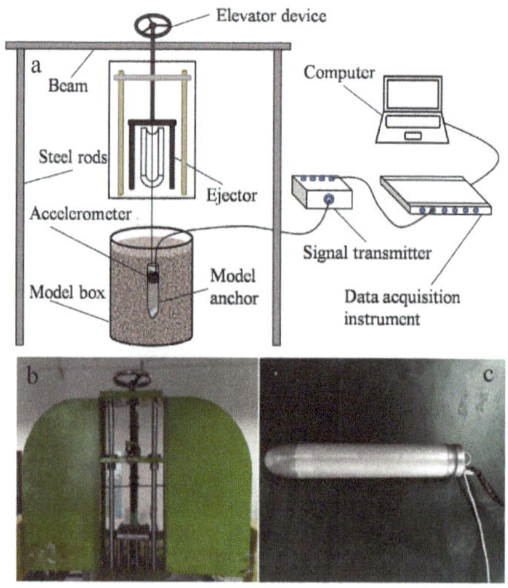

Figure 4. Schematic of the model test setup (**a**) layout of penetration test; (**b**) ejection device; (**c**) torpedo anchor.

The plexiglass tube with an inner diameter of 220 mm, a height of 600 mm and a thickness of 10 mm was used as a model box. The hole was located in the center of the bottom, and a layer of geotextile was laid to ensure the uniformity of saturated samples. The self-designed ejection device consists of a movable support frame and an ejector. The ejector is mainly composed of a tension screw, a steel strand, an ejection rod, a model torpedo anchor bracket, a velocity control pawl, a trigger, and a velocimetry system with an ejection velocity of 15 to 30 m/s. Real-time data acquisition with the MEMS accelerometer and photogate signals was performed using the TWD dynamic data acquisition instrument (Beijing Taize Technology Development Co., Ltd., Beijing, China), with an acquisition frequency of 10 kHz, which met the test requirements. The penetration depth of the model anchor was measured by the flexible rope connected to the anchor tail.

3.2. Test Methods

The height of the samples was 500 mm. According to the designed density and loading height, the required sand mass was calculated and then loaded four times and compacted in layers. After compaction, the water was slowly supplied to the sand sample from the hole at the bottom, and back pressure was pumped on the surface of the sand to make the water rise slowly from the bottom to top. When the water level was more than 2 cm above the soil sample, the water supply was stopped and left to stand for enough time to saturate the sand sample. Subsequently, a suction ball was used to absorb the excess water on the sand surface before the penetration test.

Due to the limitations of the test conditions, it was difficult to measure the locked density of the sand. Therefore, we used the maximum dry density of the sand ρ_{dmax} to replace ρ^*. In the tests, sands with four different relative densities (0.80, 0.70, 0.65, and 0.60) were labeled as A, B, C, and D, respectively, and each was assigned three velocity levels, for a total of 12 sets of tests. During the test, the model box was placed directly under the ejector, the accelerometer was preloaded into the torpedo anchor, which was installed at the bracket of the ejector, and the signal lines of the accelerometer and photogate were connected to the data acquisition instrument. The steel strand was pulled to the corresponding gear via the screw at a predesigned speed. When the signal acquisition instrument was ready, the trigger was pressed to launch the model torpedo anchor.

4. Comparison between Experimental, Analytical and Semi-Analytical Results

The strength parameters of the saturated sand with different densities were obtained by triaxial tests (Table 1). The parameters of the model torpedo anchor are shown in Table 2. According to the parameters listed in Tables 1 and 2 and the initial penetration velocity V_0 measured in the test, theoretical calculations and analyses for each penetration test were conducted. Equations (23)–(26) show the relationship between the penetration velocity, acceleration, and penetration depth of the torpedo anchor. Based on the penetration depth in the model tests, the parameters in Equations (32)–(34) were approximately determined through preliminary calculations to be $A_0 = 0.2579$, $B_0 = -0.8825$, and $C_0 = 0.9872$. The friction coefficient η_1 between the anchor tip and the soil and the friction coefficient η_2 between the anchor rod and the soil were both experimentally determined to be approximately 0.35 [28]. The passive earth pressure coefficient was adopted. For an additional comparative analysis, the semi-analytical solutions for Equations (17) and (18) were obtained using MATLAB software.

Table 1. Strength parameters of saturated sand.

No.	D_r	τ_0/kPa	λ	ρ/(g/cm^3)	E/MPa
A	0.60	22.641	1.40	1.747	100
B	0.65	25.656	1.41	1.767	
C	0.70	26.094	1.43	1.789	
D	0.80	31.549	1.47	1.834	

Table 2. Dimensions of the torpedo anchor.

L/mm	d/mm	m/g	ψ	θ_0
175	25	235	1.69	0.78

4.1. Curves of Acceleration with Depth

The median filter method was employed to filter out the high-frequency components from the acceleration time-history signal using the built-in filter program in MATLAB. To facilitate comparison, the measured acceleration time-history curve was integrated twice to obtain the variation curve of acceleration with depth. Figure 5 shows the variation curves of acceleration with depth for saturated sands with different densities and model torpedo anchors with different initial velocities. It can be observed that the acceleration decreased with increasing depth and gradually decreased to zero. Both the analytical solutions and the semi-analytical solutions are in good overall agreement with the measured results, indicating that the cavity expansion theory can be used to analyze the dynamic penetration process of torpedo anchor. In addition, there is still some deviation between the predicted results and the measured results in the later section of the variation curve of acceleration with depth.

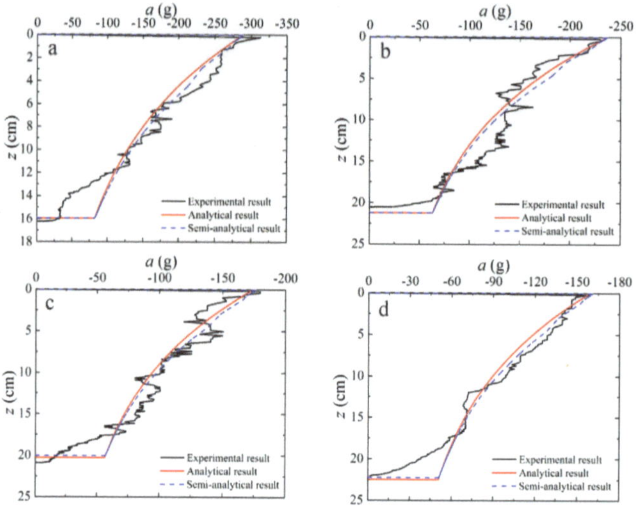

Figure 5. Curves of depth vs. acceleration (**a**) D_r = 0.80, V_0 = 22.5 m/s; (**b**) D_r = 0.70, V_0 = 23 m/s; (**c**) D_r = 0.65, V_0 = 20.1 m/s; (**d**) D_r = 0.60, V_0 = 20 m/s.

4.2. Curves of Velocity with Depth

Figure 6 shows the variation curves of velocity with depth for saturated sands with different densities and torpedo anchors with different initial velocities. As shown in the figures, the velocity decreased with increasing depth, and the analytical solution and semi-analytical solution both align well with the measured results.

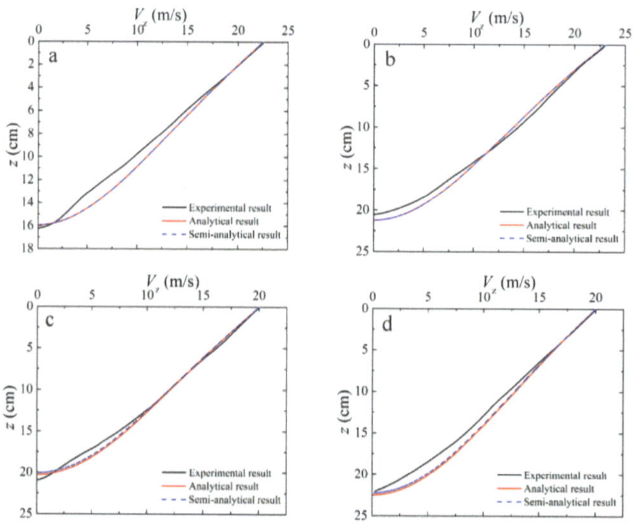

Figure 6. Curves of depth vs. velocity (**a**) D_r = 0.80, V_0 = 22.5 m/s; (**b**) D_r = 0.70, V_0 = 23 m/s; (**c**) D_r = 0.65, V_0 = 20.1 m/s; (**d**) D_r = 0.60, V_0 = 20 m/s.

4.3. Final Penetration Depth

Table 3 shows the test values of the final penetration depth of the model torpedo anchor, the integral values of the measured acceleration curve, the semi-analytical solution, and the analytical solution values under different sand densities and initial penetration velocities. As shown in Table 3, under different sand densities, the measured penetration depth is generally consistent with the calculation results based on the test acceleration time-history curve, indicating that the accelerometer used in the test has a relatively high accuracy. The analytical solution and semi-analytical solution are largely consistent with the measured results, indicating that the established equation for torpedo anchor penetration can describe the penetration process of a torpedo anchor.

Table 3. Comparison of final penetration depths.

No.	V_0/m/s	Z_0/cm	Z_1/cm	Z_2/cm	Z_3/cm
A1	16.7	12.3	12.7	11.4	11.4
A2	20.1	14.5	14.3	14.1	14.1
A3	22.5	16.2	16.3	15.9	15.9
B1	16.8	16.3	16.4	16.4	16.4
B2	20.1	18.6	18.5	18.6	18.6
B3	23.0	20.7	20.5	21.0	21.0
C1	16.6	16.9	17.0	16.8	16.8
C2	20.1	20.5	20.9	20.2	20.2
C3	22.3	21.8	21.6	22.4	22.4
D1	16.6	18.9	18.6	18.7	18.7
D2	20.0	22.5	22.2	22.4	22.4
D3	22.0	24.2	24.1	24.4	24.4

Z_0 = measured depth; Z_1 = penetration depth obtained by integrating the measured acceleration curve; Z_2 = semi-analytical solution of penetration depth; Z_3 = analytical solution of penetration depth.

5. Conclusions

By assuming that the soil is a pressure-locked material that follows the Mohr-Coulomb criterion, we analyzed the dynamic response characteristics of compressible soil based on the spherical cavity expansion theory. Under the plane strain assumption, the penetration equations of torpedo anchors were established by applying Newton's second law. The analytical solutions for acceleration vs. depth, velocity vs. depth, and penetration depth and the approximate analytical solution for penetration depth as a function of time were given. A comparison between analytical, semi-analytical, and test results showed that the theoretical solutions are in good alignment with the measured results, indicating that the analytical solution established in this study can be used to analyze the penetration process of torpedo anchors.

Author Contributions: Conceptualization, J.Z. and G.L.; methodology, J.Z. and G.L.; validation, J.L., J.Z. and G.L.; writing—original draft preparation, G.L. and Y.Y.; writing—review and editing, J.N. and J.Z.; funding acquisition, G.L. All authors have read and agreed to the published version of the manuscript.

Funding: This research was funded by the Natural Science Basic Research Program of Shaanxi Province, grant number 2021JM-535 and the Special Fund for Scientific Research by Xijing University, grant number XJ18T01.

Institutional Review Board Statement: Not applicable.

Informed Consent Statement: Not applicable.

Data Availability Statement: Not applicable.

Acknowledgments: The Youth Innovation Team of Shaanxi Universities is acknowledged.

Conflicts of Interest: The authors declare no conflict of interest.

References

1. Yu, G.L.; Wang, W.K.; Wang, C. The structure and characteristics of powered torpedo anchor. *Ocean Eng.* **2018**, *36*, 143–148.
2. Medeiros, C.J. Low cost anchor system for flexible risers in deep waters. In Proceedings of the Offshore Technology Conference, Houston, TX, USA, 6–9 May 2002.
3. Chen, X.H.; Zhang, M.X.; Yu, G.L. A self-penetration torpedo anchor with vibrational shearing. *Ocean Eng.* **2021**, *236*, 109315. [CrossRef]
4. Ads, A.; Iskander, M.; Bless, S.; Omidvar, M. Visualizing the effect of Fin length on torpedo anchor penetration and pullout using a transparent soil. *Ocean Eng.* **2020**, *216*, 108021. [CrossRef]
5. Zhang, N.; Evans, T.M. Discrete numerical simulations of torpedo anchor installation in granular soils. *Comput. Geotech.* **2019**, *108*, 40–52. [CrossRef]
6. Kim, Y.H.; Hossain, M.S.; Lee, J.K. Dynamic installation of a torpedo anchor in two-layered clays. *Can. Geotech. J.* **2018**, *55*, 446–454. [CrossRef]
7. Kim, Y.H.; Hossain, M.S.; Wang, D.; Randolph, M.F. Numerical investigation of dynamic installation of torpedo anchors in clay. *Ocean Eng.* **2015**, *108*, 820–832. [CrossRef]
8. Hossain, M.S.; Kim, Y.H.; Gaudin, C. Experimental investigation of installation and pullout of dynamically penetrating anchors in clay and silt. *J. Geotech. Geoenviron. Eng.* **2014**, *140*, 04014026. [CrossRef]
9. Lai, Y.; Zhu, B.; Chen, C.; Huang, Y.H. Dynamic installation behaviors of a new hybrid plate anchor in layered marine clay. *China Ocean Eng.* **2021**, *35*, 736–749. [CrossRef]
10. Fernandes, A.C.; de Araujo, J.B.; de Almeida, J.C.L.; Machado, R.D.; Matos, V. Torpedo anchor installation hydrodynamics. *J. Offshore Mech. Arct. Eng.* **2006**, *128*, 286–293. [CrossRef]
11. Liu, H.X.; Xu, K.; Zhao, Y.B. Numerical investigation on the penetration of gravity installed anchors by a coupled Eulerian-Lagrangian approach. *Appl. Ocean Res.* **2016**, *60*, 94–108. [CrossRef]
12. Raie, M.S.; Tassoulas, J.L. Installation of torpedo anchors: Numerical modeling. *J. Geotech. Geoenviron. Eng.* **2010**, *135*, 1805–1813. [CrossRef]
13. Sabetamal, H.; Nazem, M.; Carter, J.P.; Sloan, S.W. Large deformation dynamic analysis of saturated porous media with applications to penetration problems. *Comput. Geotech.* **2014**, *55*, 117–131. [CrossRef]
14. Sabetamal, H.; Carter, J.P.; Nazem, M.; Sloan, S.W. Coupled analysis of dynamically penetrating anchors. *Comput. Geotech.* **2016**, *77*, 26–44. [CrossRef]
15. True, D.G. Rapid penetration into seafloor soils. In Proceedings of the Offshore Technology Conference, Houston, TX, USA, 6–8 May 1974.
16. True, D.G. *Penetration of Projectiles into Seafloor Soils*; Civil Engineering Laboratory (Navy): Port Hueneme, CA, USA, 1975.
17. True, D.G. Undrained Vertical Penetration into Ocean Bottom. Soils.Ph.D. Thesis, University of California, Berkeley, CA, USA, 1976.
18. Boguslavskii, Y.; Drabkin, S.; Juran, I.; Salman, A. Theory and practice of projectile's penetration in soils. *J. Geotech. Eng.* **1996**, *122*, 806–812. [CrossRef]
19. O'Loughlin, C.D.; Richardson, M.D.; Randolph, M.F. Centrifuge tests on dynamically installed anchors. In Proceedings of the ASME 28th International Conference on Ocean, Offshore and Arctic Engineering, Honolulu, Hawaii, USA, 31 May–5 June 2009.
20. O'Loughlin, C.D.; Richardson, M.D.; Randolph, M.F.; Gaudin, C. Penetration of dynamically installed anchors in clay. *Géotechnique* **2013**, *63*, 909–919. [CrossRef]
21. O'Beirne, C.; O'Loughlin, C.D.; Gaudin, C. A release-to-rest model for dynamically installed anchors. *J. Geotech. Geoenviron. Eng.* **2017**, *143*, 04017052. [CrossRef]
22. Nazem, M.; Carter, J.P.; Airey, D.W.; Chow, S.H. Dynamic analysis of a smooth penetrometer free-falling into uniform clay. *Géotechnique* **2012**, *62*, 893–905. [CrossRef]
23. Bishop, R.F.; Hill, R.; Mott, N.F. The theory of indentation and hardness tests. *Proc. Phys. Soc.* **1945**, *57*, 147–149. [CrossRef]
24. Forrestal, M.J.; Luk, V.K. Penetration into soil targets. *Int. J. Impact Eng.* **1992**, *12*, 427–444. [CrossRef]
25. Shi, C.C.; Wang, M.Y.; Li, J.; Li, M.S. A model of depth calculation for projectile penetration into dry sand and comparison with experiments. *Int. J. Impact Eng.* **2014**, *73*, 112–122. [CrossRef]
26. Chian, S.C.; Tan, B.C.V.; Sarma, A. Projectile penetration into sand: Relative density of sand and projectile nose shape and mass. *Int. J. Impact Eng.* **2017**, *103*, 29–37. [CrossRef]
27. Fang, J.C.; Kong, G.Q.; Yang, Q. Group performance of energy piles under cyclic and variable thermal loading. *J. Geotech. Geoenviron. Eng.* **2022**, *148*, 04022060. [CrossRef]
28. Sun, J.Z.; Wang, Y.; Wang, R. Study on the behaviors of calcareous sand-structure interface and its theoretical model. *Site Investig. Sci. Technol.* **2004**, *5*, 7–9.

Journal of *Marine Science and Engineering*

MDPI

Article

Analytical Penetration Solutions of Large-Diameter Open-Ended Piles Subjected to Hammering Loads

Wei Qin [1,2,3,4,5,6,*], Xue Li [1,3,4,5], Guoliang Dai [2,6] and Pan Hu [7]

1 College of Architecture and Civil Engineering, Wenzhou University, Wenzhou 325035, China; 21461544037@stu.wzu.edu.cn
2 School of Civil Engineering, Southeast University, Nanjing 211189, China; daigl@seu.edu.cn
3 Key Laboratory of Engineering and Technology for Soft Soil Foundation and Tideland Reclamation of Zhejiang Province, Wenzhou 325035, China
4 Collaborative Innovation Center of Tide and Reclamation and Ecological Protection, Wenzhou 325035, China
5 Zhejiang Engineering Research Center of Disaster Prevention and Mitigation for Coastal Soft Soil Foundation, Wenzhou 325035, China
6 Key Laboratory of C&PC Structures, Ministry of Education, Southeast University, Nanjing 211189, China
7 School of Engineering, Design and Built Environment, Western Sydney University, Sydney 2751, Australia; p.hu@westernsydney.edu.au
* Correspondence: wei.qin@wzu.edu.cn

Abstract: This paper proposes the penetration displacement solutions of large-diameter open-ended steel pipe piles (LOSPs) with the diameter exceeding 2 m subjected to hammering load. The ultimate forcing equilibrium relationships between LOSP and soil are analyzed, and the calculated formula for self-sinking depth is derived. Next, a partial differential equation of pile hammering by single blow in soft soil is developed based on wave equation incorporating the kinematic method. A dynamic coefficient of frictional resistance (DCFR) is implemented in the process of derivation, and then the displacement Fourier analytical expression of LOSP under hammering load is presented. The parameters sensitivity of the analytical solution is investigated, and the displacement curve is compared with the numerical result. The new method presented in this paper could be used to assess the penetration development of driven piles under impact loading to predict the punching through or hammer refusal during penetration.

Keywords: large-diameter open-ended steel pipe pile (LOSP); hammering penetration; displacement solution

Citation: Qin, W.; Li, X.; Dai, G.; Hu, P. Analytical Penetration Solutions of Large-Diameter Open-Ended Piles Subjected to Hammering Loads. *J. Mar. Sci. Eng.* **2022**, *10*, 885. https://doi.org/10.3390/jmse10070885

Academic Editors: José A.F.O. Correia and Fraser Bransby

Received: 19 May 2022
Accepted: 24 June 2022
Published: 27 June 2022

1. Introduction

Large-diameter open-ended steel pipe piles (LOSPs) with diameter exceeding 1 m are often used in offshore engineering and are generally installed by the driving hammers. In terms of driven piles, i.e., LOSP, hammer refusal and punching through may unfortunately occur during driving, and these may result in accidents or failures during installation [1]. Therefore, drivability analysis is implemented in the design stage, and typical drivability analysis methods are usually implemented based on one-dimensional wave theory, such as using the GRLWEAP software [2,3]. During dynamic penetration, the pile generates inertial forces, and the soil is compressed in a way that produces transverse deformation. The stress wave generated by the hammering load is transmitted along the shaft to penetrate the pile into the soil. When it encounters the soil inside the pile, it generates a transverse Poisson effect on the magnitude of the shear stress between the pile and soil [4]. If the difference in impedance between the pile and the soil inside the pile is neglected, and the displacement and deformation compatibility between the pile and the soil are perfectly maintained, the transverse shear wave can also be completely neglected [5]. Therefore, shear waves are commonly neglected in studying the hammering penetration of LOSPs. This kind of ignorance provides an implementable framework for the application of the one-dimensional wave equation during drivability analysis for open-ended piles. However, the theoretical basis of these approaches is the discretisation of the object and the empirical value of the associated dynamical parameters, which often causes systematic errors and consequently inaccurate analysis results [6–14]. For open-ended piles, the soil-plugging effect is another key concern, but it tends to occur relatively infrequently for LOSPs dynamically driven into the soft soil due to the inertia forces during driving. Therefore, the focus of the drivability analysis is on the transfer of stress waves along the pile shaft during driving and the resistance of the soil around the pile which is called soil resistance during driving, while the displacement behavior of the pile under hammering load is not the main focus of this study.

On the other hand, load transfer method is a common way to analyze the pile subjected to axial loads [15], which could predict the load-displacement response and ultimate resistance of the pile [16]. Recently, this method has been improved to determine various types of pile [17–20]. For instance, the University of Western Australia (UWA) and the Norwegian Geotechnical Institute (NGI) have compiled a database of high-quality pile load tests in sandy and clay soils, and it is now known as the "unified" database [21]. Lehane et al. [19,21] used this "unified" load-displacement curve database to calibrate the relationship of load and settlement based on cone penetration test (CPT), which can be used to estimate pile displacements under dynamic axial loads. In addition, Xu et al. [20] indicated a steep decrease in the vertical load-displacement curve in the 1 g model test while driving an open-ended pile in soft soil. When the vertical load, P_t, is applied on the pile top at time t, the response resistances of the pile shaft and pile end are P_s and P_b, respectively [22]. And then, the piles tend to overcome the soil resistance to penetrate downward. Generally, the penetration displacement of the pile can consist of the shaft displacement (w_s) and pile end displacement (w_b), respectively. If pile compression is not considered under axial load at a given moment, the total displacement of the pile is $w_t = w_b = w_s$, where w_t is the displacement of the pile top. In this situation, the pile is assumed to be rigid (no pile shaft compression occurs), and therefore, at any given moment, the pile tip load P_t is equal to the sum of the loadings transferred to the pile shaft (P_s) and the pile tip (P_b). However, the load-displacement relationship of LOSPs under hammering loads has not been thoroughly discussed.

This paper analyses the forcing limit equilibrium relationships for self-sinking and hammering loads during the installation of the LOSPs. A solution for estimating the self-sinking is proposed. The inertia forces are then introduced to formulate the ultimate equilibrium partial differential equation for the LOSPs under hammering loads in the framework of wave theory. Finally, the Fourier analytic solution for the displacement of the LOSPs under single hammering blow is solved. For LOSPs, the sensitivity of the parameters in the solution expression was investigated and discussed. The calculated displacement curve of LOSP is compared with the result of numerical simulation. This study provides a simplified approach to determine the penetration to provide a simple method for estimating hammer refusal and punching through of LOSPs during driving.

2. Self-Sinking Penetration of LOSPs

At the beginning of installation, the driven pile will sink into the soil for a certain depth due to its own weight until it reaches equilibrium (a process known as self-sinking equilibration). The accurate assessment of the self-sink depth is critical for safe attachment of piles and the following driving operations. Assuming this process does not cause soil plug effect for LOSPs, the force distribution of the pile is shown in Figure 1. When the pile self-sinks until it comes to a standstill, the pile-soil static ultimate equilibrium relationship could be expressed as

$$f_{s\_outer}\pi Dh_{p0} + f_{s\_inner}\pi dh_{s0} + q_{tip}\left(\frac{\pi D^2}{4} - \frac{\pi d^2}{4}\right) = \gamma_p\left(\frac{\pi D^2}{4} - \frac{\pi d^2}{4}\right)L \quad (1)$$

where f_{s_outer} and f_{s_inner} are the shaft frictional resistance of the inside and outside of the LOSPs, respectively; q_{tip} is the pile tip resistance; γ_p is the volume weight of the pile; D

and d are the external and internal diameter of the piles, respectively; h_{p0} and h_{s0} are the heights of the soil inside and outside the pile after it self-sinks, respectively. By solving the above equation, the self-sinking formula of LOSPs could be obtained as follows:

$$h_{p0} = \left(\frac{1}{4}(\gamma_p L - q_{tip})\left(D^2 - d^2\right) - f_{s\_inner} d h_{s0}\right) / (f_{s\_outer} D) \tag{2}$$

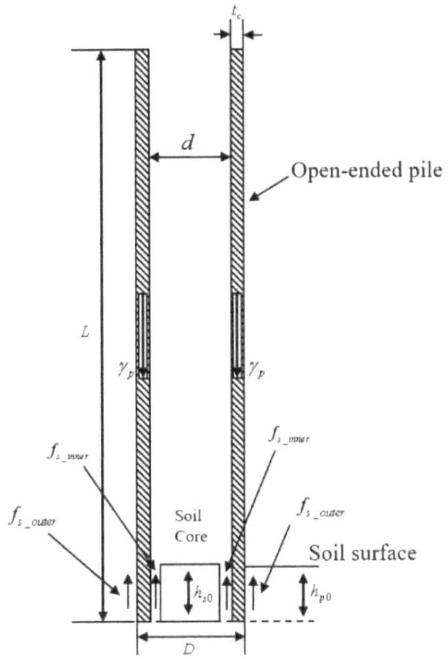

Figure 1. Schematic diagram of self-sinking force equilibrium of an open pile.

Assuming that $f_{s\_outer} = f_{s\_inner}$ and $h_{p0} = h_{s0}$, the above formula could be simplified as

$$h_{p0} = \left(\frac{1}{4}(\gamma_p L - q_{tip})\left(D^2 - d^2\right)\right) / (f_{s\_outer}(D + d)) \tag{3}$$

If the wall thickness of the pile is denoted as t_c, then $D = d + 2t_c$, which gives

$$h_{p0} = \frac{\gamma_p L - q_{tip}}{2 f_{s\_outer}} t_c \tag{4}$$

Hence, the self-sinking depth can be approximated using a few easier-to-obtained parameters, i.e., wall thickness, pile length, pile volume weight, and the pile tip and shaft resistances. By ignoring the pile tip resistance, a more simplified formula is given as follows:

$$h_{p0} = \frac{\gamma_p L}{2 f_{s\_outer}} t_c. \tag{5}$$

3. Partial Differential Analysis of LOSPs under Hammering Load

When a pile hammer strikes a pile at its top, the stress wave is generated and transmitted longitudinally along the pile body as shown in Figure 2, resulting in the penetration of the pile to the soil. One-dimensional wave equation is a suitable approach to analyze this process. Assuming isotropic property of the pile body, the basic wave-governing equation for a pile subjected to impact loading is listed as follows [23]:

$$\frac{\partial^2 u}{\partial t^2} = c^2 \nabla^2 u, \tag{6}$$

where $u(x, y, z)$ is the displacement in the x-, y-, and z-directions; $\nabla^2 = \partial^2/\partial x^2 + \partial^2/\partial y^2 + \partial^2/\partial z^2$ is the Laplace operator; c is the wave speed if one-dimensional wave propagation is assumed, in which $c = \sqrt{\lambda + 2G_p/\rho}$ for the compressional waves and $c = \sqrt{G_p/\rho}$ for the shear waves; G_p is the shear modulus of pile; ρ is the density of pile; $\lambda = \mu E/(1+\mu)(1-2\mu)$. The elastic wave velocity is calculated as $c = \sqrt{E/\rho}$. Thus, the above equation can be simplified in the framework of one-dimensional wave theory as follows:

$$\frac{\partial^2 u}{\partial t^2} = c^2 \frac{\partial^2 u}{\partial z^2}. \tag{7}$$

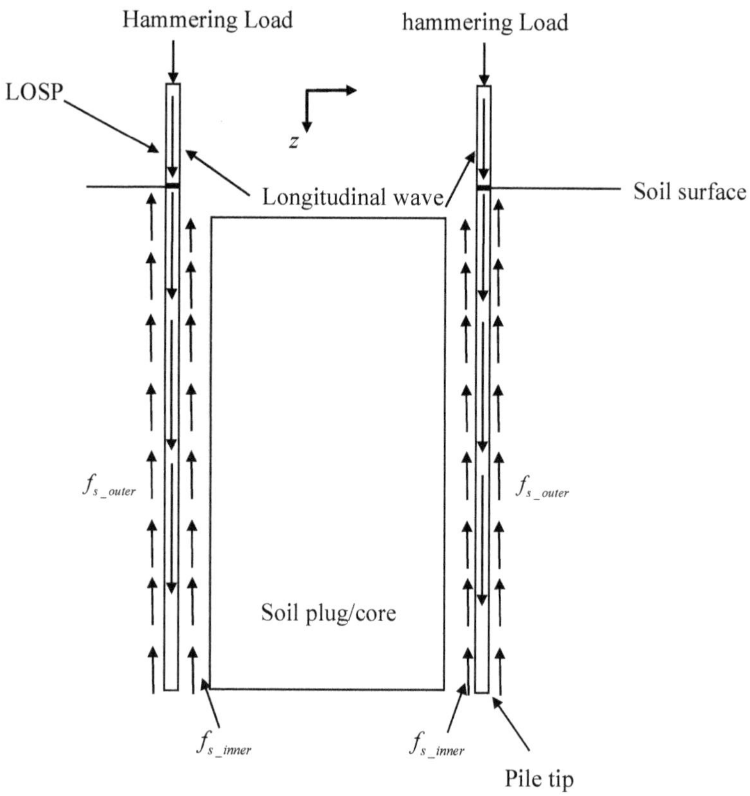

Figure 2. Load transfer of LOSP under hammering load.

The pile tip resistance is disregarded for the open-ended piles [24], and the dynamic equilibrium differential equation of the pile under a hammering load is presented as follows:

$$\rho \frac{\partial v(z,t)}{\partial t} - \frac{\partial \sigma_z(z,t)}{\partial z} = 0 \tag{8}$$

where $v(z, t)$ is the stress wave velocity along the pile shaft, and $\sigma_z(z,t)$ is the axial stress.

According to the elasticity theory,

$$\sigma_z(z,t) = E\varepsilon_z(z,t) \tag{9}$$

where $\varepsilon_z(z,t)$ is the axial strain of the pile. Substituting Equation (9) into Equation (8) results in

$$\rho\frac{\partial v(z,t)}{\partial t}-E\frac{\partial \varepsilon_z(z,t)}{\partial z}=0 \tag{10}$$

In addition, $v(z,t)=\partial u(z,t)/\partial t$ and $\varepsilon_z(z,t)=\partial u(z,t)/\partial z$, respectively. Therefore, Equation (10) can be expressed as

$$\frac{\partial u^2(z,t)}{\partial t^2}-E\frac{\partial u^2(z,t)}{\partial z^2}=0. \tag{11}$$

The Dirichlet boundary conditions are $u(0,t)=A\Delta l ln(t+1)/ln(t_0+1)$ and $u(L,t)=B\Delta w ln(t+1)/ln(t_0+1)$, where Δw is the pile penetration; Δl is the sum of the pile compression and penetration; $t_0 = L/c$, and A and B are constants. The Neumann boundary conditions are $u(z,0)=0$ and $\partial u(z,0)/\partial t=0$.

When the LOSP is loaded by the hammering loading, the longitudinal waves are applied in the pile body, and the transverse shear waves and longitudinal constrained waves exist in the soil inside the LOSP. To build a simplified analytical model, the effect of transverse shear waves in the soil inside the LOSP is ignored regardless of whether the soil plug effect occurs, at which point only longitudinal compressional stress waves exist in both the pile and the soil inside the pile under the hammering load. During hammering penetration of LOSP with thin wall, the pile tip resistance is a minor contributor to the total resistance. Thus, the tip resistance was ignored in the process of establishing the wave equation of LOSP under a single hammering blow. Alternatively, the physical mechanisms of the LOSP soil under hammering load in the case of no soil plugging and full soil plugging need to be considered separately in the theoretical analysis. As illustrated in Figure 2, under the hammering load, stress waves are transferred along the pile body and the inner and outer sides of the pile are subjected to frictional resistance. The inertial effect of soil could be considered in a dynamic situation. Thus, an inertial factor $\xi=a_{soil}/g$ was formulated in this partial differential derivation, where a_{soil} is the acceleration of the soil inside the pile, and this magnitude increases with soil plugging. When full soil plugging occurs, the displacement between the pile and the plug is completely coordinated, and now the inertia factor [25] is $\xi=a(t)/g$, where $a(t)$ is the acceleration of the pile at moment t. Hence, the wave equation that accounts for the shaft resistance is

$$\rho\frac{\partial^2 u(z,t)}{\partial t^2}-E\frac{\partial^2 u(z,t)}{\partial z^2}=\sum_{i=1}^{n}\xi_{inner_i}DCFR_{inner(i)}f_{s\_inner\_i}\frac{\pi(D-d)}{A_pL}h_{s_i}+\sum_{i=1}^{n}\xi_{outer\_i}DCFR\_outer(i)f_{s\_outer\_i}\frac{\pi D}{A_pL}h_{p\_i}, \tag{12}$$

where ξ_{inner_i} and ξ_{outer_i} are the inertia factors of the i-th soil layer, and the addition of the character i to the parameter symbols is an indication of the i-th layer of the soil; $DCFR$ is the dynamic coefficient of frictional resistance; and A_p is the cross-sectional area of the pile. The right-hand side of the above equation can be calculated as

$$M=\frac{1}{\rho}\left(\sum_{i=1}^{n}\xi_{inner\_i}DCFR\_inner(i)f_{s\_inner\_i}\frac{\pi(D-d)h_{s\_i}}{A_pL}+\sum_{i=1}^{n}\xi_{outer\_i}DCFR\_outer(i)f_{s\_outer\_i}\frac{\pi D h_{p\_i}}{A_pL}\right). \tag{13}$$

Then the partial differential equation is simplified as

$$\frac{\partial^2 u(z,t)}{\partial t^2}-c^2\frac{\partial^2 u(z,t)}{\partial z^2}=M \tag{14}$$

Thus, the problem is transformed into an inhomogeneity linear partial differential equation with the same boundary conditions as Equation (11). An auxiliary function, $w(z,t)=(B\Delta wz-A\Delta l(z-L))ln(t+1)/(Lln(t_0+1))$, is introduced which satisfies the Dirichlet boundary conditions.

Introducing the function $v(z,t)$ and letting

$$u(z,t)=v(z,t)+w(z,t) \tag{15}$$

a new format of wave equation can be obtained as

$$\frac{\partial^2 v(z,t)}{\partial t^2} - c^2 \frac{\partial^2 v(z,t)}{\partial z^2} = M - \frac{A\Delta l(z-L) - B\Delta wz}{Lln(t_0+1)} \frac{1}{(t+1)^2} \tag{16}$$

For the Dirichlet boundary, $v(0,t) = v(L,t) = 0$. For the Neumann boundary, $v(z,0) = 0$. Thus,

$$\frac{\partial v(z,0)}{\partial t} = \frac{A\Delta l(z-L) - B\Delta wz}{Lln(t_0+1)}. \tag{17}$$

The problem is then transformed into a problem of mixed inhomogeneity partial differential equations with flush boundary. The Fourier series expression for this problem is

$$u(z,t) = \sum_{n=1}^{\infty} \left(\varphi_n cos\frac{nc\pi t}{L} + \frac{L}{nc\pi}\psi_n sin\frac{nc\pi t}{L} + \frac{1}{nc\pi}\int_0^t f_n(\tau) sin\frac{nc\pi(t-\tau)}{L} d\tau \right) sin\frac{n\pi z}{L} \tag{18}$$

According to the boundary conditions and nonsimultaneous nature of the wave equation, $\varphi_n = 0$. Therefore,

$$\psi_n = \frac{2}{Lln(t_0+1)} \left(\left(A\Delta l - \frac{A\Delta l - B\Delta w}{n\pi} L \right) cos\ n\pi - A\Delta l \right) \tag{19}$$

and

$$f_n(t) = \frac{2}{n\pi} \left(\frac{A\Delta l - B\Delta w}{ln(t_0+1)} \frac{1}{(t+1)^2} cos\ n\pi + \left(M + \frac{A\Delta l}{ln(t_0+1)} \frac{1}{(t+1)^2} \right) (1 - cos\ n\pi) \right) \tag{20}$$

Equations (19) and (20) represent an analytical solution to the displacement of the open-ended pile stress wave under a hammering load, where A and B are constants. When the pile is assumed to be a rigid body, $\Delta l = \Delta w$. The vertical displacement of the pile accompanying stress wave propagation in the presence of resistance inside and outside of the pile can be determined by substituting the corresponding resistance into the analytical solution.

4. Parameters Sensitivity

Assuming that no soil plugging effect occurs during LOSPs driving and the pile is a rigid body, and the soil inside and outside the pile are at the same height. The pile displacement under a hammering load is calculated according to the analytical formula (Equation (18)). Sensitivity analyses of the parameters in Equation (18) are conducted based on LOSPs for internal diameters of 2, 4 and 6 m, separately. The specific calculation parameters are shown in Table 1. The effect of pile compression is not considered during the hammering process. Therefore, the effect of constant B does not need to be considered in Equations (19) and (20). The sensitivity analyses of the proposed displacement equation for LOSPs are shown in Figures 3–5.

Table 1. Parameters of open-ended piles installed by impact loading.

Elastic modulus of the pile, E (kPa)	2×10^8	Inner diameter of the pile, d (m)	2, 4, 6
Density of the pile (kg/m^3)	7.63×10^3	Pile length, L (m)	10 d
Wave velocity, c (m/s)	5119.7	Ratio between thickness of pile wall and inner diameter, t/d	0.015
Maximum number of summations of Fourier solutions, n	1000	*DCFR*	0.1, 0.5, 1.0
Constant A in Equation (19)	1, 1000, 1,000,000	Shaft friction inside and outside the pile, f_s (kPa)	40.0
Single hammer penetration depth, Δl (m)	0.01	Average acceleration of soil inside and outside the pile, a (m/s^2)	1 (Outside the pile), 10 (Inside the pile)

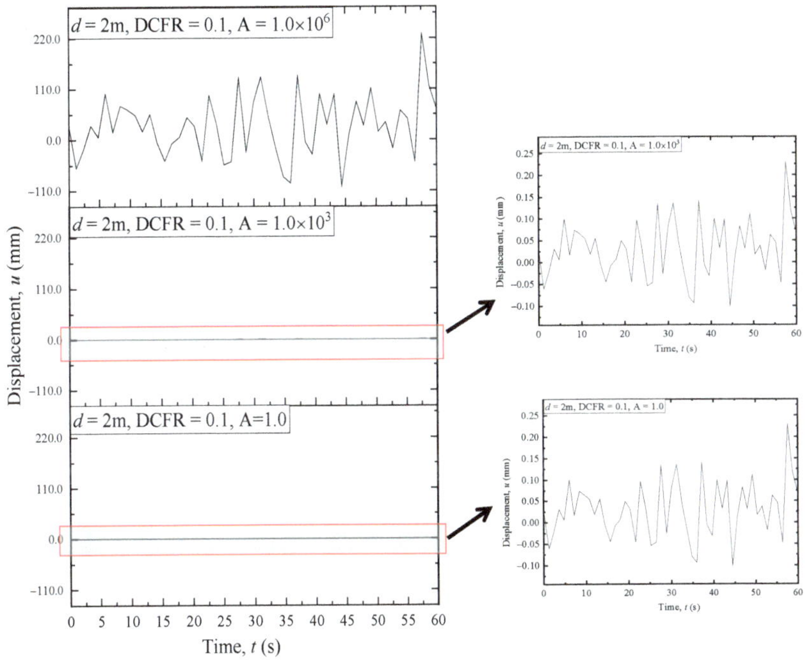

Figure 3. Sensitivity analysis of the constant A.

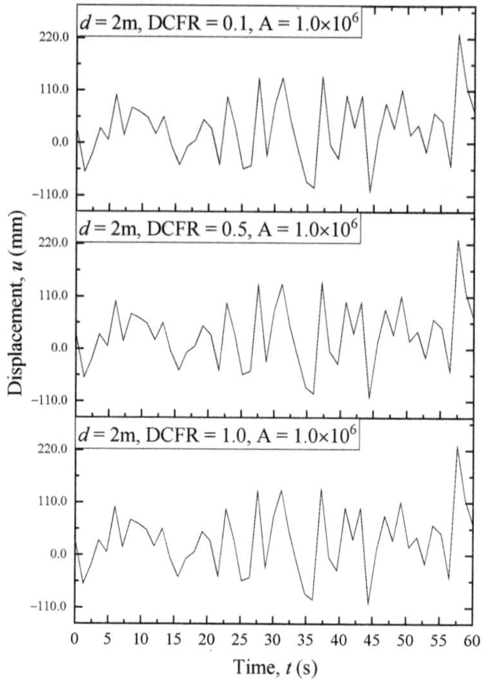

Figure 4. Sensitivity analysis of DCFR.

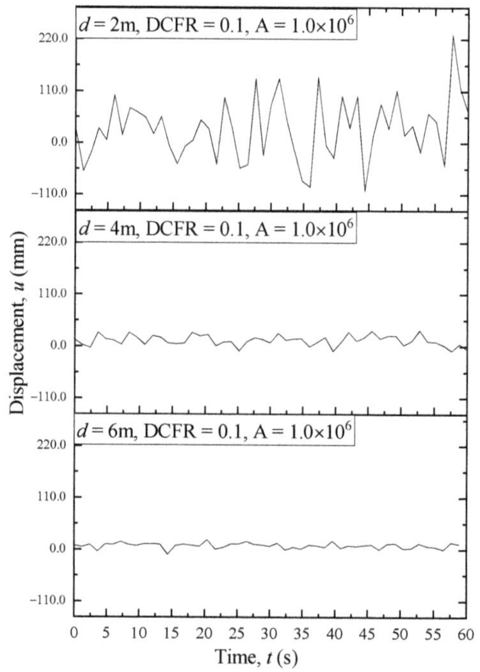

Figure 5. Sensitivity analysis of the diameter of LOSP.

Although the type of curve oscillation did not vary, the decrease in the magnitude of the constant A resulted in a significant jump down in the scale of the displacement amplitude for LOSP under hammering load (for the case of D = 2 m). The constant A is an empirical parameter and therefore requires more field data to determine the value. As illustrated in Figure 4, the analytical solution is insensitive to the *DCFR* value. With no soil plugging effect, the value of the DCFR for the LOSP during driving is relatively small and therefore has less influence on the Fourier solution results. The increasing diameter of the pile will reduce the penetration and will also change the shape of the vibration curve under the effect of identical single hammering energy. The other parameters are the same, indicating that for the identical hammering energy, the larger the pile diameter, the smaller the penetration.

5. Comparison between Numerical Method and Fourier Solution

A finite element (FE) numerical model in ABAQUS/Standard incorporated with the remeshing and interpolation technique with a small strain (RITSS) was built [26], which is carried out on the effective stress analysis for axisymmetric pile impact driven into saturated soil, as shown in Figure 6. The soil model had a width of $10d$ and height of $2h_p$ (where h_p being the target penetration depth) to eliminate the boundary effects. The soil within a distance of $2d$ longitudinally down from the pile tip and $2d$ laterally away from the outside pile was finely meshed, and the element size gradually increased away from the pile. In the region of penetration, the minimum size of the element was equal to the wall thickness, and the maximum size was 30 cm. In the far region, the minimum size was 30 cm, and the maximum size was $1.5d$. The boundary conditions were free vertically and constrained laterally on the axisymmetric side and constrained in both directions in the far boundary of the model. The element type of this numerical model is CAX4P. The essence of the RITSS technique is to map the variables of the integration points of the deformed elements into the remesh element in the finite element framework [27,28], thus enabling a method for high precision analysis of large deformation problems, i.e., the pile-driving process and spudcan penetration stimulation [29]. The numerical results were compared to the solution curve calculated using Equation (18). The specific parameters values are listed in Table 2. The numerical model was used to obtain the last penetration curve. The comparative analysis is shown in Figure 7, which indicates that the two penetration curves are similar in magnitude. However, the oscillation of the analytical solution curve is more significant for two main reasons. (1) The single hammer load in the numerical model is artificially divided into several steps to achieve convergence. (2) The number of time points for extracting penetration data in the numerical model is not as high as in the analytical solution because of computational cost constraints to reduce the cost of the calculation without affecting the accuracy and overall trend of the displacement curve. In fact, the first displacement peak of the numerical and theoretical curves is essentially the pile penetration, as the theoretical results do not take into account the rebound of the stress wave from the pile bottom and the corresponding resistance of the soil to the pile kickback, which is why the larger displacement fluctuations of the pile occur after the first displacement peak in the figure. In general, the theoretical Equation (18) can be used for rigid pile displacement prediction under one blow. Considering the transmission of stress waves along the pile body during driving, the curves of the first 2 s in the figure are the focus of interest.

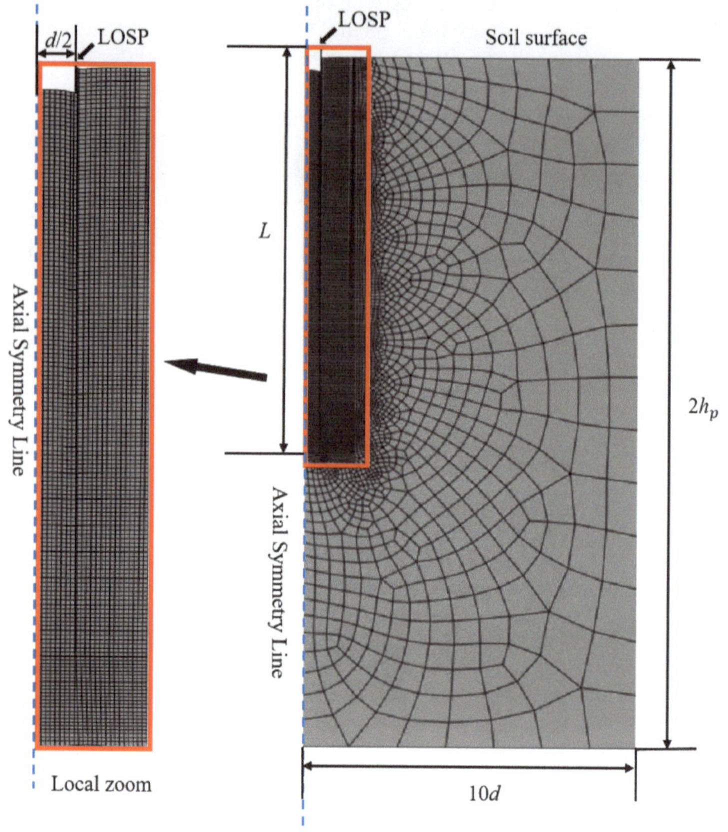

Figure 6. A typical finite element model in this study.

Table 2. Adopted parameters values for comparison analysis.

Parameters	Values
Internal diameter of pile, d (m)	2.0
Wall thickness, t_c (m)	0.03
Pile length, L (m)	20.5
Elastic modulus of pile, E (kPa)	2×10^8
Density of pile (kg/m^3)	7.63×10^3
Average frictional resistance inside and outside the pile, f_s (kPa)	35
Average acceleration of soil inside the pile (m/s^2)	17.0
Average acceleration of soil outside the pile (m/s^2)	12.1
Pile penetration by single blow, Δl (m)	1.50
Pile penetration depth, h_p (m)	20.02
Shaft resistance dynamic coefficient, $DCFR$	0.1
Duration of single blow (s)	19.79
Constant A in Equation (18)	65
Maximum number of summations of Fourier solutions, n	1000

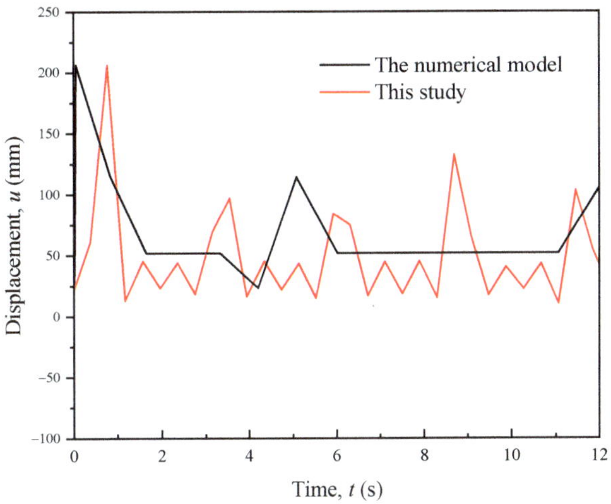

Figure 7. Comparison of displacement curves predicted from this study and the numerical result.

6. Conclusions

The LOSPs are widely used in offshore engineering and their driveability analysis has been one of the focuses among the academia and industry. This study proposes a simple approach to calculate the self-sinking depth of LOSPs under self-weight. The concepts of inertia force and inertia factor are employed based on the one-dimensional wave theory to develop a partial differential equation considering the shaft resistance. Further, a Fourier solution for the displacement of the open-ended pile under one blow strike is proposed, and a parametric sensitivity analysis is carried out for the application of the equation in LOSPs. The analytical solution is compared with the numerical result to confirm the validity of the proposed methodology. The results of this study indicate that the simple calculation approach has the potential for predicting the condition of hammer refusal or punching through under single hammer blow for LOSPs driven in offshore soft soil. It should be noted that when LOSP is driven into soft soils, full soil plugging rarely occurs. If it occurs, the acceleration of the soil within the pile should be aligned with the pile during the calculation. With the proposed displacement calculation method for LOSP with a single hammer, the maximum downward displacement in the curve can be considered as the penetration of the single hammer, and the displacement curve of LOSP during the whole driving process can be determined by accumulating and iterating the displacements of the single hammer.

Author Contributions: Conceptualization, W.Q. and G.D.; methodology, W.Q.; software, W.Q.; validation, W.Q.; formal analysis, W.Q.; investigation, W.Q.; resources, G.D. and W.Q.; data curation, W.Q.; writing—original draft preparation, W.Q. and X.L.; writing—review and editing, W.Q., X.L. and P.H.; visualization, W.Q. and X.L.; supervision, G.D. and W.Q.; project administration, G.D. and W.Q.; funding acquisition, W.Q. and G.D. All authors have read and agreed to the published version of the manuscript.

Funding: This research was funded by the National Natural Science Foundation of China (grant number: 52008318(granter: W.Q.) and 51878160(granter: G.D.), Education Department of Zhejiang Province (grant number: Y202146717(head of funding: W.Q.; participant: X.L.)), and by the Graduate Innovation Fund Project of Wenzhou University (grant number: 316202101061(head of funding: W.Q.; participant: X.L.)).

Data Availability Statement: The authors declare that they have no known competing financial interests or personal relationships that could have appeared to influence the work reported in this paper.

Acknowledgments: Thanks to Mark Randolph of The University of Western Australia for his valuable advice during the conceptualization of this study. Thanks to Erchuan Zhang of a former Ph.D. student at the University of Western Australia for his advice on the derivation of partial differential equations in this paper.

Conflicts of Interest: The authors declare no conflict of interest.

References

1. Nguyen, T.; McVay, M.; Herrera, R. Case Studies of Rebounds on Long, Slender Piles. In Proceedings of the 10th International Conference on Stress Wave Theory and Testing Methods for Deep Foundations, Lowell, MA, USA, 1–3 June 2016; pp. 640–650.
2. Song, Y.P.; Sun, Y.F.; Cao, C.L.; Li, S.L. Study on Pile-Soil Interaction during Pile Sinking and Drivability of Pile in Offshore Platform. *Appl. Mech. Mater.* **2012**, *164*, 137–141. [CrossRef]
3. Milewski, H.; Kennedy, J. Application of Friction Fatigue Pile Driving Models in GRLWEAP. In Proceedings of the ASME 2019 38th International Conference on Ocean, Offshore and Arctic Engineering, Glasgow, Scotland, 9–14 June 2019; OMAE2019-95944.
4. Xiao, S.; Wang, K.; Gao, L.; Wu, J. Dynamic characteristics of a large-diameter pile in saturated soil and its application. *Int. J. Numer. Anal. Methods Geomech.* **2018**, *42*, 1255–1269. [CrossRef]
5. Dean, E.T.R.; Deokiesingh, S. Plugging criterion for offshore pipe pile drivability. *Geotechnique* **2013**, *63*, 796–800. [CrossRef]
6. Coyle, H.M.; Sulaiman, I.H. Skin friction for steel piles in sand. *J. Soil Mech. Found. Div.* **1967**, *93*, 261–277. [CrossRef]
7. Holloway, D.M.; Clough, G.W.; Vesic, A.S. The effects of residual driving stresses on pile performance under axial loads. In Proceedings of the 10th Offshore Technology Conference, Houston, TX, USA, 8–10 May 1978; pp. 2225–2236.
8. Lee, J.H.; Salgado, R. Determination of pile base resistance in sands. *J. Geotech. Geoenviron. Eng.* **1999**, *125*, 673–683. [CrossRef]
9. Kim, S.; Jeong, S.; Cho, S.; Park, I. Shear load transfer characteristics of drilled shafts in weathered rocks. *J. Geotech. Geoenviron. Eng.* **1999**, *125*, 999–1010. [CrossRef]
10. Guo, W.D.; Randolph, M.F. Vertically loaded piles in non-homogenous media. *Int. J. Numer. Anal. Methods Geomech.* **1996**, *21*, 507–532. [CrossRef]
11. Lehane, B.M.; Gavin, K.G. Base resistance of jacked pipe piles in sand. *J. Geotech. Geoenviron. Eng.* **2001**, *127*, 473–480. [CrossRef]
12. Nicola, A.D.; Randolph, M.F. The plugging behaviour of driven and jacked piles in sand. *Géotechnique* **2015**, *47*, 841–856. [CrossRef]
13. Paik, K.; Salgado, R. Determination of Bearing Capacity of Open-Ended Piles in Sand. *J. Geotech. Geoenviron. Eng.* **2003**, *129*, 46–57. [CrossRef]
14. Paik, K.; Salgado, R.; Lee, J.; Kim, B. Behavior of Open and Closed-Ended Piles Driven into Sands. *J. Geotech. Geoenviron. Eng.* **2003**, *129*, 296–306. [CrossRef]
15. Goel, S.; Patra, N.R. Prediction of load displacement response of single piles under uplift load. *Geotech. Geol. Eng.* **2007**, *25*, 57–64. [CrossRef]
16. O'Neill, M.W.; Raines, R.D. Load transfer for pipe piles in highly pressured dense sand. *J. Geotech. Eng.* **1991**, *117*, 1208–1226. [CrossRef]
17. Liu, Y.; Vanapalli, S.K. Load displacement analysis of a single pile in an unsaturated expansive soil. *Comput. Geotech.* **2019**, *106*, 83–98. [CrossRef]
18. Xu, H.F.; Zhang, J.X.; Liu, X.; Geng, H.-S.; Li, K.-L.; Yang, Y.-H. Analytical Model and Back-Analysis for Pile-Soil System Behavior under Axial Loading. *Math. Probl. Eng.* **2020**, 1369348. [CrossRef]
19. Lehane, B.M.; Li, L.; Bittar, E.J. CPT-based load-transfer formulations for driven piles in sand. *Géotech. Lett.* **2020**, *10*, 568–574. [CrossRef]
20. Xu, X.B.; Zhang, T.Y.; Wang, J.C.; Hu, M.Y.; Chen, K.L. Model tests on the bearing capacity of precast open-ended micro pipe piles in soft soil. *Geotech. Eng.* **2020**, *173*, 500–518. [CrossRef]
21. Lehane, B.M.; Lim, J.K.; Carotenuto, P.; Nadim, F.; Lacasse, S.; Jardine, R.J.; Van Dijk, B.F.J. Characteristics of Unified Databases for Driven Piles. In Proceedings of the 8th International Conference of Offshore Site Investigation and Geotechnics (OSIG 2017), London, UK, 12–14 September 2017.
22. Randolph, M.F.; Wroth, C.P. Analysis of Deformation of Vertically Loaded Piles. *J. Geotech. Geoenviron. Eng.* **1978**, *104*, 1465–1488. [CrossRef]
23. Graff, K.F. *Wave Motion in Elastic Solids*; Ohio State University Press: Columbus, OH, USA, 1975.
24. Paikowsky, S.G. Use of Dynamic Measurements to Predict Pile Capacity under Local Conditions. Master's Thesis, Israel Institute of Technology, Haifa, Israel, July 1982.
25. Paikowsky, S.G.; Chernauskas, L.R. Dynamic analysis of open-ended pipe piles. In Proceedings of the 8th International Conference on the Application of Stress-Wave Theory to Piles, Lisbon, Portugal, 8–10 September 2008.
26. Qin, W.; Cai, S.; Dai, G.; Wang, D.; Chang, K. Soil resistance during driving (SRD) of offshore large-diameter open-ended thin-wall pipe piles driven into clay by impact hammers. *Comput. Geotech.* 2022, *submitted*.

27. Hu, Y.; Randolph, M.F. H-adaptive FE analysis of elasto-plastic non-homogeneous soil with large deformation. *Comput. Geotech.* **1998**, *23*, 61–83. [CrossRef]
28. Hu, Y.; Randolph, M.F. A practical numerical approach for large deformation problems in soil. *Int. J. Numer. Anal. Methods Geomech.* **1998**, *22*, 327–350. [CrossRef]
29. Wang, D.; Randolph, M.F.; White, D.J. A dynamic large deformation finite element method based on mesh regeneration. *Comput. Geotech.* **2013**, *54*, 192–201. [CrossRef]

Journal of
Marine Science and Engineering

MDPI

Article

Model Test of the Pullout Bearing Capacity of End-Bearing Torpedo Anchors

Gang Li [1], Jinli Zhang [2,*] and Jia Liu [3]

1 Shaanxi Key Laboratory of Safety and Durability of Concrete Structures, Xijing University, Xi'an 710123, China; t_bag945@126.com
2 State Key Laboratory of Coastal and Offshore Engineering, Dalian University of Technology, Dalian 116024, China
3 School of Geological Engineering and Geomatics, Chang'an University, Xi'an 710054, China; 15929935077@163.com
* Correspondence: jlzhang@dlut.edu.cn; Tel.: +86-139-4094-9667

Abstract: Torpedo anchors (TAs) are regarded as one of the most efficient mooring solutions for taut mooring systems, and end-bearing TAs are a new type of TA that primarily relies on end-bearing plates at the tail to generate a pile-end resistance to improve their pulloutbearing capacity (P). Therefore, the estimation of the pullout capacity of the end-bearing TA is vital for the design of offshore floating facilities. In this study, pullout model tests were conducted to investigate the P of conventional and end-bearing TAs and examine the effects of factors such as the embedment depth (h), the relative density (D_r), the pullout angle (α), and the area (A) of the bearing plates on P. The test results show that, under oblique pullout loading, the P on the conventional TA increased slowly as displacement increased, while there was a peak on the load–displacement curve of each end-bearing TA with a relatively large A. The end-bearing TAs considerably outperformed the conventional TAs in terms of the P. In addition, increasing h, D_r, and A significantly increased the P of the end-bearing TAs. However, increasing h and D_r slightly decreased the ability of the bearing plates to increase the P of the end-bearing TAs. These research results can provide a guideline for TA installation in deep-sea engineering.

Keywords: torpedo anchor; pullout capacity; embedment depth; relative density; pullout angle

Citation: Li, G.; Zhang, J.; Liu, J. Model Test of the Pullout Bearing Capacity of End-Bearing Torpedo Anchors. *J. Mar. Sci. Eng.* **2022**, *10*, 728. https://doi.org/10.3390/jmse10060728

Academic Editor: Dracos Vassalos

Received: 24 April 2022
Accepted: 23 May 2022
Published: 26 May 2022

1. Introduction

Human demand for environmental resources is increasing with rapid social and economic development. Exploitation and effective utilization of marine resources has become a primary means of addressing the imbalance between the supply of, and demand for, limited resources [1–3]. Marine oil and gas resources are generally extracted on fixed or floating platforms. The effective anchorage of oil platforms is a central issue in marine oil and gas exploitation. Oil platforms can be fixed to the seafloor through structures such as tension legs, piles, suction caissons, and drag anchors in shallow-sea environments, and through structures such as plate anchors, suction caissons, and drag anchors in deep-sea environments [4,5]. However, the installation of various anchorage systems in deep-sea environments is extremely difficult and costly due to relatively deepwater depths and the significant effects of external loads (e.g., waves and winds). In recent years, taut-leg mooring systems composed of anchors (often torpedo anchors (TAs)) and anchor chains have been used to fix oil platforms for deep-sea oil extraction. TAs rely primarily on gravity to penetrate the seafloor. As a result, their installation is fast, easy, and relatively inexpensive, and does not require special installation vessels [6–8].

The pullout bearing capacity (P) of TAs is vitally important for the reliability of a mooring system, and is affected primarily by factors such as the properties of the seafloor soil, the penetration speed and depth, and the TA parameters [9,10]. Chen et al. [11]

conducted tests to investigate the penetration and pullout behaviors of torpedo anchors under vibration, and found that the anchoring force reached more than 83.2% of holding capacity after 7 d in the silty bed. Wang et al. [12] studied the penetration depth into cohesive soil, considering the effect of torpedo anchor shape and size, and a formula to calculate the penetration depth was established based on energy conservation principles and measurements. Ads et al. [13] indicated that the fin length of a torpedo anchor correlated negatively with penetration depth, and correlated positively with extraction resistance. With increasing penetration depth, soil displacement initiallyincreased and then keptconstant. Based on centrifuge tests, Hossain et al. [14] studied the dynamic installation and monotonic pullout of torpedo anchors into calcareous silt, and a total energy expression was proposed to predict the embedment depth, accounting for anchor geometry, anchor mass, and impact velocity. Lai et al. [15] assessed the free-fall penetration behavior of dynamically penetrating anchors into marine clay, and an energy-based model and a force-based model were established to predict the penetration depth of anchors with different geometries and impact velocities. Takinginto account the effect of density, aspect ratio, scale ratio, and fin size, Hasanloo et al. [16] investigated the falling velocity and drag coefficient of torpedo anchors during acceleration, and a mathematical model for anchor motion was proposed. Wang et al. [17] studied the relationship between pullout capacity, anchor properties, and soil type, and a formula to calculate the monotonic holding capacity of torpedo anchors was established, accounting for embedded depth, weight, geometry, and soil properties. Wang et al. [18–20] investigated the effect of pullout velocity on the pullout behavior of torpedo anchors by numerical analysis and model testing, and formulas to predict the maximum vertical and inclined pullout force were established. Meanwhile, an empirical formula to predict the penetration depth was established and validated, based on energy analysis and test results.Santiago et al. [21] conducted centrifuge tests on torpedo piles installed in loose sand, and found that the interactions between the adjacent piles depended on the load direction and pile spacing. Based on the test results, the optimal pile spacing for maximum cluster efficiency was proposed. During the installation process, Fernandes et al. [22,23] concluded that a torpedo anchor have a minimum directional stability, with a less-than-three degree vertical angle required for control.Compared to four-fin torpedo anchors, Chang et al. [24] noted that the vertical and horizontal bearing capacities of the novel installed 'fish' anchors were 4.0 and 5.6 times greater. Raaj et al. [25] found that the hydrodynamic performance of torpedo anchors followed the order of hemispheric nose, ellipse, and ogive, and that the conical nose showed inferior performance. Kunitaki et al. [26] compared the Monte Carlo method and a fuzzy arithmetic method to treat uncertainties during torpedo anchor installation, and concluded that the fuzzy arithmetic method was an effective design tool due to its outstanding computational efficiency. O'Beirne et al. [27] validated the new release-to-rest model for torpedo anchors, based on test results, and found that the model-calculated results were consistent with field data of over 100 anchor installations, with an accuracy of $\pm10\%$.Coupling finite element of deformation and fluid flow in porous media, Raie and Tassoulas [28] indicated that the dissipation rate of excess pore-water pressure and recovery of soil strength were critically important for predicting the pull-out capacity of torpedo anchors at different times. Based on the statistical model uncertainty evaluated using test results and finite element simulation results, Sagrilo et al. [29] found that the single reliability safety factor was significantly lower than the value used in the design of torpedo anchors, considering the working stress design method. DeSousa et al. [30] indicated that the total contact area of a torpedo anchor is a necessary parameter for determining the load capacity, and that the load capacity of an anchor can significantly increase with an increase in undrained shear strength and fluke number. Based on a large deformation finite element approach, Kim et al. [31–34] investigated the installation and pullout process of a torpedo anchor. It was concluded that embedment depth was significantly decreased with increasing strain rate dependency of a soil, and the pullout capacity of a finless anchor under 45° was 29% of that of a four-fin anchor. Based on the results, an improved analytical embedment model considering shearing resistance and drag resistance was established. Sabetamal et al. [35,36] observed that a smooth discretization of the contact interface between soil and structure is a crucial factor for avoiding oscillations of soil when simulating the installation of torpedo anchors.

Current studies have primarily focused on investigating the P of conventional TAs through model tests, combined with numerical simulations, while end-bearing TAs have received relatively less attention. In this study, pullout model tests were performed to investigate the P of conventional and end-bearing TAs, and the effects of factors such as the embedment depth (h), the relative density (D_r), the pullout angle (α), and the area (A) of the bearing plates on the P. The results have a certain reference value for guiding the design and installation of TAs in deep-sea projects.

2. Materials and Methods

2.1. Test Materials

Fujian standard sand was used in the tests, and the specific gravity test, relative density test, and particle size analysis test were conducted according to the Geotechnical Test Method Standard (GBT 50123-2019). Table 1 summarizes the physical properties of the sand sample. Figure 1 shows the particle-size distribution curve of the test sand. Particles with a diameter of less than 2 mm accounted for 100% of the total weight, and particles with a diameter larger than 0.5 mm exceeded 50% of the total weight. Therefore, the test sand used in this study can be classified as coarse sand. Based on the values of the coefficient of uniformity (3.623) and the coefficient of curvature (1.099), the test sand was determined to be poorly graded.

Table 1. Physical parameters of the test sand.

G_s	e_{max}	e_{min}	C_u	C_c
2.650	0.802	0.349	3.623	1.099

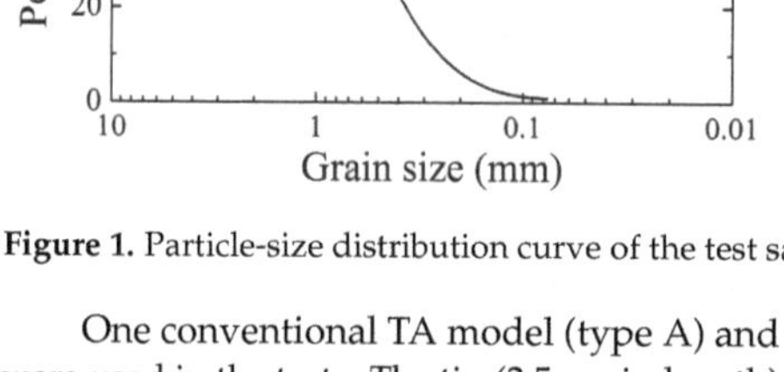

Figure 1. Particle-size distribution curve of the test sand.

One conventional TA model (type A) and five end-bearing TA models (type B1–B5) were used in the tests. The tip (2.5 cm in length) and bearing plates of each end-bearing TA were made of stainless steel, while the shank (17.5 cm in length (L), 2.5 cm in diameter (d), and 207.0 g in mass (m)) was made from a hollow aluminum tube. The bearing plates in the type B1–B5 TAs had external diameters of 3.53, 4.33, 5.00, 5.59, and 6.02 cm, respectively, areas that were 2–6 times the cross-sectional area of the shank, and masses of 9.3, 16.9, 26.3, 33.6, and 39.2 g, respectively, as shown in Figure 2. A poly(methyl methacrylate) bucket with a diameter of 45 cm, a height of 75 cm, and a wall thickness of 2 cm was used in the tests.

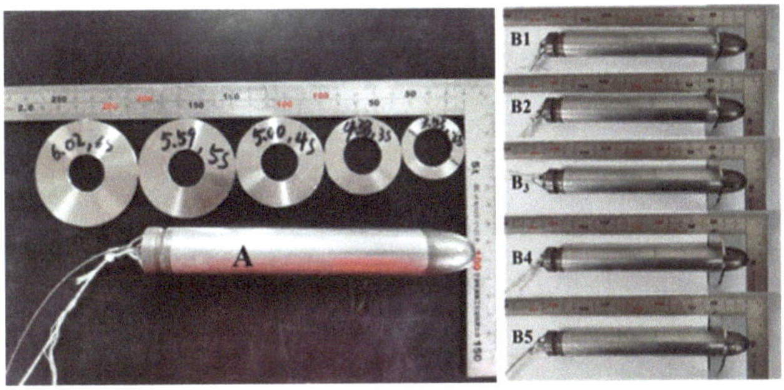

Figure 2. Model of torpedo anchors.

2.2. Model Arrangement

Figure 3 shows the TA pullout test setup, which consisted mainly of a support frame, a fixed pulley, a sample barrel, a drive motor, a tension sensor, and a data acquisition system. The drive motor of a triaxial shear apparatus was used to apply a load, and met the requirement for loading at a constant rate within its loading capacity. During loading, the back of the anchor was connected via a thin steel strand to an "S"-shaped tension/compression sensor with a measuring range of ±200 kg, which was, in turn, connected via a signal magnifier to a TWD information acquisition box using data cables. Data were measured at a frequency of 200 Hz. The sensor was calibrated by graded loading, and the results indicated that the performance of a tension sensor ranging from 0 to 50 kg could meet the test requirements. The measured data were smoothed by MATLAB 2020a software (MathWorks Software (Beijing) Co., Ltd., Beijing, China). The data acquisition system was used to measure and record the data in real time during the test process, in which the model was pulled out. The loading rate was set to 1 mm/s. The corresponding displacement (S) was calculated based on the test time.

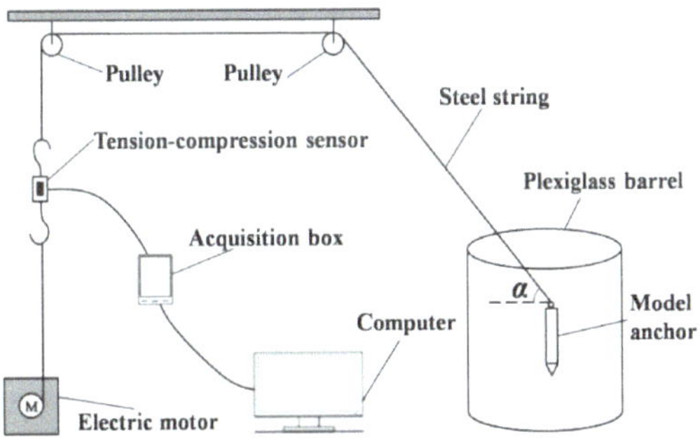

Figure 3. Schematic of the pullout test setup.

2.3. Test Methods

Based on the American Petroleum Institute (API) standard [37] and the similarity principle listed in Table 2, a total of 42 tests were conducted on the type A and B TAs for different values of D_r, h, and α, and Table 3 summarizes the test schemes. Sand samples were prepared through compaction in the test process. Specifically, sand was first weighed according to the predesigned height and D_r of the sample, and subsequently placed layer-by-layer in the barrel until the desired height was reached. Each layer of sand was compacted before the next layer was placed. A 10 cm-tall layer of sand was first placed at the bottom of the barrel. Then, the TA was installed in the center of the barrel via a fixed thin steel strand, the length of which was adjusted to place the TA at a certain depth. When placing sand near the TA, the TA was fixed with a sleeve to prevent this process from affecting its verticality. Subsequently, the water was filled on the surface of the sand bed to ensure sand saturation, and 10 cm of free water was maintained above the sand surface during the model test. Because each TA had a diameter of 2.5 cm, and the barrel had an internal diameter of 41 cm, the boundary effect was effectively eliminated.

Table 2. Similarity principle between prototype and model.

Parameter	Symbol	Scale (Prototype/Model)
Length	L	λ
Depth	h	λ
Area	A	λ^2
Mass	m	λ^3
Gravity	G	λ^3
Water density	ρ_w	1
Soil density	ρ_s	1
Time	t	$\lambda^{0.5}$
Drag resistance	P	λ^3
Displacement	S	λ

Table 3. Test schemes.

D_r	h (cm)	α (°)
0.6	2.0L	0°, 45°, 90°
0.6	2.5L, 3.0L	90°
0.7	2.0L	90°
0.8	2.0L	90°

During the test process, the embedded TA was first connected to a tension transducer via a steel strand, while α was adjusted through a fixed pulley. Then, the steel strand was connected to the rotating shaft of a triaxial apparatus. The strain-control mode was used. The loading rate was set to 1 mm/s. Data were acquired at a frequency of 200 Hz. The test was terminated when a peak appeared on the P–S curve, or when the preset value was reached. To ensure reliable test results, parallel tests were also conducted.

3. Results and Discussion

3.1. Analysis of Vertical Load–Displacement Curves

The main factors affecting the P value of a TA include the soil properties, h, the type of TA, α, and the interaction between the TA and the soil. To examine the P of different types of TAs, the P–S curves of the TAs in sand with different D_r at h = 2.0L were obtained, as shown in Figure 4. At h = 2.0L, as displacement increased, the P value of the conventional TA increased slowly to a certain value, and then remained approximately at this value, with no notable peak on its curve. In comparison, the P on each end-bearing TA increased rapidly as S increased, with a peak appearing, and then sharply decreased. The main reason for this phenomenon is that the P of the TA increased gradually with the increasing of S. When the P reached a peak, the end-bearing TA loosened the upper soil due to the end-bearing resistance of the bearing plate. If the vertical pullout loading is continuously applied, the P will sharply decrease with increasing of S. Under vertical pullout loading, the P of the end-bearing TA was much higher than that of the conventional TA and increased as A increased. This finding suggests that under vertical pullout loading, the bearing plates of an end-bearing TA compress the soil above them, thereby improving the P of the TA. For an end-bearing TA, a relatively large S is required to reach its P. Therefore, in-depth research is warranted to improve the P of TAs through innovative end-bearing designs.

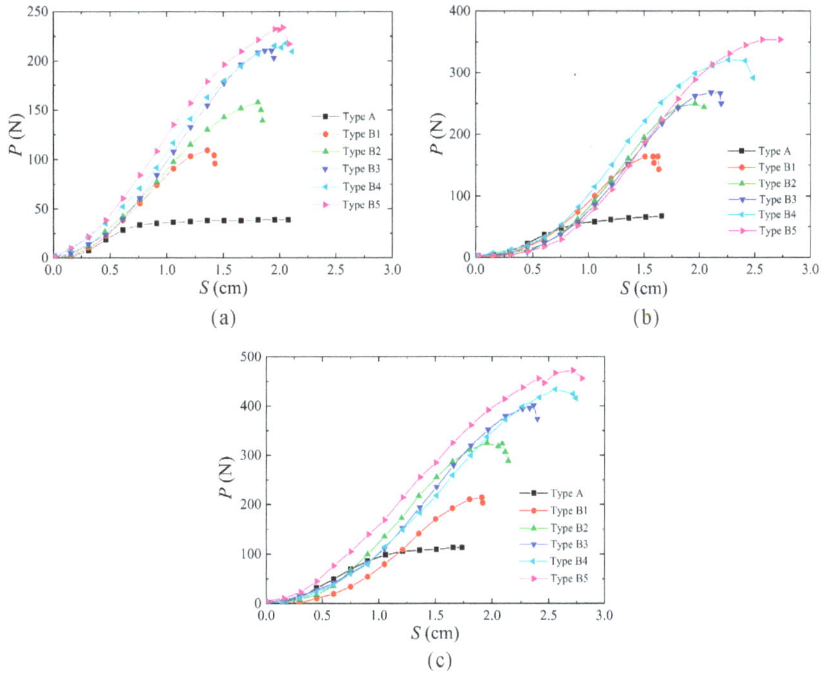

Figure 4. Load–displacement curves of the TAs under vertical pullout loading (**a**) $D_r = 0.6$; (**b**) $D_r = 0.7$; (**c**) $D_r = 0.8$.

3.2. Analysis of Oblique and Horizontal Load–Displacement Curves

A TA in normal operation is generally subjected to an oblique pullout load. To examine the effects of α on the P value, tests were conducted at $D_r = 0.6$, $h = 2.0L$, and $\alpha = 0°$ and $45°$. Figure 5 shows the test results. As shown in Figure 5a, the horizontal P–S curves of the two types of TAs were similar, and each contained a peak. However, under horizontal pullout loading, the P value of the end-bearing TAs was slightly higher than that of the conventional TA. Under horizontal pullout loading, a TA primarily moves through rotation and is subjected to passive earth pressure at both sides of the center of rotation, and its P depends on the passive earth pressure. The bearing plates of end-bearing TAs are horizontal, the same as the pullout load, and therefore play a relatively non-significant role. As a result, the above test results were obtained. Figure 5b reveals a significant difference between the P–S curves of the two types of TAs at $\alpha = 45°$. For the conventional TA, P increased slowly as S increased, similar to the trend observed under vertical pullout loading. In contrast, each end-bearing TA with a relatively large A exhibited a significantly higher P, with a peak on its curve. The curve of the conventional TA was similar to that of each end-bearing TA with a relatively small A, but the end-bearing TAs had a slightly higher P. Under oblique pullout loading, a TA not only moves vertically, but also rotates horizontally, and its P depends on both the lateral earth pressure and the frictional resistance between it and the soil. The presence of a vertical S allows the end resistance provided by the bearing plates of an end-bearing TA to play its role. Therefore, the P of end-bearing TAs should be higher than that of conventional TAs. The test results show that the P values of the type B1–B5 TAs were 16%, 30%, 64%, 83%, and 118% higher than that of the type A TA, respectively.

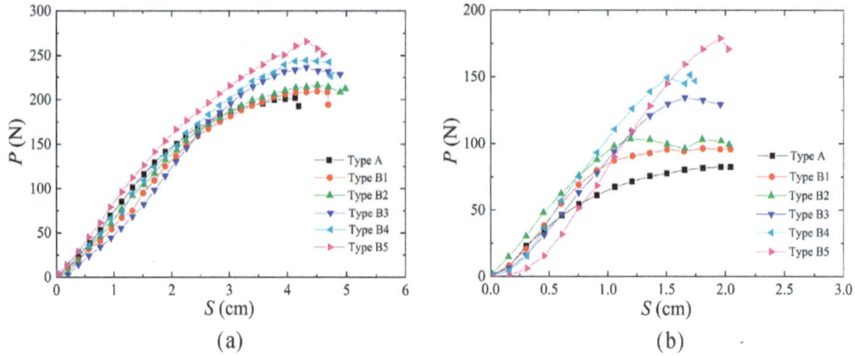

Figure 5. Load–displacement curves of the TAs under oblique and horizontal pullout loading (**a**) $\alpha = 0°$; (**b**) $\alpha = 45°$.

3.3. Effects of *h* on *P*

h is a major factor that affects the *P* of a TA. Therefore, the *P*–*S* curves of the type A and B TAs at $h = 2.0L$, $2.5L$, and $3.0L$ under vertical pullout loading were obtained (Figure 6). *P* and *S* were non-dimensionalized based on the self-weight of the TA and the diameter of the shank of the TA, respectively. The *P* of each type of TA increased as *h* increased. As *S* increased, the *P* of the type A TA increased considerably at $h = 3.0L$, and slowly at $h < 3L$. Therefore, maximizing *h* is crucially important for improving the *P* of type A TAs. For the type B TAs, similar *P* curves were obtained at different *h* values.

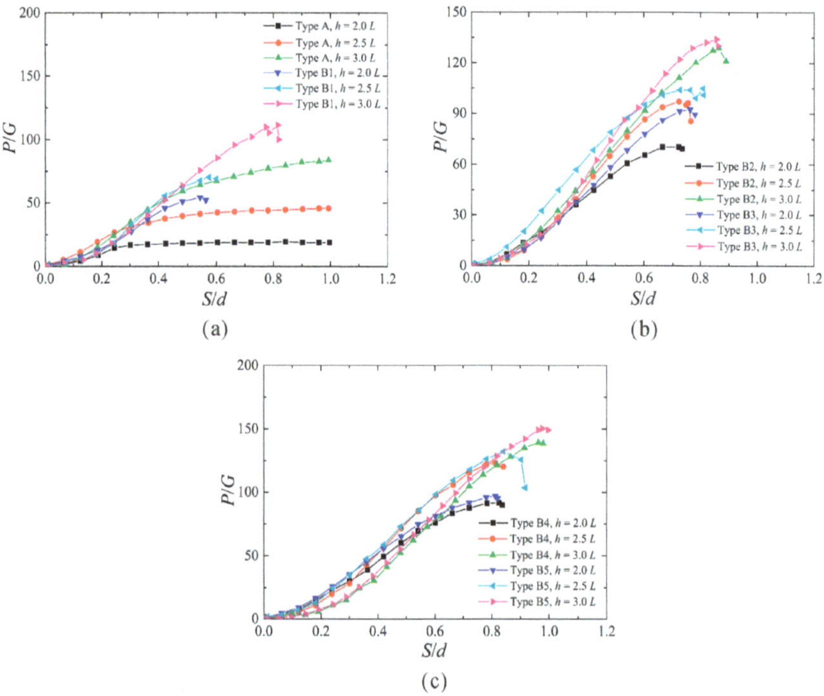

Figure 6. Effects of *h* on the load–displacement curves of the TAs under vertical pullout loading (**a**) Types A and B1; (**b**) Types B2 and B3; (**c**) Types B4 and B5.

Figure 7 shows the *P–h* curves of the two types of TAs at $D_r = 0.6$. At the same *h*, the *P* of the end-bearing TA was significantly higher than that of the conventional TA, and increased as *A* increased. At $h = 2.0L$, the *P* values of the type B1–B5 TAs were 220%, 345%, 515%, 530%, and 575% higher than that of the type A TA, respectively. At $h = 3.0L$, the *P* values of the type B1–B5 TAs were 93%, 130%, 149%, 169%, and 197% higher than that of the type A TA, respectively. These results show that increasing *h* reduces the effectiveness of bearing plates in improving *P*.

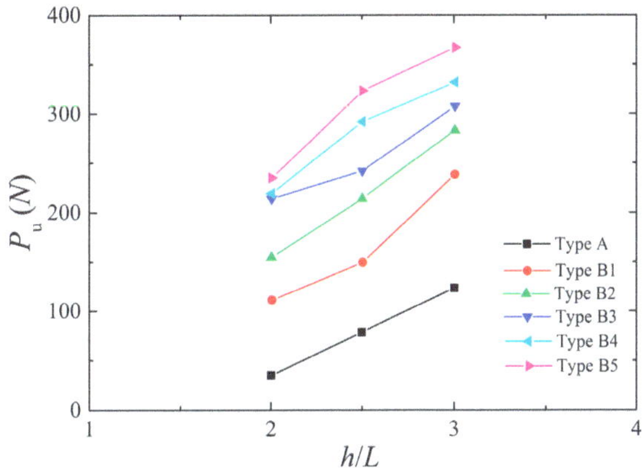

Figure 7. *P–h* curves of the TAs under vertical pullout loading.

3.4. Effects of D_r on *P*

The effects of D_r on the *P* values of the two types of TAs were analyzed. Figure 8 shows the *P–S* curves of the TAs at $h = 2.0L$ under vertical pullout loading. It can be seen that as D_r increased, the *P* of the type A and B TAs increased. At the same *h*, the *P* of each type B TA was significantly higher than that of the type A TA. At $D_r = 0.7$, the *P* values of the type B1–B5 TAs were approximately 169%, 309%, 337%, 424%, and 475% higher than that of the type A TA, respectively. Figure 9 shows the *P–*D_r curves of the two types of TAs at $h = 2.0L$. Under vertical pullout loading, the *P* of the type A and B TAs gradually increased as D_r increased. At each D_r, the *P* of each type B TA was significantly higher than that of the type A TA. At $h = 2.0L$, the *P* of the type A TA increased by 68.4% and 71.9% as D_r increased from 0.6 to 0.7, and from 0.7 to 0.8, respectively. Transducer failure caused errors in the measurement of the *P* of the type B3 TA. For the remaining type B TAs, the vertical *P* values of the type B1, B2, B4, and B5 TAs increased by 48.3%, 62.5%, 47.8%, and 50.8% as D_r increased from 0.6 to 0.7, respectively; their vertical *P* values increased by 30.2%, 30.8%, 35.9%, and 34.2% as D_r increased from 0.7 to 0.8, respectively. This finding suggests that increasing D_r can reduce the effectiveness of bearing plates in increasing the *P* of a TA.

3.5. Effects of *A* on *P*

The *P* of an end-bearing TA depends on *A*. The type B1–B5 TAs used in this study had *A* values of 4.9, 9.8, 14.7, 19.6, and 24.5 cm^2, respectively. Figure 10 shows the *P–A* curves at different *h* and D_r values. At $D_r = 0.6$, the vertical *P* of the type B TA increased as *A* increased, and increased significantly as *h* increased. At $h = 2.0L$, the vertical *P* of the type B TA increased as *A* increased, and increased considerably as D_r increased. These findings are consistent with those obtained from earlier analyses.

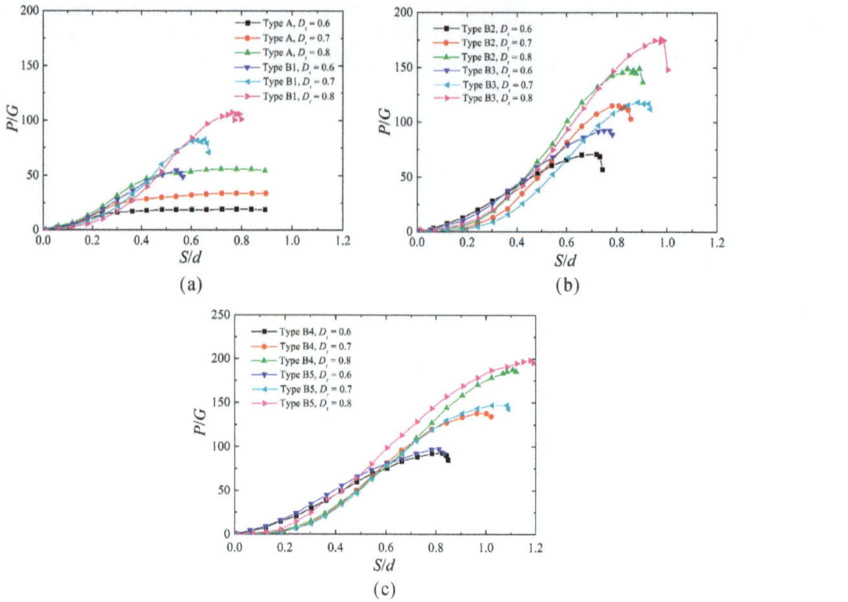

Figure 8. Effects of D_r on the load–displacement curves of the TAs under vertical pullout loading (**a**) Types A and B1; (**b**) Types B2 and B3; (**c**) Types B4 and B5.

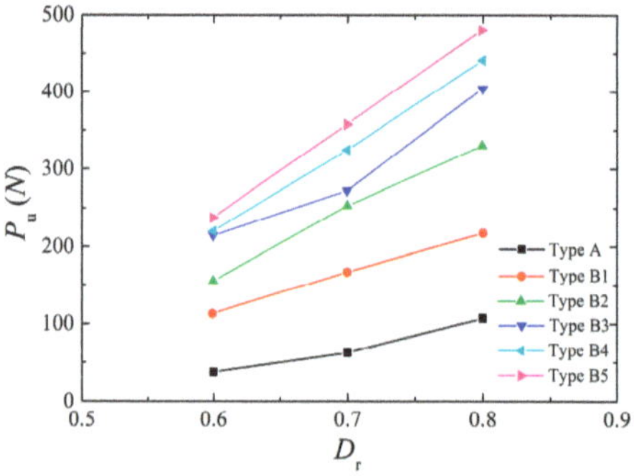

Figure 9. P–D_r curves of the TAs under vertical pullout loading.

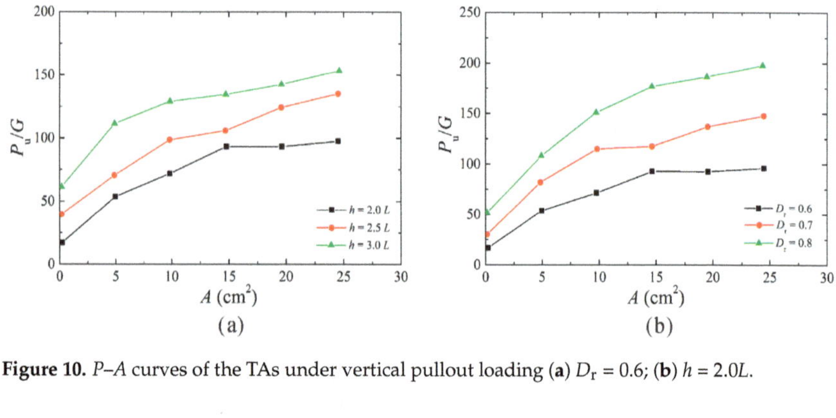

Figure 10. P–A curves of the TAs under vertical pullout loading (**a**) D_r = 0.6; (**b**) h = 2.0L.

4. Conclusions

Based on the pullout model tests, this study examined the bearing capacity of conventional and end-bearing torpedo anchors and investigated the effects of factors such as embedment depth, relative density, pullout angle, and area of the bearing plates on the bearing capacity. The main findings of this study are summarized as follows:

(1) When embedment depth reached 2 times the shank length, for the conventional torpedo anchor, the vertical bearing capacity increased slowly as displacement increased, with no peak on the bearing capacity–displacement curve. In comparison, for each end-bearing torpedo anchor, the vertical bearing capacity increased rapidly as displacement increased, with a notable peak on the bearing capacity–displacement curve. Under vertical pullout loading, the bearing capacity of each end-bearing torpedo anchor was much higher than that of the conventional torpedo anchor, and increased as area increased.

(2) Under horizontal pullout loading, similar bearing capacity-displacement curves were obtained for the conventional and end-bearing torpedo anchors, and each had a peak. In addition, these two types of torpedo anchors displayed essentially similar bearing capacity values. Under pullout loading at a pullout angle equal to 45°, area had a relatively significant impact on the bearing capacity of the end-bearing torpedo anchor. The bearing capacity increased relatively considerably as area increased beyond three times the cross-sectional area of the shank, with a notable peak on the bearing capacity–displacement curve. In contrast, there was no peak on the bearing capacity–displacement curve when the area was smaller than three times the cross-sectional area of the shank.

(3) Both embedment depth and relative density were found to have a major impact on the bearing capacity. The bearing capacity of the conventional and end-bearing torpedo anchors increased as embedment depth and relative density increased. In addition, increasing embedment depth and relative density had a significantly greater impact on the bearing capacity of the conventional torpedo anchor than that of the end-bearing torpedo anchor. At fixed embedment depth and relative density, the bearing capacity of the end-bearing torpedo anchor was significantly higher than that of the conventional torpedo anchor, and increased as area increased.

(4) Based on the research results, it can be found that the bearing capacity of torpedo anchors under horizontal pullout loading is greater than that of oblique and vertical pullout loading. The bearing capacity of the end-bearing torpedo anchor significantly increased as shank area, embedment depth, and relative density increased, and it reached maximum when embedment depth reached 3 times the shank length, relative density reached 0.8, and shank area reached 6 times the cross-sectional area. These research results can provide a guideline for installation of torpedo anchors, which need to be validated in further studies.

Author Contributions: Conceptualization, J.Z. and G.L.; methodology, J.Z. and G.L.; validation, J.L., J.Z. and G.L.; writing—original draft preparation, G.L.; writing—review and editing, J.Z.; funding acquisition, G.L. All authors have read and agreed to the published version of the manuscript.

Funding: This research was funded by the Natural Science Basic Research Program of Shaanxi Province, grant number 2021JM-535 and the Special Fund for Scientific Research by Xijing University, grant number XJ18T01.

Institutional Review Board Statement: Not applicable.

Informed Consent Statement: Not applicable.

Data Availability Statement: Not applicable.

Acknowledgments: The Youth Innovation Team of Shaanxi Universities is acknowledged.

Conflicts of Interest: The authors declare no conflict of interest.

References

1. Zhou, L.L.; Bao, X.W.; Yu, J.; Zhang, Y.; Feng, R.Y. Recent initiatives and developments in the ecological utilization of marine resources (EUMR) in China. *J. Coast. Res.* **2019**, *93*, 443–449. [CrossRef]
2. Zereshkian, S.; Mansoury, D. A study on the feasibility of using solar radiation energy and ocean thermal energy conversion to supply electricity for offshore oil and gas fields in the Caspian Sea. *Renew. Energy* **2021**, *163*, 66–77. [CrossRef]
3. Tay, J.H.; Show, K.Y.; Lee, D.J.; Hong, S.Y. Reuse of wastewater sludge with marine clay as a new resource of construction aggregates. *Water Sci. Technol.* **2004**, *50*, 189–196. [CrossRef] [PubMed]
4. Spagnoli, G.; Doherty, P.; Murphy, G.; Attari, A. Estimation of the compression and tension loads for a novel mixed-in-place offshore pile for oil and gas platforms in silica and calcareous sands. *J. Pet. Sci. Eng.* **2015**, *136*, 1–11. [CrossRef]
5. Yan, W.; Zhang, G.X.; Xia, B.; Zhang, L.; Yang, Z.; Lei, Z.Y.; Yao, H.Q. Seismic characteristics and development patterns of miocene carbonate platform in the Beikang basin, southern South China Sea. *Acta Geol. Sin.* **2020**, *94*, 1651–1661. [CrossRef]
6. Zhang, N.; Evans, T.M. Discrete numerical simulations of torpedo anchor installation in granular soils. *Comput. Geotech.* **2019**, *108*, 40–52. [CrossRef]
7. Kim, Y.H.; Hossain, M.S.; Lee, J.K. Dynamic installation of a torpedo anchor in two-layered clays. *Can. Geotech. J.* **2017**, *55*, 446–454. [CrossRef]
8. Soh, B.; Pao, W.; Chen, X.H. Numerical analyses for improved terminal velocity of deep water torpedo anchor. *Int. J. Numer. Methods Heat Fluid Flow* **2017**, *27*, 428–443. [CrossRef]
9. Wang, C.; Zhang, M.X.; Yu, G.L. Vertical pullout behaviour of a torpedo anchor vertically in stiff-over-soft cohesive soil bed. *IOP Conf. Ser. Earth Environ. Sci.* **2019**, *300*, 022139. [CrossRef]
10. Raie, M.S.; Tassoulas, J.L. Installation of torpedo anchors: Numerical modeling. *J. Geotech. Geoenviron. Eng.* **2009**, *135*, 1805–1813. [CrossRef]
11. Chen, X.H.; Zhang, M.X.; Yu, G.L. A self-penetration torpedo anchor with vibrational shearing. *Ocean Eng.* **2021**, *236*, 109315. [CrossRef]
12. Wang, W.K.; Wang, X.F.; Yu, G.L. Penetration depth of torpedo anchor in cohesive soil by free fall. *Ocean Eng.* **2016**, *116*, 286–294. [CrossRef]
13. Ads, A.; Iskander, M.; Bless, S.; Omidvar, M. Visualizing the effect of Fin length on torpedo anchor penetration and pullout using a transparent soil. *Ocean Eng.* **2020**, *216*, 108021. [CrossRef]
14. Hossain, M.S.; O'Loughlin, C.D.; Kim, Y. Dynamic installation and monotonic pullout of a torpedo anchor in calcareous silt. *Geotechnique* **2015**, *65*, 77–90. [CrossRef]
15. Lai, Y.; Huang, Y.H.; Chen, C.; Zhu, B. Free-fall penetration behaviors of a new dynamically installed plate anchor in marine clay. *China Ocean Eng.* **2020**, *34*, 795–805. [CrossRef]
16. Hasanloo, D.; Pang, H.L.; Yu, G.L. On the estimation of the falling velocity and drag coefficient of torpedo anchor during acceleration. *Ocean Eng.* **2012**, *42*, 135–146. [CrossRef]
17. Wang, W.K.; Wang, X.F.; Yu, G.L. Vertical holding capacity of torpedo anchors in underwater cohesive soils. *Ocean Eng.* **2018**, *161*, 291–307. [CrossRef]
18. Wang, C.; Wang, X.F.; Chen, X.H.; Yu, G.L. Maximum vertical pullout force of torpedo anchors in cohesive seabeds at different steady pullout velocities. *J. Coastal Res.* **2020**, *36*, 1068–1078. [CrossRef]
19. Wang, C.; Chen, X.H.; Yu, G.L. Maximum force of inclined pullout of a torpedo anchor in cohesive beds. *China Ocean Eng.* **2019**, *33*, 333–343. [CrossRef]
20. Wang, C.; Zhang, M.X.; Yu, G.L. Penetration depth of torpedo anchor in two-layered cohesive soil bed by free fall. *China Ocean Eng.* **2018**, *32*, 706–717. [CrossRef]
21. Santiago, P.C.; Saboya, F.; Tibana, S.; Reis, R.M.; Borges, R.G. Centrifuge modelling of a combined pile-type anchor subjected to general inclined loading. *Mar. Struct.* **2020**, *74*, 102815. [CrossRef]
22. Fernandes, A.C.; de Araujo, J.B.; de Almeida, J.C.L.; Machado, R.D.; Matos, V. Torpedoanchor installation hydrodynamics. *J. Offshore Mech. Arct. Eng.* **2006**, *128*, 286–293. [CrossRef]
23. Fernandes, A.C.; Sales, J.J.S.; Silva, D.F.C.; Diederichs, G.R. Directional stability of the torpedo anchor pile during its installation. *IES J. Part A Civ. Struct. Eng.* **2011**, *4*, 180–189. [CrossRef]
24. Chang, K.; Hossain, M.S.; Kim, Y.H. Performance of a novel dynamically installed fish anchor in calcareous silt. *J. Geotech. Geoenviron. Eng.* **2019**, *145*, 04019019. [CrossRef]
25. Raaj, S.K.; Saha, N.; Sundaravadivelu, R. Freefall hydrodynamics of torpedo anchors through experimental and numerical analysis. *Ocean Eng.* **2022**, *243*, 110213. [CrossRef]
26. Kunitaki, D.M.K.N.; de Lima, B.S.L.P.; Evsukoff, A.G.; Jacob, B.P. Probabilistic and fuzzy arithmetic approaches for the treatment of uncertainties in the installation of torpedo piles. *Math. Probl. Eng.* **2008**, *2008*, 512343. [CrossRef]
27. O'Beirne, C.; O'Loughlin, C.D.; Gaudin, C. A release-to-rest model for dynamically installed anchors. *J. Geotech. Geoenviron. Eng.* **2017**, *143*, 04017052. [CrossRef]
28. Raie, M.S.; Tassoulas, J.L. Simulation of torpedo anchor set-up. *Mar. Struct.* **2016**, *49*, 138–147. [CrossRef]
29. Sagrilo, L.V.S.; de Sousa, J.R.M.; Lima, E.C.P.; Porto, E.C.; Fernandes, J.V.V. A Study on the holding capacity safety factors for torpedo anchors. *J. Appl. Math.* **2012**, *2012*, 102618. [CrossRef]

30. De Sousa, J.R.M.; de Aguiar, C.S.; Ellwanger, G.B.; Porto, E.C.; Foppa, D.; de Medeiros, C.J. Undrained load capacity of torpedo anchors embedded in cohesive soils. *J. Offshore Mech. Arct. Eng.* **2011**, *133*, 021102. [CrossRef]
31. Kim, Y.H.; Hossain, M.S.; Wang, D. Effect of strain rate and strain softening on embedment depth of a torpedo anchor in clay. *Ocean Eng.* **2015**, *108*, 704–715. [CrossRef]
32. Kim, Y.H.; Roshera, L.T. Performance of novel dynamic installed anchors during installation and monotonic pullout. *Geomech Eng.* **2019**, *18*, 153–159.
33. Kim, Y.H.; Hossain, M.S. Numerical study on pull-out capacity of torpedo anchors in clay. *Géotechn. Lett.* **2016**, *6*, 275–282 [CrossRef]
34. Kim, Y.H.; Hossain, M.S.; Wang, D.; Randolph, M.F. Numerical investigation of dynamic installation of torpedo anchors in clay. *Ocean Eng.* **2015**, *108*, 820–832. [CrossRef]
35. Sabetamal, H.; Carter, J.P.; Nazem, M.; Sloan, S.W. Coupled analysis of dynamically penetrating anchors. *Comput. Geotech.* **2016**, *77*, 26–44. [CrossRef]
36. Sabetamal, H.; Nazem, M.; Carter, J.P.; Sloan, S.W. Large deformation dynamic analysis of saturated porous media with applications to penetration problems. *Comput. Geotech.* **2014**, *55*, 117–131. [CrossRef]
37. American Petroleum Institute. *Recommended Practice for Planning, Designing and Constructing Fixed Offshore Platforms-Working Stress Design*, 21st ed.; RP2A-WSD; American Petroleum Institute: Washington, DC, USA, 2000.

MDPI AG
Grosspeteranlage 5
4052 Basel
Switzerland
Tel.: +41 61 683 77 34

Journal of Marine Science and Engineering Editorial Office
E-mail: jmse@mdpi.com
www.mdpi.com/journal/jmse

www.ingramcontent.com/pod-product-compliance
Lightning Source LLC
Chambersburg PA
CBHW042011260926
51850CB00020B/3163